ELECTRONIC STRUCTURES IN SOLIDS

ELECTRONIC STRUCTURES IN SOLIDS

Lectures presented at the Second Chania Conference, held in Chania, Crete, June 30-July 14, 1968

Edited by

E. D. Haidemenakis

International Center for
Advanced Studies
and
Faculté des Sciences
Université de Paris
Paris, France

Springer Science+Business Media, LLC 1969

Library of Congress Catalog Card Number 71-82759

ISBN 978-1-4899-6250-8 ISBN 978-1-4899-6537-0 (eBook)
DOI 10.1007/978-1-4899-6537-0

Originally published by Plenum Press in 1969.

Softcover reprint of the hardcover 1st edition 1969

To Agni

PREFACE

The Chania Conferences, held each summer on the island of Crete, are international meetings in which each topic in a given subject area is treated first in a didactic manner with a thorough discussion of fundamental up-to-date knowledge. Then, the lectures develop rapidly into an advanced stage and include recent findings on current research. A number of panel and informal discussions also contribute to close interaction among the lecturers themselves and between the lecturers and the audience, which consists of research workers with at least four years of experience in the field.

The present proceedings include the lectures and seminars of the Chania Conference on Electronic Structures in Solids, held on the island of Crete from June 30 to July 14, 1968. Among the topics discussed were: energy band theory, reflectivity, photoemission and electroreflectance experiments, magnetooptics, tunneling spectroscopy, x-ray diffraction, many-body theory, neutron spectroscopy, quantum plasmas, transition metal oxides, superconductivity, quantum transport phenomena, the Kondo effect, nonlinear optics, light scattering, narrow band semiconductors, crystal structure and defects, impurity states, luminescence, excitons, radiative recombination, lasers, and the Gunn, acoustoelectric, and other effects.

The conference attracted more than one hundred participants from twenty countries. The general enthusiasm expressed by both lecturers and participating audience at the 1967 and 1968 Chania Conferences has led to the establishment of the International Center for Advanced Studies, the objective of which is to expand the frontiers of scientific, technological, and socio-economic learning through international exchange and cooperation.

We are deeply grateful to the sponsors who contributed to the realization and success of this meeting. These are: Public Power Corporation of Greece, Greek Office of International Meetings, Philips Industries, Olympic Airways, U.S.Office of Naval Research,

IBM World Trade Corporation, Industrial Development Bank of Greece, and also, National Bank of Greece, Boeing Scientific Laboratories, Hellenic Handicrafts Organization, Efthymiades Lines, Papastratos Cigarettes, Castel Wines, Metaxas Drinks, Viochym and Lampathakis. We would also like to thank the Governor and Mayor of Chania, the Lykion Ladies, and the people of the village of Anopolis for their warm hospitality.

In particular, I would like to thank the directors of this conference, Dr. Frank Herman and Professor Yves Rocard, for their interest and cooperation. My sincere appreciation goes also to the publication committee, Drs. Hans Frederikse and Frank Herman, for their prompt and hearty support and collaboration.

E. D. Haidemenakis
Director,
International Center for
Advanced Studies

April, 1969

CONTENTS

TUNNELING IN SOLIDS

Leo Esaki*

IBM Watson Research Center

INTRODUCTION

According to quantum mechanics, an electron represented by the wave function penetrates into classically forbidden region, thus tunnels through a thin potential barrier without acquiring enough energy to pass over the top. If our discussion is limited to the macroscopic potential barriers in the bulk of solids rather than those on the surface or in the atomic or crystal field, tunneling barriers, which we are familiar with, are classified into two categories:

1) interband tunneling in one and the same semiconductor
2) tunneling in conductor-insulator-conductor systems, such as MIM (metal-insulator-metal), MIS (metal-insulator-semiconductor or -semimetal), Schottky junctions, etc.

The former effect unambiguously demonstrated in the degenerate semiconductor p-n junctions--the tunnel diodes in 1957.[1] The effect was originally proposed by Zener as an explanation for dielectric breakdown in insulators.[2] Although electrons here tunnel across an energy gap rather than across a potential barrier, it was shown that the energy gap could be treated in the manner of the potential barrier. The Zener mechanism in dielectric breakdown has never been proved to be important in reality. If a high electric field is applied to the bulk crystal, hot electron effects

* Sponsored in part by the Army Research Office, Durham, N.C., under Contract DA-31-124-ARO-D-478.

such as impact ionization, avalanche, Gunn effect, etc., always precede tunneling, thus the field never reaches a critical value for tunneling.

In the latter systems, the observation of electron tunneling between superconductor films by Giaever in 1960[3] and Cooper pair tunneling by Josephson in 1962[4] stimulated extensive, experimental as well as theoretical investigations, of which results significantly contributed to understanding of the superconductors in the last several years.

INTERBAND TUNNELING IN ONE AND THE SAME SEMICONDUCTOR

As mentioned before, it has been known that interband tunneling can be seen most easily and almost exclusively in the high built-in field region of narrow semiconductor p-n junctions.[1]

The characteristics of the diode were analyzed in terms of interband tunnel current, excess (tunnel) current and conventional thermal current, as shown in Fig. 1. The Fermi levels in those heavily doped semiconductors penetrate deeply inside the bands. One can see that, as the bias is increased, the bands, which overlap each other at zero bias, become uncrossed. Since the energy

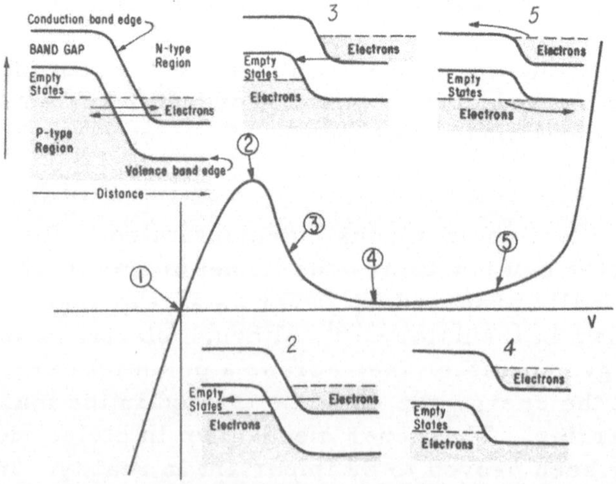

Fig. 1 Schematic illustration of the change in the energy diagram, as the bias voltage is increased, as well as the current-voltage characteristic, in the tunnel diode.

should be conserved in tunneling, the current obviously should decrease as the bands become uncrossed. The excess current could be explained in terms of tunneling through mid-gap levels.[5] Measurements of the junction width indicate that the built-in field strength is as high as 5×10^5 to 10^6 volt/cm.

The interband tunneling current for $I_{c \to v}$ and $I_{v \to c}$ at a bias voltage V were formulated as follows:

$$I_{c \to v} = A \int_{-\infty}^{\infty} f_c(E) \rho_c(E) |M|^2 \{1 - f_v(E+eV)\} \rho_v(E+eV) dE,$$

$$I_{v \to c} = A \int_{-\infty}^{\infty} f_v(E+eV) \rho_v(E+eV) |M|^2 \{1 - f_c(E)\} \rho_c(E)\, dE$$

and the resultant current,

$$I = I_{c \to v} - I_{v \to c} = A \int_{-\infty}^{\infty} \{f_c(E) - f_v(E+eV)\} |M|^2 \rho_c(E) \rho_v(E)\, dE \qquad (1)$$

where f_c and f_v are the Fermi-Dirac distribution functions, ρ_c and ρ_v are the density-of-states in the conduction band and valence bands, respectively, and M is the matrix element for the transition across the gap. Figure 2 is the comparison of current-voltage curves calculated with the measured points at

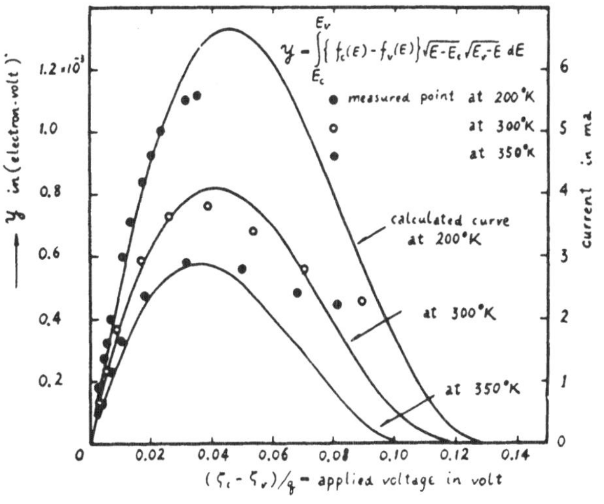

Fig. 2 Comparison of calculated current-voltage curves with measured points at 200°K, 300°K, and 350°K.

200°K, 300°K, and 350°K, which was shown in our first paper on the tunnel diode.[1] In this case, Z was considered to be almost constant in the voltage range involved. Although the energy conservation was taken into consideration here, there was no concomitant momentum conservation law. This is probably true only when tunneling electrons interact (incoherently) with individual impurities in the junctions. However, in the case of "elastic tunneling" or the coherent excitation of collective modes of the junction by tunneling electrons, it will be required to apply a conservation law on the component of momentum parallel to the junction plane. If this is taken into consideration, the current for elastic tunneling is given by

$$j = \frac{4\pi e}{\hbar} \sum_{k_t} \int_{-\infty}^{\infty} |M|^2 \rho_c \rho_v (f_c - f_v)\, dE \tag{2}$$

Now, M vanishes unless the transverse wave number k_t is the same for the initial and final states and ρ_c and ρ_v are density of states for fixed k_t. The integral over energy should be taken at fixed k_t also.

Because of the reciprocal relation between the particle velocity and the density of states, we obtain

$$j = \frac{2e}{h} \sum_{k_t} \int_{-\infty}^{\infty} \exp\left(-2\int |k_x| dx\right)(f_c - f_v)\, dE \tag{3}$$

without the density of states factor in the integrand. This is a consequence of the independent-particle model.[6] The above equation can be simplified by replacing the sum over k_t by an integral over the projection of a constant energy surface onto the plane of the barrier, as follows:

$$j = \frac{e}{2\pi^2 h} \int_{-\infty}^{\infty} dE(f_c - f_v) \int d^2k_t \exp\left(-2\int |k_x| dx\right). \tag{4}$$

k_x here is an imaginary wave number in the forbidden gap which could be obtained by applying the k·p perturbation approach.[7] In the simplest two-band formulation, the dispersion of states in the forbidden gap is given by

$$E(\vec{k}) \sim \left[\left(\frac{E_g}{2}\right)^2 + \frac{\hbar^4 \vec{k}^2 P^2}{m_o}\right]^{1/2}$$

where E_g is the energy gap, m_o is the free electron mass and P is the momentum matrix element involved in the perturbation.

In the wake of the Ge tunnel diode, the silicon tunnel diode[8] was fabricated with similar techniques. The characteristics are shown in Fig. 3. Other semiconductor materials also quickly were explored to obtain the tunnel diodes; InSb,[9] GaAs,[10] InAs,[11] PbTe,[12] GaSb,[13] SiC,[14] etc. Table I shows the material constants and Fig. 4 shows a comparison among those at the normalized peak current. The tunneling characteristics (including the current density) depend on the "band structure" and other material constants, thus detailed analyses of refined data have yielded a great deal of information about the properties of heavily doped semiconductors as well as a detailed understanding of the tunneling process itself.

When the silicon tunnel diode was cooled down to 4.2°K, a significant fine structure was found on the current-voltage curve, as shown in Fig. 5.[15] We were surprised that four characteristic values almost exactly agreed with the four phonon energies obtained from the optical absorption spectra of pure silicon or analysis of intrinsic recombination radiation in 1958.[16] (Their assignment was not correct at that time), indicating the existence

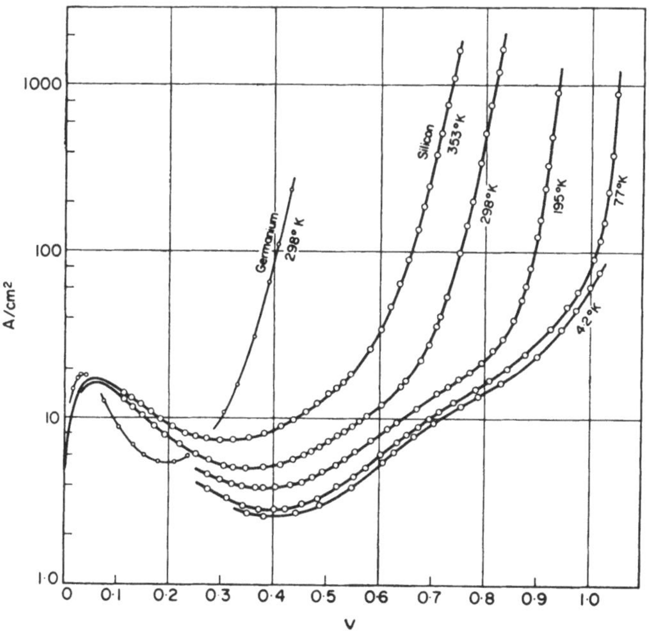

Fig. 3 Current-voltage characteristics in the forward direction of Ge and Si tunnel diodes at various temperatures.

TABLE I

	energy gaps (eV)		effective mass ratios		static dielectric constants
	$0^{o}K$	$300^{o}K$	electrons	holes	
PbSe	0.16	0.27	m_t 0.040	m_t 0.034	250
			m_ℓ 0.070	m_ℓ 0.068	
PbTe	0.185	0.30	m_t 0.024	m_t 0.024	412
			m_ℓ 0.27	m_ℓ 0.27	
InSb	0.235	0.17	0.0155	ℓ 0.02	17.9
				h 0.4	
PbS	0.28	0.41	m_t 0.080	m_t 0.075	175
			m_ℓ 0.105	m_ℓ 0.105	
InAs	0.43	0.35	0.025	ℓ 0.025	14.6
				h 0.41	
Ge	0.785	0.66	m_t 0.082	ℓ 0.04	16.0
			m_ℓ 1.64	h 0.3	
GaSb	0.81	0.72	0.047	ℓ 0.052	15.7
				h 0.35	
Si	1.21	1.08	m_t 0.19	ℓ 0.16	12.0
			m_ℓ 0.98	h 0.5	
InP	1.41	1.27	0.073	---	12.4
GaAs	1.53	1.40	0.07	ℓ 0.12	13.1
				h 0.68	
GaP	2.4	2.25	0.35	---	10.2

ℓ: light holes, h: heavy holes
m_t: transverse mass, m_ℓ: longitudinal mass

of the indirect (phonon-assisted) tunneling[17] due to the band structure of silicon. Namely, the phonon-assisted tunneling occurs when the energy band extrema are not located at the same point in the k-space. In this case, since the transverse components of momentum on both sides of the junctions are no longer equal, the difference in momentum can be supplied by phonons. There are two mechanisms in the phonon-assisted tunneling: i) a first order mechanism, in which an electron on one side scatters to a state on the other side with the emission of a phonon (Keldysh, Kane); ii) a second order mechanism, in which an electron tunnels to an intermediate state in a higher band on the other side and then scatters with the emission of a phonon to a final state (Kleinman).

At any rate, a number of phonon humps can be seen on the conductance-versus-voltage curve, serving as "phonon tunneling spectroscopy."

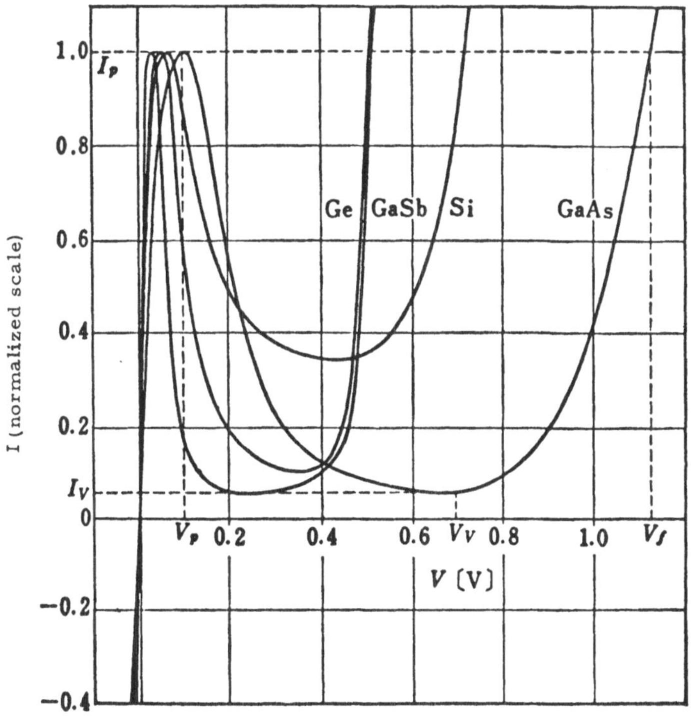

Fig. 4 Comparison of current-voltage characteristics of Ge, GaSb, Si and GaAs tunnel diodes.

The announcement of the tunnel diode has stimulated not only experimental investigation but also theoretical studies of interband tunneling.[18] The theories may be classified into two schools: one traditional, Zener[2]-Houston[2] - Keldysh[19]- Kane[7] - Price[20] school, and the other, relatively new, Fredkin, Wannier[21] - Shuey[22]- Takeuti, Funada[23] school. The former assumes a constant and modestly high electric field, under which electrons cycle repeatedly through the Brillouin zone according to the equation of motion; $d\vec{k}/dt = eF/\hbar$, the cycle time $\tau = 2\pi\hbar/eFa$, where a is the lattice constant. The rate of leaking into the adjacent band is greatest when the k vector is at the band edge. In the latter theory, the tunneling probability is essentially defined in terms of the formalism of the scattering theory. Incident and out-going electrons are represented by the Bloch waves in respective bands and the energy conservation in "elastic" tunneling is to be applied. Therefore, one may say that the former treats tunneling inside the junction region or the high field region, whereas the latter looks at tunneling from the outside regions.

In the simple theory of interband tunneling, Bloch electrons interact with a "steady electric field: E, which contributes to the Hamiltonian a term $eE\vec{r}$. This can be compared with the interband optical absorption process in which Bloch electrons interact with an "alternating electric field" of the vector

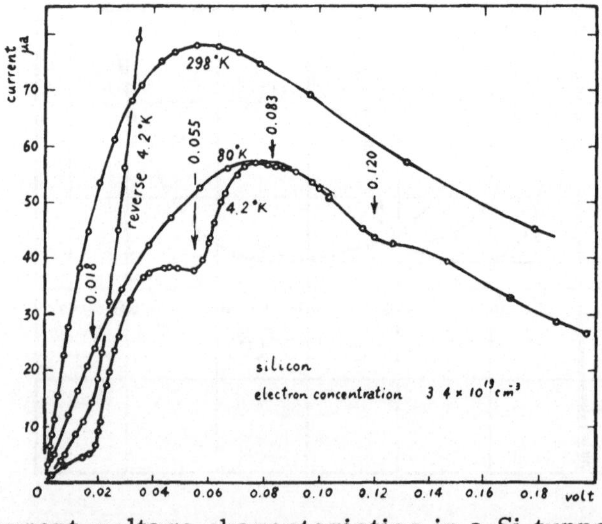

Fig. 5 Current-voltage characteristics in a Si tunnel diode at 4.2°K, 80°K and 298°K. Notice structure in the curve at 4.2°K.

potential A. Usually the additional term (e/m) $\vec{A} \cdot \vec{p}$ to the Hamiltonian is treated as a perturbation in the simple radiation theory. In both cases, one ought to derive the matrix element governing the transition probability. The phonon bumps observed in the tunneling current will correspond to phonon effects in the absorption spectra or recombination radiation spectra. The magneto-tunneling and the pressure effect on tunneling will also be considered to be the counterparts of the magneto-optical effect and the pressure dependence of the optical effects, respectively. The Franz-Keldysh effect[24] and the photon-assisted tunneling[25] are worth mentioning here, because they are really a combination of tunneling and photon absorption or emission under a steady electric field.

As described before, the excess current is attributed to band-to-band tunneling via deep impurity states in the forbidden gap. The current often shows structure or humps, particularly if doped with deep impurities[26] or irradiated by high energy particles.[27] Since the study provides energy positions of these levels, one might call it "excess-current tunneling spectroscopy."

ELECTRON TUNNELING INTO OR THROUGH DISSIMILAR SOLIDS

A considerable number of studies have been devoted to electron tunneling between two metals separated by a thin insulating layer, particularly since the observation of the superconducting energy gap in 1960.[3] The superconducting metal tunnel junction, which should be discussed from the many-particle point of view,[28] is not given here in detail, although it is an extremely useful tool in measuring the superconducting energy gap.[29] It should be noted that, in this case, the transition matrix element, M, in Eqs. 1 and 2 is rather insensitive to the wave number k and is considered to be constant, thus the superconducting density of states can be obtained with tunneling conductance measurements. Recently, the Schottky tunnel junctions have been used for this purpose.[30] One might call the field "superconducting tunneling spectroscopy."

Theoretical calculations for the normal metal tunnel junctions with the WKB approximation have led to analytical expressions of the current-voltage characteristics,[31] and experimentally, the $Al-Al_2O_3-Al$ junctions have been extensively studied.[32] Recently tunneling studies have been further extended to metal-insulator-semiconductor[33, 34, 35] or semimetal[36, 37] (MIS) systems as well as the Schottky junctions.(38) One direction taken in the studies is to observe electronic energy states in semiconductors or semimetals: band edge energies,(34, 36, 37) electron affinities,(34, 35) density of states,(39) and localized states and their changes with magnetic fields.(35) It appears that this type of tunneling spectroscopy is an extremely versatile tool, applicable to a variety of materials, which will be discussed in detail later. The other direction of intensive investigation is measurements and interpretations for rapidly varying tunnel conductance in the neighborhood of zero-bias at low temperatures; the so-called zero-bias anomalies, which was initiated by the observation of narrow conduction minima in III-V compound tunnel diodes.(40) This effect is believed to be due to the process of inelastic tunneling, where the tunneling electron excites some internal degree of freedom, thus serves as a spectroscopic probe.(41) A typical example is the observation of vibrational spectra for molecules contained in the junction region.(42, 43) Another example is the observation of conductance maxima(44) or minima(45) and an intriguing interpretation invoking the resonant excitation of spin-flip transitions of magnetic impurities in the insulation by tunneling electrons.(46) The previously mentioned "phonon tunneling spectroscopy" now may well be classified in this category.(47) Indeed, a recent observation of optical phonon structure in GaAs Schottky junctions has been interpreted in terms of bulk polar electron-phonon coupling from the many-particle point of view.(39) (Another interesting aspect of this study in GaAs Schottky junctions is that the tunneling current is claimed to depend upon the density of states of the semiconductor.)

TUNNELING SPECTROSCOPY FOR GROUP V SEMIMETALS AND IV-VI SEMICONDUCTORS

Introduction

Tunneling spectroscopy is a technique to use tunneling electrons of known energy distribution as a spectroscopic probe, instead of photons of known frequency in optical spectroscopy. The technique can be applied to the following two investigations: one, the study of electronic energy states in solids (semiconductors, semimetals, insulators and superconductors); the other, the observation of the excitation of modes in the junction; phonons, magnons, molecular vibrations, etc., as shown in Fig. 6. We often see both effects; transitions to higher electronic energy states and mode excitations, simultaneously. This can be compared with the optical measurements, where photons either excite electrons to higher levels or interact some kind of modes in solids, or both.

During the past few years, we have been applying the technique to group V semimetals(36, 37) such as Bi, Bi-Sb alloys, etc., and IV-VI compound semiconductors(34, 35) such as SnTe, GeTe, PbTe, PbSe, etc. The study is primarily aimed at the observation of energy bands as well as localized levels, and possibly their changes with applied magnetic fields. All of these materials are isovalent (average valence five), of which the crystal structure is either cubic or deviates in some small way from that by the addition of distortion along the trigonal axis, and therefore from a closely related group of materials.(48) A theoretical consideration on the band structure indicates that they are, in general, narrow direct-gap semiconductors or semimetals, thus many band edges are expected to be located in a relatively small energy range near the Fermi energy.(49)

Both Bi and Sb are the most common semimetals and they

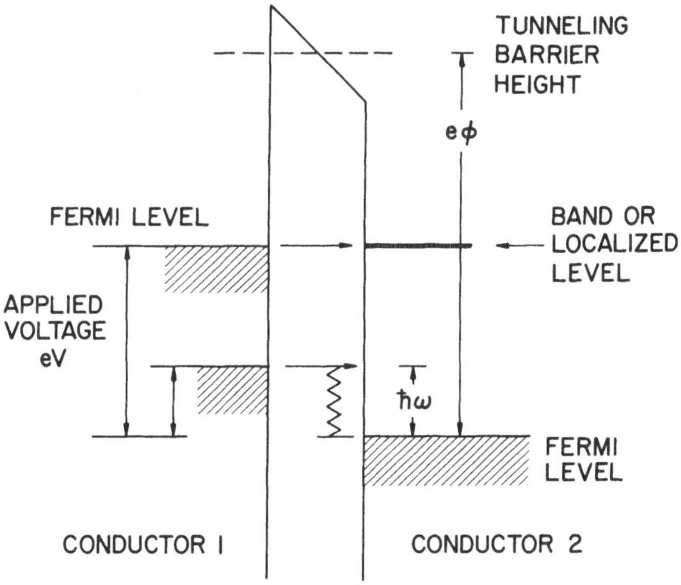

Fig. 6 Schematic illustration of the tunnel junction for tunneling spectroscopy.

make a continuous solid solution. Galvanomagnetic measurements for some BiSb alloys indicate that they are semiconductors with energy gaps from zero to ~.03eV, depending on composition, from about 5 to about 40 atomic percent Sb in Bi. Both SnTe and GeTe are usually nonstoichiometric and highly p-type owing to high concentrations of Sn or Ge vacancies. There had been uncertainty as to whether they were semiconductors or semimetals. The present studies have given a clear answer that they indeed have a thermal energy gap. The lead salts, PbTe, PbSe and PbS are the most well-known semiconductors among the materials involved here.

Preparation of Junctions

In carrying out tunneling spectroscopy, it is of utmost importance to obtain proper tunnel junctions. Two techniques have been employed to fabricate metal-insulator-semiconductor or semimetal (MIS) tunnel junctions.

The first involves depositing the material under investigation over an oxidized metallic layer, which is commonly used in the metal-insulator-metal (MIM) system. This provides a simple method to fabricate tunnel junctions, although all materials including semiconductors and semimetals are polycrystalline. It has proved possible to deposit on Al_2O_3, films of SnTe and GeTe whose electrical, X-ray and optical properties are comparable to bulk properties.

The second technique which requires a sophisticated evaporation system, involves the introduction of a thin insulating layer on a single crystal surface and then evaporating a metal counter electrode to complete the tunnel junction.

Figure 7 shows a schematic fabrication process as well as a finished device of PbTe, PbSe or Bi. We used either bulk-grown crystals or epitaxially grown monocrystalline films on a (100) surface of NaCl substrates.[50] The tunnel junctions were formed by successive evaporation of pure sapphire (Al_2O_3) and a counter-electrode metal (Al, Au or Pb) with an electron-beam gun in the following way; we placed a 10 mil wide mask a few mils away from the surface and then evaporated the sapphire up to 1000~1500 angstroms at a rate of 100 Å/min., as shown in Fig. 7. This process was repeated after the mask was rotated by 90 degrees, thus we obtained a window surrounded by thick Al_2O_3. By proper adjustment of the variable involved, the window could be covered with a thin pinhole-free Al_2O_3 layer of appropriate thickness for tunneling.

Our evaporation system, with a mask changer, enables us to make several junctions on a small piece of single crystal in one operation without breaking vacuum. Each junction has an area of about $5 \times 10^{-4} cm^2$. Although we have little knowledge about mechanism of formation of the thin Al_2O_3 layer underneath the mask, we find the technique works for most materials investigated except for PbS, where sulphur possibly prevents the thin oxide layer formation. In the case of use of a bulk-grown crystal, it is obvious that considerable precaution should be taken in order to obtain clean and damageless surfaces, whereas the problem does not exist in the case of formation of tunnel junctions on epitaxially grown films.

The characteristics of semiconductor films, polycrystalline or monocrystalline, are obviously dependent upon the evaporation conditions, particularly, the substrate temperature. The substrate was kept at 100 $\sim$300°C for SnTe, and 150 $\sim$300°C for GeTe and 350 $\sim$400°C for the monocrystalline PbTe and PbS. In the cases of SnTe and GeTe, higher hole concentrations were

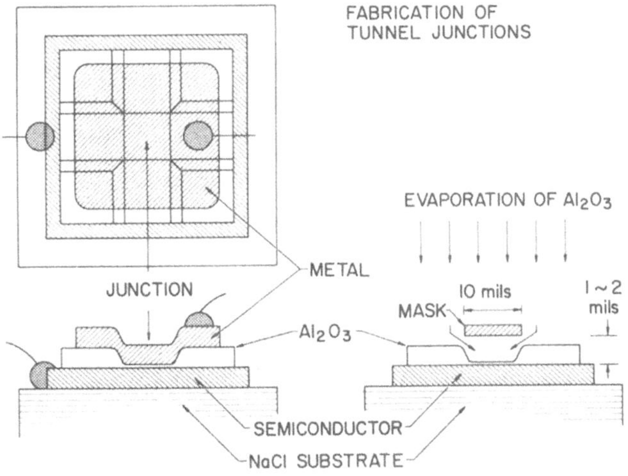

Fig. 7 Fabrication process and structure of the tunnel junction.

obtained at lower substrate temperatures ($10^{20} \sim 10^{21}$ cm^{-3}). However, at very low temperatures (below 150°C), GeTe becomes amorphous and the devitrification temperature of the material is probably in the neighborhood of 130°C, where the resistivity changes the order of a million. The epitaxial PbTe and PbSe films, a few thousand angstroms thick, have hole concentrations in the range of $1 \sim 8 \times 10^{18}$ cm^{-3} and Hall mobilities up to 150,000 cm^2/volt· sec at 4.2°K.

Analyses

Assuming "elastic tunneling" (the conservation of energy E and transverse momentum k_t) and the WKB approximation, the tunneling current density along x direction between two conducting regions 1 and 2 through a forbidden region is given on the independent-particle model[6] by

$$j = \frac{e}{2\pi^2 h} \int dE(f_1 - f_2) \int d^2k_t \; \exp(-2\int |k_x| dx) \qquad (5)$$

as is the case of interband tunneling in Eq. 4. Here f_1 and f_2 are the Fermi distribution function in the two conducting regions and the integral of k_x in the tunneling exponent should be over the tunneling region. For the MIS junction under consideration, the constant energy surface in k space for electrons in the semiconductor or in the semimetal is, in general, considerably smaller than that in the metal. As a result, the tunneling of electrons to the metal is always assumed to be allowable. If it is further assumed that the energy band of the solids involved are parabolic with an isotropic electron mass m, Eq. 5 at zero temperature can be reduced to

$$j = \frac{me}{2\pi^2 \hbar^3} \iint \exp(-2\int |k_x| dx) \; dE_t dE, \qquad (6)$$

where E_t and E are now the transverse and total kinetic energy of electrons in the semiconductor or the semimetal. The limits of integration for E_t are zero and E; and for E, are simply the two Fermi levels because of the assumption of zero temperature.

Although we have mentioned Eqs. 5 and 6 based on the WKB approximation, primarily, because of simplicity, it should be noted that there are two extreme cases: the graded boundary and the sharp boundary. In the former, the potential changes slowly and hence the fractional change in the wavelength is small over a distance of a wavelength, where the WKB method provides a good approximation. In the latter, electron waves are reflected by such a discontinuity, just as light waves are, where the wave function as well as its derivative should be matched. It is important to point out that, in the sharp boundary condition, $|M|^2\rho_1\rho_2$ (see Eq. 2) is not independent of energy. Actual experimental specimens probably lie somewhere in between.

Experimental Results

We have measured usually the current as well as the conductance dI/dV with a small ac signal (peak-to-peak less than 1 meV) of a few kilo-cycles as a function of applied voltages at liquid helium temperatures.

i) Bi and BiSb alloys. In search of band edge effects in monocrystalline Bi and Bi-Sb alloy tunnel junctions, prominent structure was observed in our early studies.[36] The structure of peaks in the conductance curves seemed to be considerably stronger than theoretical predictions.[6,51] For instance, assuming the exponent in Eq. 6 to be essentially constant because of relatively low applied voltages, one obtains the tunneling conductance proportional to $(V-V_b)$, where eV_b is the band edge energy with respect to the Fermi level. The situation is not saved much, even if the sharp boundary approximation is taken. Namely, in such a case, the conductance is given to be proportional to $(V-V_b)^{3/2}$. These are obviously too small to account for the observed structure.

The puzzle, we believe, has been solved by taking into account "inelastic tunneling," whereas Eq. 6 is derived in view of "elastic tunneling." Recently, we have observed fine structure in Bi,[37] of which analysis clearly indicates that such structure arises from a combination of phonon-assisted tunneling and energy band edges.

In carefully examining the structure near zero bias of more than fifty units (most trigonal), we found that, if any bumps were detected at all, they appeared at only four distinct voltages: 3 ± 1, 6 ± 1, 10 ± 1, and 14 ± 1 mV, regardless of the polarity of the applied voltage. These four voltages, designated as A, B, C, and D in Fig. 8, correspond closely to the measured phonon energies of the four modes (TA, LA, TO, and LO) in the trigonal direction.

In Bi tunnel junctions, however, not only the partially filled conduction and valence bands but also other bands are expected to contribute to the tunneling current, and we have indeed found the phonon-assisted process to be equally significant for tunneling into these higher and lower bands. This is clearly seen in Fig. 8 near 40 and 90 mV. There is a band just below each of these two voltages. The bumps F and G for one band, and J, K, L, and M for the other band are due to phonon-assisted tunneling. It is very important to note that, choosing the higher band-edge energy to be 88 meV, the voltage differences from the band edge to J, K, L, and M are almost identical to the voltages of A, B, C, and D. With this method of determining band edge energies with considerable accuracy, we find four band edges: +90 ± 5, +38 ± 2, -35 ± 3, and -60 ± 5 meV; and less reliably, nine additional band edges: +650, +145, -80, -110, -140, -200, -600, -800, and -1600 meV, where + and - refer to below and above the Fermi level, respectively. It should be added that, in polycrystalline Bi tunnel junctions, the phonon process is difficult to resolve, in contrast to the marked structure of single crystal units. This is probably due to the larger number of scattering centers in poorer-quality materials.

In BiSb alloy tunnel junctions, we have found a large dip of half-width of 10 ~ 15 mV, as shown in Fig. 9, resulting in the semiconductor energy gap of 20 ~ 30 meV. The value is in good agreement with the previous estimate.[52]

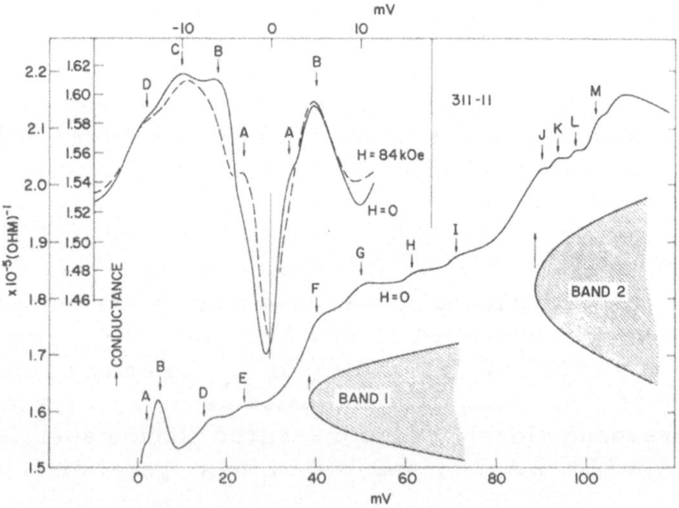

Fig. 8 Conductance versus applied voltage curve in the Bi tunnel junction, showing band edge effects due to the bands 1 and 2 below the Fermi level.

ii) SnTe and GeTe. It is shown in calculating of the tunneling current in Eq. 6 that, under certain conditions, the current-voltage characteristic of an insulating barrier tunnel junction between a metal and a degenerate semiconductor exhibits a negative resistance, similar to that of the tunnel diode. This type of negative resistance arises from the voltage dependence of the tunneling probability of the barrier. We have observed this negative resistance in the polycrystalline tunnel junctions of Al-Al_2O_3- SnTe and -GeTe systems,[34] as shown in Fig. 10, in the course of studying the electronic band structure of SnTe. The characteristic is of significance since this is the first observation of the "thermal" energy gaps, Eg, in SnTe as well as GeTe.

It is possible to obtain analytical expressions for the tunneling current in Eq. 6 by the use of the effective barrier approach. This procedure involves expanding the terms in the integrand of the tunneling exponent λ. Figure 11 shows the schematic energy diagrams at zero bias and at an applied voltage V of the SnTe-Al_2O_3-Al system. Now, assuming that k_x at given E and E_t

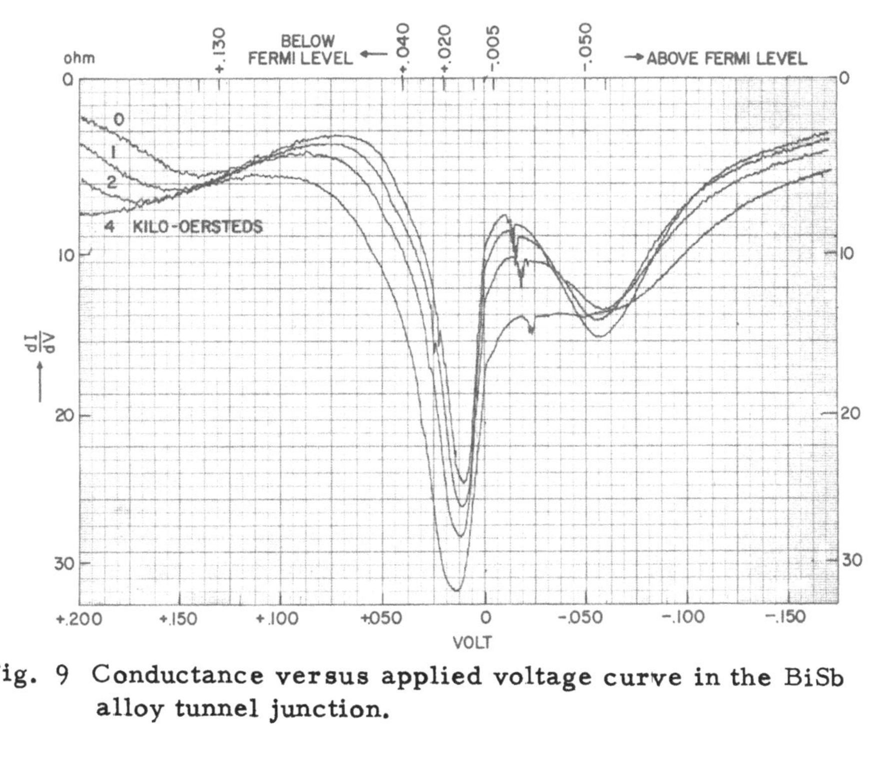

Fig. 9 Conductance versus applied voltage curve in the BiSb alloy tunnel junction.

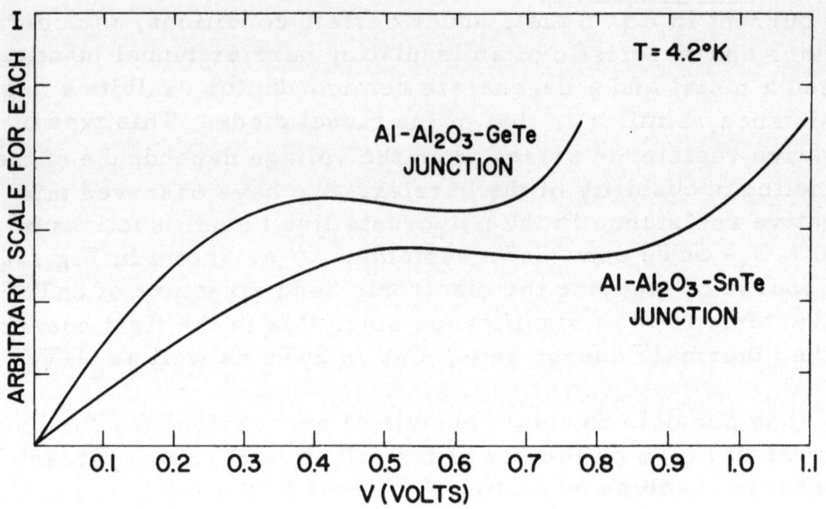

Fig. 10 Current-voltage characteristics of Al-Al_2O_3-GeTe and -SnTe tunnel junctions.

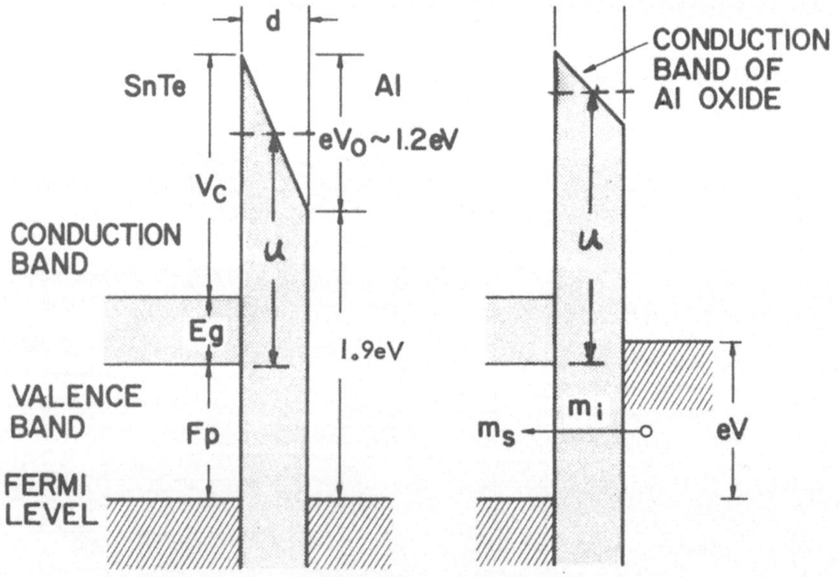

Fig. 11 Energy diagrams in the SnTe junction at zero bias and at an applied voltage V (right). The GeTe junction is also similar to this.

is determined only by the conduction band edge in the Al_2O_3, the exponent λ for the valence band of SnTe (or GeTe) is given by

$$\lambda = 2(2m)^{\frac{1}{2}}/\hbar \int_0^d \left[eV_c + E_g + E - E_t + \frac{e(V - V_o)x}{d}\right]^{\frac{1}{2}} dx \qquad (7)$$

where eV_c is the energy difference between the conduction band edges in SnTe or GeTe and in Al_2O_3 at the boundary, V_o is the built-in voltage across the oxide due to the work function difference between two conductors and d is the thickness of Al_2O_3. The applied voltage V is taken to be positive if the semiconductor side is positively biased with respect to metal. It should be mentioned that the tunneling probability decreases with increase in positive applied voltage.

The current, I_v, due to the valence band, is readily derived as a function of applied voltage from Eqs. 6 and 7 as follows (the effective mass, m, to be kept constant in all regions);

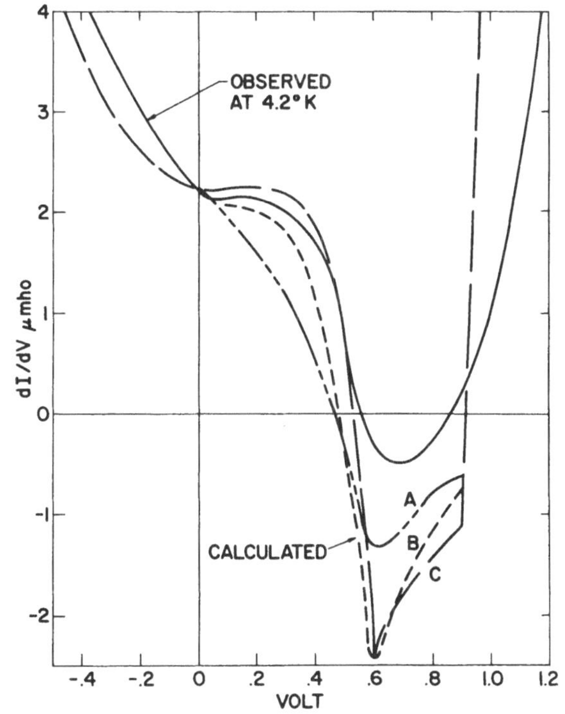

Fig. 12 Observed and calculated conductance-versus-applied voltage curves.

for $eV < F_p$

$$I_v = C \exp(-\tfrac{1}{2}\gamma eV) \times$$

$$[\exp(\gamma eV) - 1 - \tfrac{1}{2}\exp(\gamma(2eV - F_p)) + \tfrac{1}{2}\exp(-\gamma F_p)] \quad (8)$$

for $eV > F_p$

$$I_v = C \exp(-\tfrac{1}{2}\gamma eV) \times$$

$$[\tfrac{1}{2}\exp(\gamma F_p) - 1 + \tfrac{1}{2}\exp(-\gamma F_p)], \quad (9)$$

where F_p is the Fermi energy, C is a slowly varying function of the applied voltage and

$$\gamma = \frac{d}{\hbar}(2m)^{\frac{1}{2}}(eV_c + E_g - \tfrac{1}{2}eV_o)^{-\frac{1}{2}}.$$

The current I_v shows a maximum around $eV = F_p$ and then decreases with increase in V, namely, it illustrates a negative resistance as mentioned before. When the applied voltage eV reaches $F_p + E_g$, the current I_c due to the conduction band starts and soon offsets the negative resistance. The current-voltage relationship can be calculated numerically in the sharp boundary condition.

Observed (solid line) and calculated (dotted lines) conductance curves for the SnTe junction are shown in Fig. 12, where the calculated curve C (WKB approximation) is derived from derivatives of Eqs. 8 and 9 and $\lambda = 5(eV)^{-1}$ and the curves A and B (sharp boundary) are obtained by exact mating of wave functions and $\lambda = 5$ and $7.5(eV)^{-1}$, respectively. $E_g = 0.3$eV and $F_p = 0.6$eV for 8×10^{20}cm^{-3} at 4.2°K have been selected to get the closest fit of the calculated one to the observed one in all cases. The same technique was applied to GeTe, yielding $E_g = 0.2$eV and $F_p = 0.4$eV for 2.8×10^{20}cm^{-3} at 4.2°K.

Further extending the effective barrier approach with the WKB approximation, one can obtain the theoretical current-voltage relationship from the low-voltage negative-resistance region up to the Fowler-Nordheim region. The results are compared with the experimental data at 4.2°K for three SnTe junctions in Fig. 13 and for two GeTe junctions in Fig. 14. The Fowler-Nordheim plots (applied voltages are higher than the barrier heights) are shown in Fig. 15 for thick-oxide SnTe and GeTe junctions. It should be noted in these figures that the data are seen to follow the theoretical

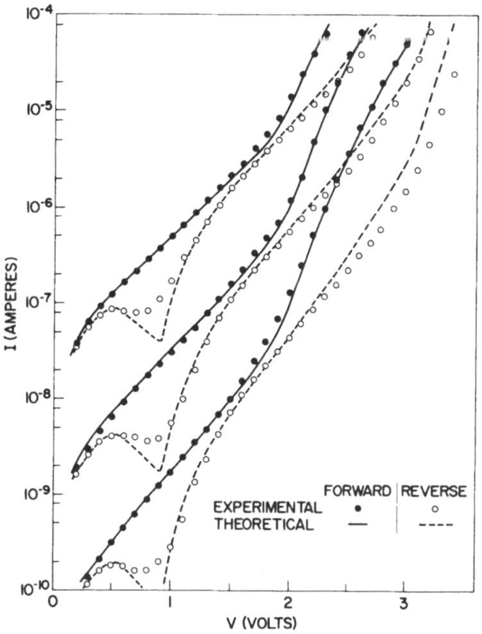

Fig. 13 Current-voltage characteristics for Al-Al_2O_3-SnTe junctions at 4.2°K. Three sets of curves are shown corresponding to samples with various oxide thickness.

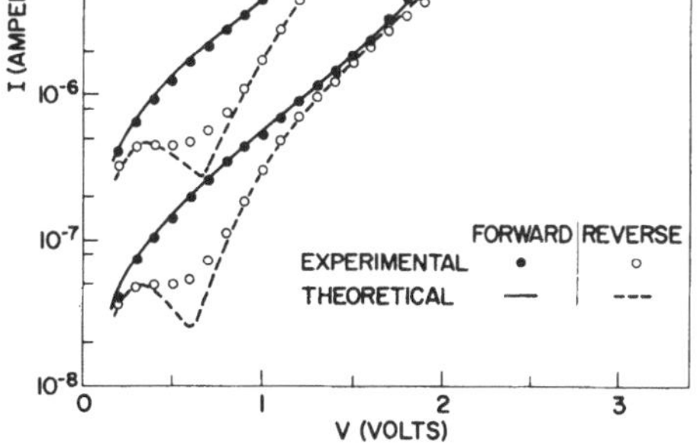

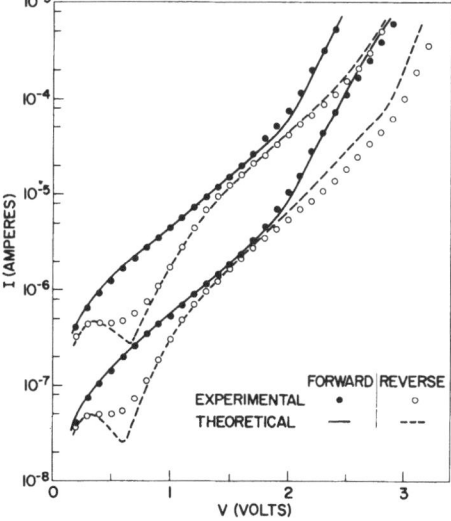

Fig. 14 Current-voltage characteristics for Al-Al_2O_3-GeTe junctions at 4.2°K. Two sets of curves are shown corresponding to samples with various oxide thickness.

expressions over several decades. We have used here above-described values of E_g and F_p and have derived that the work function is 5.1 eV for SnTe and 4.8 eV for GeTe, and the electron affinity of both materials is 4.2 eV, using 4.2 eV for the Al work-function.

We have also measured the temperature dependence of the energy gap below 100°K. The experiments were performed by passing a constant current through the tunnel junction and measuring the voltage drop across it. Here $V > F_p + E_g$, and the temperature dependence of the voltage is due to that of E_g, as seen in Fig. 16. In these junctions, the current for $V < F_p$ is independent of the temperature, indicating that the barrier remains unchanged. Figure 16 is a plot of the results for GeTe and two types of SnTe units. The results for all SnTe units tested, except for SnTe #111, were similar to those of SnTe #89. The anomalous increase in E_g at low temperatures, seen in most of the SnTe specimens, suggests the possibility of a phase transition.

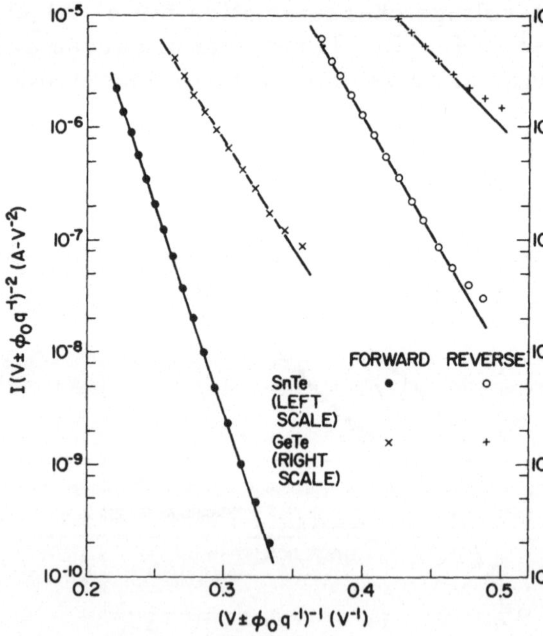

Fig. 15 Fowler-Nordheim plots for Al-Al_2O_3-SnTe and -GeTe junctions at 4.2°K on samples with an extremely thick oxide.

We have carried out a preliminary investigation of tunneling on a superconducting film of GeTe[53] with a hole concentration of $1.2 \times 10^{20}/cm^3$. Figure 17 shows the conductance as a function of applied voltages at different temperatures. At temperatures above 0.5°K the structure is entirely due to the superconducting energy gap in the Al film. At temperatures below 0.5°K, additional structure is observed which is due to the superconducting energy gap of GeTe. This was the first observation of the superconducting energy gap in a semiconductor. The superconducting energy gap extrapolated to $T = 0^\circ$K is 0.14 meV, and T_c is about 0.4°K. It is possible to carry out a program of energy gap versus temperature for varying concentrations for a system such as SnTe or GeTe.

iii) PbTe and PbSe. We have observed an atom-like energy spectrum and spin-splitting, arising from imperfections in the vicinity of the surface of epitaxial PbTe and PbSe films. In plotting conductance as a function of applied voltages for monocrystalline PbTe and PbSe tunnel junctions at liquid helium temperatures,

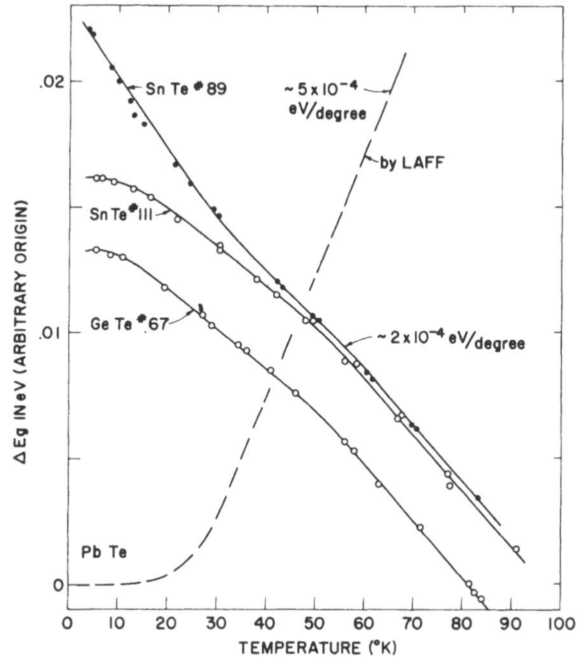

Fig. 16 Temperature dependence of energy gaps in SnTe and GeTe, measured by the tunneling technique, comparing with that of PbTe.

we have seen a series of strikingly sharp peaks, as shown in Figs. 18 and 19 for PbTe and PbSe, respectively, at energies in the forbidden gap. (Plus voltage corresponds to semiconductor positive.)

The general shape of the conductance curve is primarily determined by the energy gap and the Fermi energy, as described previously in SnTe and GeTe tunnel junctions. However, as the carrier concentration is not so high in the present case ($1 \sim 8 \times 10^{18} cm^{-3}$), surface band-bending should be taken into account. Our analysis indicates that the PbTe surface is depleted and its Fermi level at the interface is located in the neighborhood of $40 \sim 60$ meV above the valence band edge at zero bias for doping

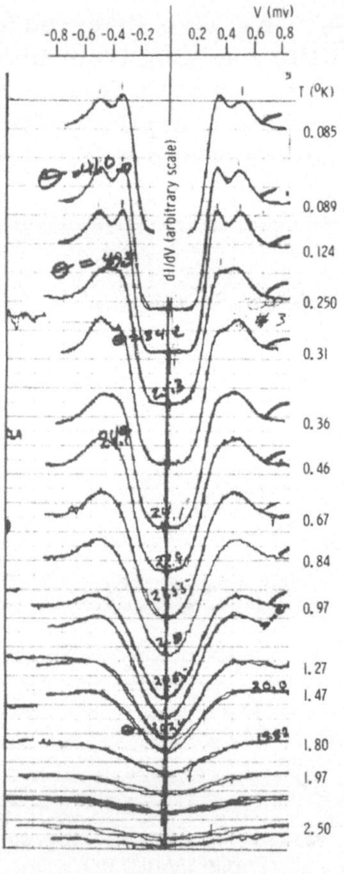

Fig. 17 Conductance versus applied voltage curves for a superconducting GeTe junction at temperatures between 0.085°K and 2.5°K.

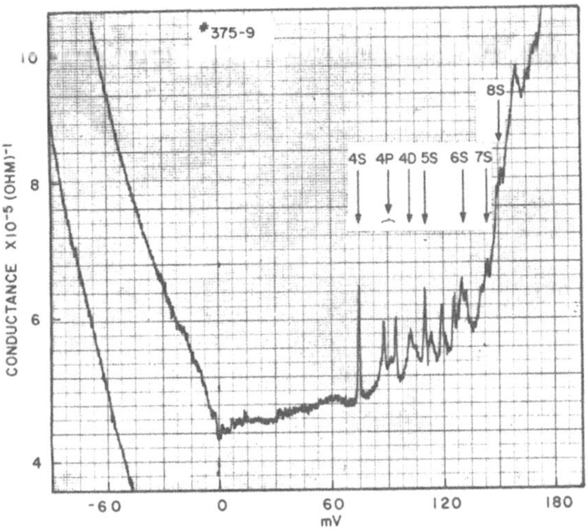

Fig. 18 Conductance versus applied voltage at 1.3°K for a PbTe junction, illustrating the general shape and the structure, S, P, and D. The far left-hand curve was plotted with the conductance level offset by 2.1 x 10^{-5} ohm^{-1}.

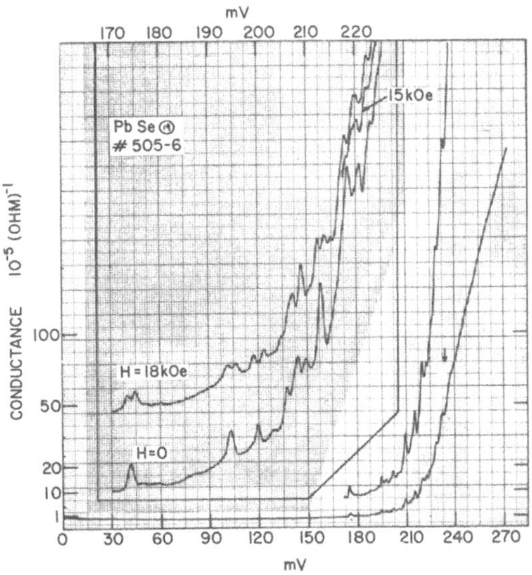

Fig. 19 Conductance versus applied voltage at 2°K for a PbSe junction. The insert shows curves at H = 0, H = 15 kOe, and H = 18 kOe on an expanded scale.

levels involved, as schematically shown in Fig. 20. With an increase in applied voltage, however, the amount of band-bending does not change appreciably. We have observed, in some junctions, a negative resistance, arising from the voltage dependence of tunneling probability.

When the Fermi level in the metal is raised by an applied voltage to an energy equal to one of the localized states, the current shows a step, resulting in a peak in conductance. We have noticed that the energy separation between successive peaks decreases in going toward the conduction band edge. The location of the band edge is clearly seen in some unit, as shown with an arrow in Fig. 19. The structure is seen to cease to exist beyond the arrow point.

Now, we would like to discuss mostly the results of PbTe which were studied in detail. The band edge, which usually can be identified within $\pm$ meV, is used here as an energy reference. It is interesting to see that application of magnetic fields up to 90 kOe

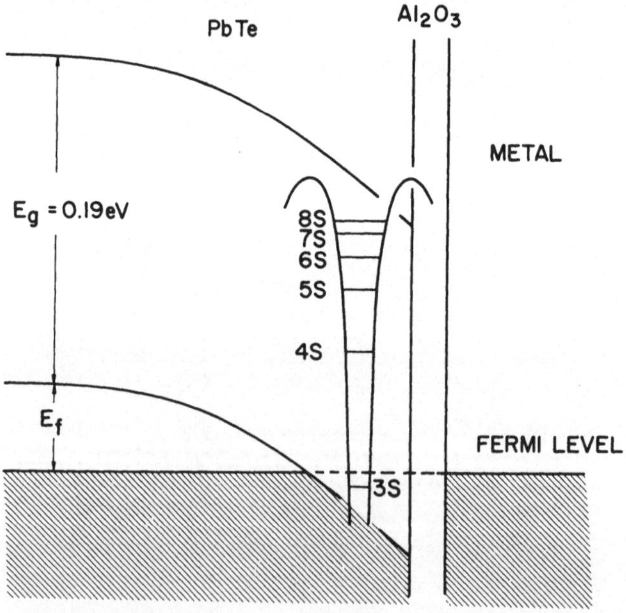

Fig. 20 Energy diagram of the PbTe junction at zero bias, including the energy levels of a deep center.

in the (100) direction perpendicular to the film clearly splits almost all peaks with no appreciable shift of the centers, as shown in Fig. 21 and in the insert of Fig. 19. The amount of the splitting is verified to be proportional to the magnetic field, yielding g values from 2 up to as high as 18. The results for five of the units are summarized in Table II, where the positions of peaks are measured in meV downwards from the conduction band edge and observed g values are shown in the parentheses. The clearest peak, corresponding to the 4S state, however, can be seen in most units and always is located at 100 $\pm$ 10 meV below the conduction band edge; thus, it serves as another energy reference.

We have analyzed the experimental results as follows: First of all, the "sharpest" peaks are identified and designated by S as in Fig. 18 and 21. This can be done without uncertainty since those are not only the "sharpest" but also have the "largest" g values. Corresponding energies for peaks agree with the following formula, if n = 3 is set for the deepest one (the closest

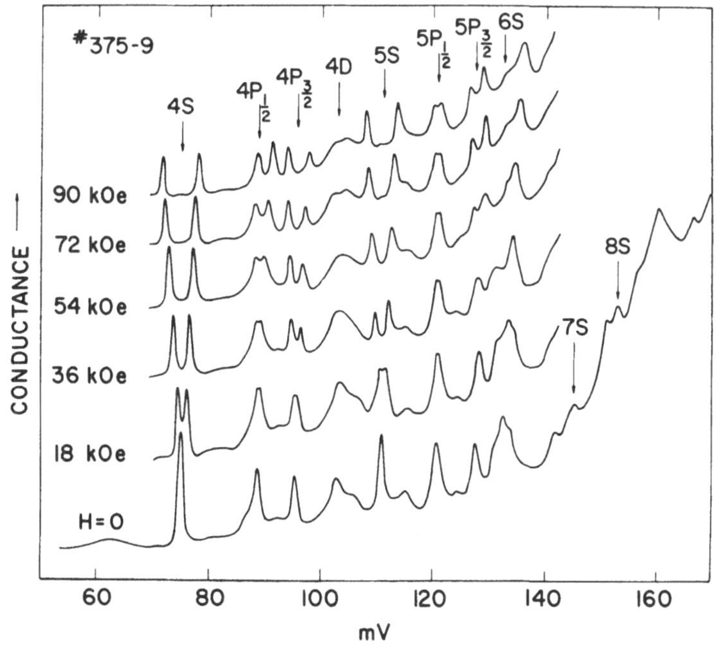

Fig. 21 Conductance curves under magnetic fields at 1.3°K on an expanded scale, offset for clarity for increments of 18 kOe.

Table II Observed energies in meV and g values in the parentheses for five of the experimental units together with assigned levels and calculated energies.

Assigned levels	375-9	446-5	446-11	430-12	438-3	Calculated energies
3S		167(8.2)	167	171		178
$3P_{1/2}$		117(∼6)	152(∼8)			
$3P_{3/2}$		110(∼7)	145(∼10)			
4S	100(12.4)	100(14.5)	100(11.5)	100(8.2)	100	100
$4P_{1/2}$	86.5(4.8)	81(∼6)	84(∼5)		86.5	
$4P_{3/2}$	80 (7.7)	78(∼7)	80(∼5)		80	
4D	72 (∼2)		72		72	
5S	64 (10.5)	66(17.2)	65(15.5)		64	64
$5P_{1/2}$	54 (2.0)				54	
$5P_{3/2}$	47 (3.8)				47	
6S	43 (∼10)	44	43(12.9)	43(18.4)	43	44.5
	41.5				40	
					37	
	33.5				32	
7S	30	30		30(12.3)	30	32.6
					28	
					25	
	25				24	
8S	23			20	22	25

one to the valence band),

$$E_n = C/n^2 \tag{10}$$

where n is an integer and C is a constant. The energy values calculated from the above equation are shown in the last column of Table II, choosing a value of the constant C for the best fit. There does not seem to be much arbitrariness in assigning the n numbers.

Secondly, the "principal" characteristic seen between the sharpest peaks consists of doublets which are usually separated by 4 ~7 meV and are designated by P in the figures and the table. A comparison of measured g values with Landé g factors for doublets in LS coupling(54) seems informative, assuming that scaling is allowed. Taking the measured g values of 4S and 4P states in unit 375-9, the ratio for the P states is 4.8/7.7 = 0.63 and the ratio for the S state to the sum of the P states is 12.4/(7.7 + 4.8) = 1, while, according to the Landé formula, the former is 0.5 and the latter is 1. (For different units, the former varies from 0.6 to 1, while the latter is always ~1.) This correlation suggests that the sharpest one is $S_{1/2}$ state and the doublet is composed of $P_{1/2}$ and $P_{3/2}$ states. We simply designate "diffused" peaks as D states.

If one takes a conventional approach for shallow donors and acceptors,(55) the constant C in Eq. 10 is given, as follows:

$$C = E_H m^* Z^2 / m_o \kappa^2 \tag{11}$$

where E_H: the hydrogenic IS ionization energy (13.6 eV), m^*/m_o: the ratio of the effective mass to the free electron mass, Ze: the charge of a center and κ: the dielectric constant. The value of C, we have obtained, is equal to 0.12 E_H. It is obviously too large if one simply uses the effective mass for conduction electrons, $m^* < 0.1\ m_o$ in PbTe, the static dielectric constant κ_o, 412 ± 40,(56) (or even the optical dielectric constant, κ_{opt}, 32.6),(57) and Z = 1. (κ_o, 250; κ_{opt}, 23 for PbSe.(58))

If the localized states are coupled to either the conduction band or the valence band states, one could expect large g values. Reported experimental $g_{//}$ components are as large as 45(59) and 57(60) for electrons, and 51(59) for holes. For PbSe the reported values are 27 and 32 for electrons and holes, respectively.(59) Therefore, if the ellipsoids are mixed,(61)

$g \doteq \frac{1}{3} g_{//}$, which generally coincides with the experimental values of 10 ~ 15 obtained here.

The epitaxial layers of PbTe and PbSe are (100) films. We have noticed that the spin-splitting is a function of direction of the magnetic field as shown in a pseudo-three-dimensional plot, Fig. 22, of conductance - vertical angle into the paper and applied voltage - horizontal. Each curve here was taken by rotating the magnetic field at 10 degree increment. When rotating the magnetic field in four different planes, (100) $(1\bar{1}0)$ $(0\bar{1}0)$ and $(\bar{1}\bar{1}0)$, g values change as shown in four curves, A, B, C, and D of Fig. 23. The results indicate that the g value does not have the full symmetry of the crystal and has a nearly ellipsoidal g tensor and the maximum at one of the <111> directions. If, in fact, we are observing a single type of center, the present observation may suggest that the centers consist of clusters of two or more Pb or Te vacancies, orientated in one of the <111> directions.

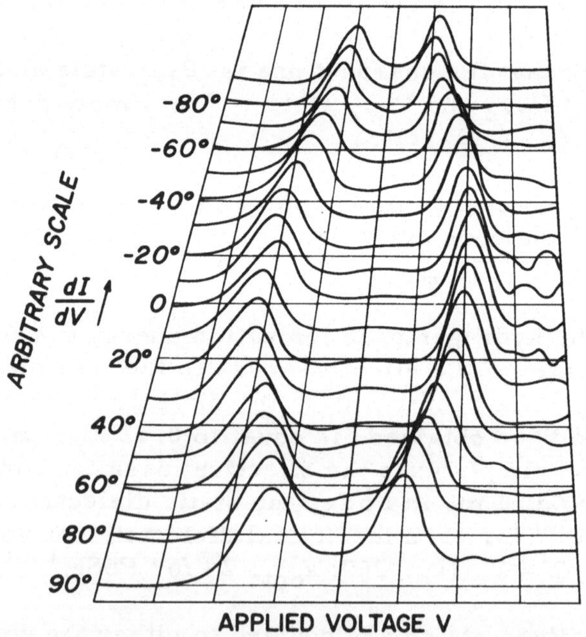

Fig. 22 Orientation effect of the spin-splitting in PbTe, indicated by a pseudo-three-dimensional plot. Conductance(vertical) and applied voltage(horizontal), where each curve was taken by rotating the magnetic field at ten degree increment in a certain plane.

The ratio of the maximum to the minimum g values in the rotation data ranges from 2 to 4 for PbTe, and 1 to 1.3 for PbSe. Although the anisotropy of the g tensor in these materials is little known, it is expected to be related to that of the effective mass. The larger ratio for values of g in PbTe which also has a larger mass anisotropy thus indicates the coupling of the observed levels to the bands. Deep centers, in general, have not yet been a subject of theoretical investigation even in well-known semiconductors, while extensive studies on shallow impurities have been carried out.[55] It has been known that PbTe (or PbSe), rock-salt crystal, has Pb as well as Te (or Se) vacancies, providing acceptor and donors, respectively.[62] Therefore, it is plausible that there exist complex clusters of those centers as in the alkali halides, as above-described with the rotation data. At any rate, we believe that further studies of the rotation effect would be very fruitful for detailed understanding of structure of the centers.

We are quite certain that the center is not situated in the Al_2O_3 because the structure is only seen for PbTe and PbSe. The sharpness and intensity of each peak in the conductance may depend on the average as well as the distribution in depth of the clusters from the surface.

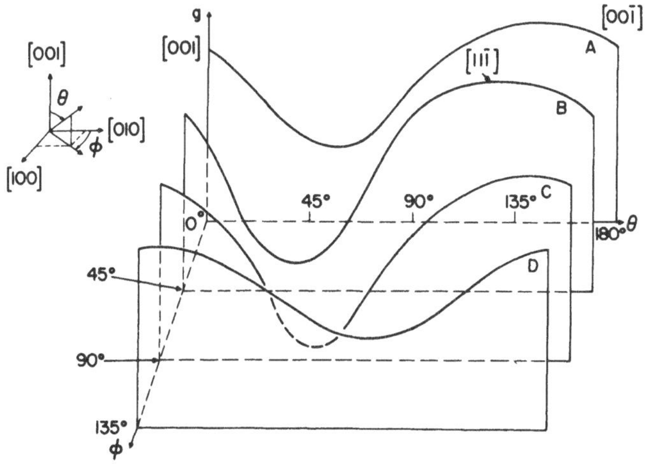

Fig. 23 The g values in PbTe as the magnetic field is rotated in four different planes, (100) at $\phi = 0$, $(1\bar{1}0)$ at $\phi = 45^\circ$, $(0\bar{1}0)$ at $\phi = 90^\circ$, and $(\bar{1}\bar{1}0)$ at $\phi = 135^\circ$.

TUNNELING SPECTROSCOPY FOR Eu CHALCOGENIDES

We have observed an entirely new magneto-tunneling in tunnel junctions of magnetically ordered insulators such as Eu chalcogenides.(63) The study, we believe, has opened up another area in which tunneling spectroscopy provides useful information about the tunneling medium itself.

EuSe and EuS are involved here, of which the magnetic transition temperatures have been reported to be 4.7 and 16.5°K, respectively. Experimental junctions were constructed simply by successive evaporations, metal, the Eu chalcogenide, and metal, on a heated sapphire substrate. Thus the thin film sandwiched between metal electrodes provides a barrier for electrons. The thickness of the Eu chalcogenides ranges from 200 to 600 Å and the junction area is approximately $4 \times 10^{-4} cm^2$.

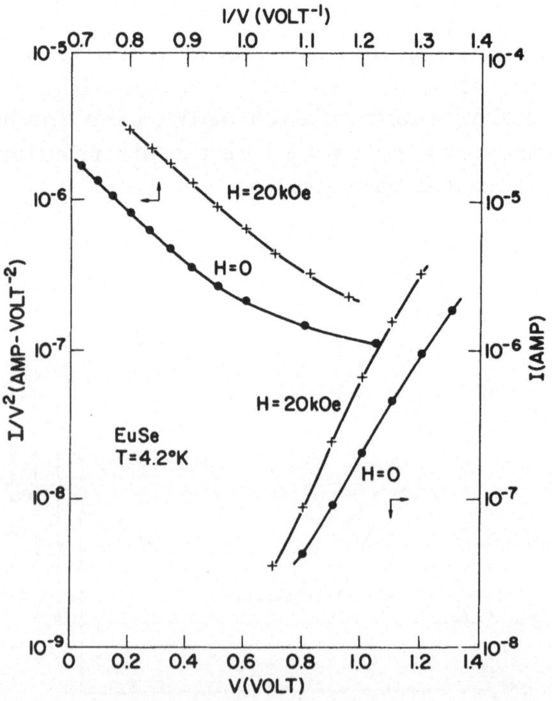

Fig. 24 Log I versus V for EuSe at H = 0 and 20 kOe in the lower right-hand side, log (I/V^2) versus I/V for EuSe at H = 0 and 20 kOe in the upper left-hand side, at T = 4.2°K.

The current-voltage characteristics on a semi-log plot for an EuSe junction at 4.2°K are shown in the lower right-hand side of Fig. 24. It is seen that the curve markedly shifts to lower voltages with a magnetic field of 20 kOe. We have found that current-voltage curves at high voltages always approach an asymptote expressed by Fowler-Nordheim tunneling. Figure 25 shows the shape of the potential barrier of the height, ψ, at a high applied voltage, V.

We have verified that the barrier height changes rather rapidly near the transition temperature and also that the application of magnetic field considerably affects the barrier height below the transition temperature, making it small with increase in field, particularly for EuSe, as follows; the voltage at a given current level (normalized to its value at 4.2°K) is plotted as a function of temperature with and without a magnetic field. The results are shown in Fig. 26, where the dashed line indicates the EuSe junction with 20 kOe. The magnetic field effect on the barrier

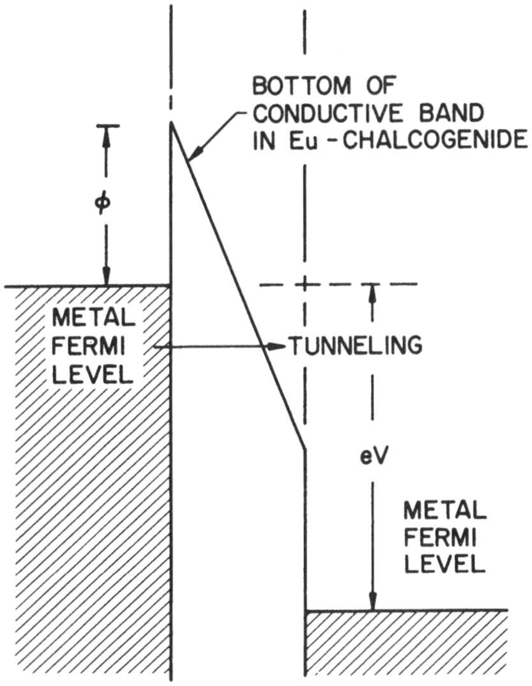

Fig. 25 The potential barrier of the junction at an applied voltage V. ϕ, barrier height.

height of EuSe is rather dramatic. This is clearly shown in Fig. 27, where the voltage at a given current level (normalized to its value at H = 0) V_H/V_0 is plotted versus H for an EuSe junction, where

$$V_H/V_0 = (\psi_H/\psi_0)^{3/2}.$$

The decrease of about 35% in voltage means that the barrier height is lowered by as much as 25% with 20 kOe.

Now, in the Eu chalcogenides, it has been suggested that the 5d states provide a conduction band and its width was recently calculated to be a few electron volts. Thus, with this model, it is believed that the barrier height ψ in our junctions is determined by the energy difference between the Fermi level in the metal and the 5d band. The magnetic field effect appears to confirm the shift, due to spin ordering of the 4f electrons, of the 5d band with respect to the metal Fermi level, thus with respect to the vacuum level. The small magnetic field effect in the EuS junctions probably comes from the fact that not only is EuS a normal ferromagnet but also the applied field is much lower than its Weiss field. The decrease in the barrier height is indeed reminiscent of a shift of all 4.2°K optical absorption curves

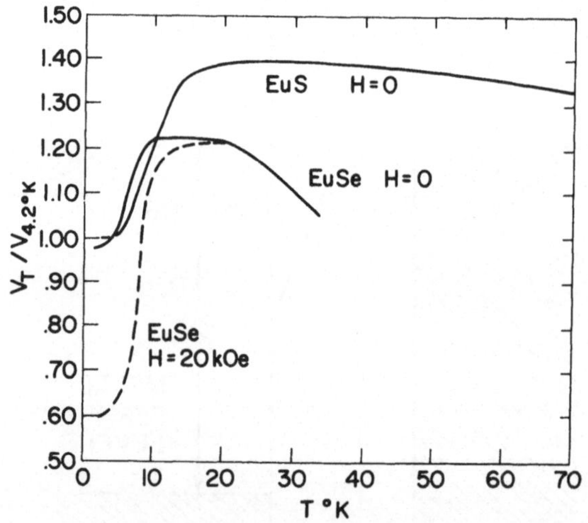

Fig. 26 $V_T/V_{4.2°K}$, the voltage at a given current level (normalized to its value at 4.2°K), versus temperature, for EuSe at H = 0 and H = 20 kOe and for EuS at H = 0.

toward longer wavelengths with applied magnetic fields. The difference between the fields applied parallel and perpendicular to the junction plane, as is shown in Fig. 27, can be interpreted as being due to the demagnetizing effect in the film. ($4\pi M$ in the bulk is as large as 13.8 kOe.)

CONCLUSIONS

We have demonstrated that the technique of tunneling spectroscopy yields not only corroborative but also entirely new information on the band edge energies, the energy gaps, the Fermi energies, the workfunctions, the magnetic levels, the deep impurity levels and their changes with applied magnetic field.

In order to carry out this spectroscopy successfully, it is of utmost importance to fabricate decent junctions of an extremely thin tunneling barrier with refined techniques. Since the tunneling current I or the conductance dI/dV (or the second derivative

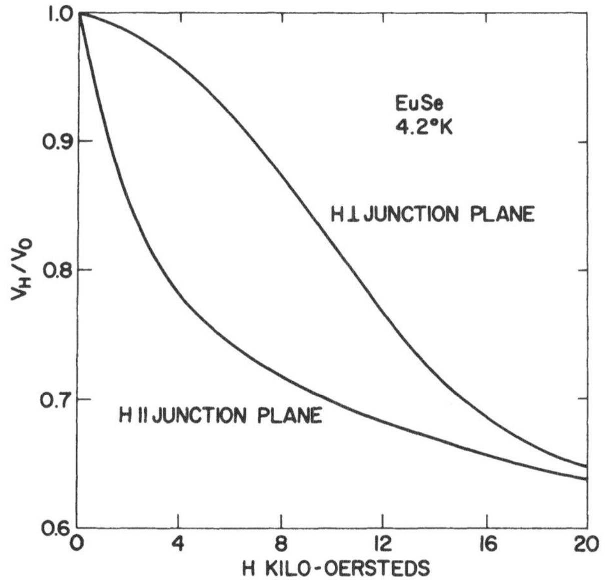

Fig. 27 V_H/V_o, the voltage at a given current level (normalized to its value at H = 0), versus magnetic field H, for EuSe at 4.2°K.

d^2I/dV^2) is plotted versus the applied voltage, obtained information is expressed as a function of energy. The Fermi level here gives an unequivocal energy reference which is usually not provided by optical spectroscopy. In this respect, the tunneling and optical techniques might be considered as dc and ac measurements, respectively.

The author wishes to express his gratitude to many colleagues at IBM for helpful discussions, particularly P. J. Stiles and L. L. Chang for their cooperation in preparing the manuscript.

REFERENCES

1. L. Esaki, Phys. Rev. 109, 603 (1958); L. Esaki, Solid State Physics in Electronics and Telecommunications, Proceedings of an International Conference held in Brussels, June 2-7, 1958, edited by M. Desirant and J. L. Michels. Volume 1, Semiconductors, Part 1 (Academic Press, New York, 1960) p. 514.

2. C. Zener, Proc. Roy. Soc. (London) 145, 523 (1934). W. V. Houston, Phys. Rev. 57, 184 (1940).

3. I. Giaever, Phys. Rev. Letters 5, 147 (1960).

4. B. D. Josephson, Physics Letters 1, 251 (1962).

5. T. Yajima and L. Esaki, J. Phys. Soc. Japan 13, 1281 (1958).

6. W. A. Harrison, Phys. Rev. 123, 85 (1961).

7. E. O. Kane, J. Phys. Chem. Solids 12, 181 (1959), and J. Appl. Phys. 32, 83 (1961).

8. L. Esaki and Y. Miyahara, Solid State Elect. 1, 13 (1960).

9. R. L. Batdorf, G. C. Dacey, R. L. Wallace and D. J. Walsh, J. Appl. Phys. 31, 613 (1960).

10. N. Holonyak and I. A. Lesk, Proc. IRE 48, 1405 (1960).

11. H. P. Kleinknecht, Solid-State Electronics 2, 133 (1961).

12. R. H. Rediker and A.R. Calawa, J. Appl. Phys. 32, 2189 (1961).

13. L. Armstrong, IRE Trans. ED-9, 114 (1962).

14. R. F. Rutz, IBM J. Res. Develop. 8, 539 (1964).

15. L. Esaki and Y. Miyahara, unpublished and also Solid State Electronics 1, 13 (1960); N. Holonyak, I. A. Lesk, R. N. Hall, J. J. Tiemann and H. Ehrenreich, Phys. Rev. Letters 3, 167 (1959).

16. G. G. Macfarlane, T. P. McLean, J. E. Quarrington and V. Roberts, J. Phys. Chem. Solids 8, 388 (1959) and J. R. Haynes, M. Lax and W. F. Flood, J. Phys. Chem. Solids 8, 392 (1959).

17. L. V. Keldysh, Zh. eksper, teor, Fiz. 34, 962 (1958). Translation: Soviet Physics--JETP 7, 665 (1958); E. O. Kane, J. Appl. Phys. 32, 83 (1961); L. Kleiman, Phys. Rev. 140, A637 (1965).

18. See, for instance, Conference on Electron Tunneling in Solids, Philadelphia (1961), J. Phys. Chem. Solids 23, 173 (1962); Proc. Int. Conf. Semiconductor Physics, Prague (1960); NATO Summer School and Conference on Electron Tunneling in Solids, Riso, Denmark (1967), Proceedings of the Summer School, to be published, and Abstracts in the Conference, Solid State Communications 5, No. 10, i-xvi (1967).

19. L. V. Keldysh, Zh. eksper, teor, Fiz. 33, 994 (1957).

20. P. J. Price and J. M. Radcliffe, IBM J. Res. Develop. 3, 364 (1959).

21. D. R. Fredkin and G. H. Wannier, Phys. Rev. 128, 2054 (1962).

22. R. T. Shuey, Phys. Rev. 137, A1268 (1965).

23. Y. Takeuti and H. Funada, J. Phys. Soc. Japan 20, 1854 (1965).

24. W. Franz, Z. Naturforsch 13, 484 (1958); L. V. Keldysh, Zh. eksper teor. Fiz. 34, 1138 (1958). Translation: Soviet Physics-JETP 7, 788 (1958).

25. J. I. Pankove, Phys. Rev. Letters 9, 283 (1962).

26. C. T. Sah, Phys. Rev. 123, 1594 (1961).

27. R. S. Classen, J. Appl. Phys. 32, 2372 (1961).

28. J. Bardeen, Phys. Rev. Letters 6, 57 (1961).

29. For instance, Proc. 8th and 9th Int. Conf. on Low Temp. Phys. (London 1962 and Columbus, Ohio, 1964, respectively) and also Ref. 18.

30. S. von Molnar, W. A. Thompson and A. S. Edelstein, Appl. Phys. Letters 11, 163 (1967).

31. R. Stratton, J. Phys. Chem. Solids 23, 1177 (1962); J. G. Simmons, J. Appl. Phys. 34, 1793 (1963).

32. J. C. Fisher and I. Giaever, J. Appl. Phys. 32, 172 (1961); D. Meyerhofer and S. A. Ochs, J. Appl. Phys. 34, 2535 (1963); S. R. Pollack and C. E. Morris, J. Appl. Phys. 35, 1502 (1964).

33. P. V. Gray, Phys. Rev. 140, A179 (1965).

34. L. Esaki and P. J. Stiles, Phys. Rev. Letters 16, 1108 (1966); L. L. Chang, L. Esaki and F. Jona, Appl. Phys. Letters 9, 21 (1966); P. J. Stiles, L. Esaki and W. E. Howard, Proc. 10th Int. Conf. Low Temp. Phys., Moscow (1966); L. Esaki, J. Phys. Soc. Japan 21 (supplement), 589 (1966); L. L. Chang, P. J. Stiles and L. Esaki, J. Appl. Phys. 38, 4440 (1967).

35. L. Esaki, P. J. Stiles, and L. L. Chang, Phys. Rev. Letters 20, 1108 (1968).

36. L. Esaki and P. J. Stiles, Phys. Rev. Letters 14, 902 (1965). L. Esaki and P. J. Stiles, Phys. Rev. Letters 16, 574 (1966). J. J. Hauser and L. R. Testardi, Phys. Rev. Letters 20, 12 (1968).

37. L. Esaki, L. L. Chang, P. J. Stiles, D. F. O'Kane, and N. Wiser, Phys. Rev. 167, 637 (1968).

38. J. W. Conley, C. B. Duke, G. D. Mahan, and J. J. Tiemann, Phys. Rev. 150, 466 (1966); F. A. Padovani and R. Stratton,

Solid State Electron. 9, 21 (1966).

39. G. D. Mahan and J. W. Conley, Appl. Phys. Letters 11, 29 (1967); J. W. Conley and G. D. Mahan, Phys. Rev. 161, 681 (1967).

40. R. N. Hall, J. H. Racette and H. Ehrenreich, Phys. Rev. Letters 4, 456 (1960).

41. C. B. Duke, GE Report 67-C-305 (1967).

42. R. C. Jaklevic and J. Lamb, Phys. Rev. Letters 17, 1139 (1966).

43. W. A. Thompson, Phys. Rev. Letters 20, 1085 (1968).

44. A. F. G. Wyatt, Phys. Rev. Letters 13, 401 (1964).

45. J. M. Rowell and L. Y. L. Shen, Phys. Rev. Letters 17, 15 (1966).

46. J. Appelbaum, Phys. Rev. Letters 17, 91 (1966); P. W. Anderson, Phys. Rev. Letters 17, 95 (1966); J. Solyom and A. Zawadowski, KFK1-Report 14 (1966), Central Research Institute for Physics, Budapest.

47. C. B. Duke, S. D. Silverstein and Alan J. Bennet, Phys. Rev. Letters 19, 315 (1967).

48. M. Cardona and D. L. Greenaway, Phys. Rev. 133, A1685 (1964).

49. M. H. Cohen, L. M. Falicov and S. Golin, IBM J. Res. Develop. 8, 215 (1964).

50. R. B. Schoolar and J. N. Zemel, J. Appl. Phys. 35, 1848 (1964).

51. D. J. BenDaniel and C. B. Duke, Phys. Rev. 160, 679 (1967).

52. A. L. Jain, Phys. Rev. 114, 1518 (1959); D. M. Brown and S. J. Silverman, Phys. Rev. 136, A290 (1964).

53. P. J. Stiles, L. Esaki, and J. Schooley, Phys. Letters 23, 206 (1966).

54. See, for example, G. Herzberg, Atomic Spectra and Atomic Structure, (Dover, New York, 1944) p. 109.

55. W. Kohn, Solid State Phys. (Academic Press, New York, 1957) Vol. 5, p. 258.

56. W. Cockran, R. A. Cowley, G. Dolling and M. M. Elcombe, Proc. Royal Soc. London 293A, 433 (1966).

57. J. R. Dixon and H. R. Rieale, Phys. Rev. 138A, 873 (1965).

58. J. N. Zemel, J. D. Jensen, and R. B. Schoolar, Phys. Rev. 140A, 330 (1965).

59. K. F. Cuff, M. R. Ellet, C. D. Kuglin, and L. R. Williams, Proc. International Conf. Phys. Semiconductors, Paris (1964) (Academic Press, New York, 1964) p. 677.

60. C. K. N. Patel and R. E. Slusher, Bull. Am. Phys. Soc. 13, 480 (1968).

61. L. M. Roth, Phys. Rev. 118, 1534 (1960).

62. W. W. Scalon, Solid State Phys. (Academic Press, New York 1959) Vol. 9, p. 83.

63. L. Esaki, P. J. Stiles, and S. von Molnar, Phys. Rev. Letters 19, 539 (1967).

ENERGY BAND STRUCTURE OF SEMICONDUCTORS

F. Herman

Lockheed Palo Alto Research Laboratory

Palo Alto, California

During the past three years we have developed greatly improved methods for determining the electronic energy band structure of crystals. In the course of perfecting these methods, we investigated the electronic band structure and related physical properties of over 50 crystalline solids, including the diamond-type crystals; various polytypes of SiC; a large number of III-V, II-VI, and I-VII compounds having the sphalerite structure; additional II-VI and I-VII crystals having the wurtzite and rock-salt structures; several II-IV compounds (anti-fluorite structure); a number of IV-VI compounds (rock-salt structure); and various solid rare gases (face-centered cubic structure). Most of the substances studied were semiconductors with band gaps in the range from 0.5 to 5 eV; some were small band-gap semiconductors or semi-metals; the others were insulators with band gaps larger than 5 eV. All these materials are of scientific interest; many, especially Si, Ge, SiC, and various III-V, II-VI, and IV-VI compounds, are also of considerable technological interest.

In pursuing these studies, our overall objective was to determine physically reliable energy band models for a wide variety of crystals, with the expectation that these models would find wide use in the interpretation of existing experimental information, and in the planning of future experimental programs. We believe that our results do indeed provide a useful and reliable theoretical guide for present and future experimental investigations, as well as a convenient framework for more specialized theoretical studies of the electronic and optical properties of crystals.

For most of the crystals that we have examined, it is possible to obtain a qualitatively reliable energy band model by carrying out a first-principles OPW (orthogonalized plane wave) energy band calculation using a simple but physically realistic crystal potential. It is usually sufficient to use a trial crystal potential having the form of a spatial superposition of overlapping atomic potentials, but we have the option of iterating our solutions so as to obtain self-consistent crystal potentials. We have developed powerful methods for performing self-consistent OPW energy band calculations, electronic density of states calculations, and optical spectrum calculations. For most of the semiconductors that we have studied, the energy band structure calculated from first principles is usually consistent with experiment to within about 0.5 eV - on the average - over a 10 to 15 eV range. This is quite remarkable, considering the many physical and mathematical approximations that underlie a first-principles energy band calculation. The generally satisfactory nature of our results is a consequence of the care that we have taken with our numerical work, and with our choice of physical assumptions and mathematical techniques.

In practice, it is convenient to determine the electronic energy levels from first principles only at a selected set of points in the reduced zone, and then to map out the band structure in the remainder of the zone with the aid of a suitably chosen interpolation scheme. Although we have experimented with a variety of interpolation schemes, we favor a modified version of the pseudopotential interpolation scheme, since this is essentially a simplified form of the full-scale OPW method. By using an interpolated band structure which has been fitted to our first-principles energy levels, we are usually able to account for a wide variety of experimental results, including the gross features of optical and photoemission spectra.

If we wish to improve the accuracy of our first-principles energy band model, so as to be able to study the finer details of experimental spectra, or for other reasons, we modify our first-principles crystal potential in such a way that it leads to an energy band model which reproduces the best-known features of the experimental band structure (typically the direct and indirect band gaps). If this empirical modification of our first-principles crystal potential is carried out properly, the <u>empirically-refined first-principles</u> energy band model will be in closer agreement with experiment not only in those regions which have been deliberately adjusted to experiment, but in other regions as well. It is usually possible to carry out this process knowing only two or three key interband transition energies. Of course, the more experimental information at our disposal, the easier it is to obtain an energy band model that is accurate throughout the entire reduced zone over a wide energy range. The advantages of our method (which is typically about 95 percent "first principles" and only about 5 percent "empirical" in character) over a purely empirical method are discussed at length in our recent series of papers. In effect, our empirically-refined first-principles approach combines the best features of a purely first-principles approach with the best features of a purely empirical approach.

In addition to obtaining improved estimates of the band structures in those regions which are poorly understood from an experimental point of view, our work leads to electronic wave functions and other theoretical information that can be used to calculate various quantities of physical and chemical interest, such as effective mass tensors, net deformation potential differences, spin-orbit splittings, electronic charge distributions, and X-ray scattering factors.

During the past three years we have published a dozen papers dealing with our empirically-refined first-principles approach to energy band problems (see list of references at the end of this paper). In Table I we have listed the various crystals that we have studied, as well as the references in which these crystals are treated. For some materials we have already reported detailed results, such as the overall energy band structure and optical spectrum; for others we have so far reported only a few numerical results, for example, key energy levels and spin-orbit splittings. Reference 1 provides a comprehensive introduction to the present series of papers, as well as detailed illustrations of our approach to energy band problems.

We are still in the process of studying the electronic and optical properties of several of the crystals listed in Table I. (Such crystals are marked with an asterisk.) We intend to publish our results for these crystals in due course. At the present time we are preparing two additional scientific papers, one on the II-IV compounds (Mg_2Si, Mg_2Ge, Mg_2Sn, and Mg_2Pb), and another on high band gap III-V compounds (BN, BP, AlP, AlAs, AlSb, InP, and InAs). We are also working on a book in which we present a systematic account of the essential mathematical and physical features of our energy band calculations, as well as a comprehensive summary of our results.

The research reported in this paper was supported in part by the Lockheed Independent Research Fund; in part by the Air Force Cambridge Research Laboratories, Office of Aerospace Research, under Contract No. AF 19(628)-5750; and in part by the Aerospace Research Laboratories, Office of Aerospace Research, under Contract No. AF 33(615)-5072.

TABLE I

Crystal	Reference												
C, Si, Ge, αSn	1	2	3		5	6	7					12	
SiC, BN, BP			3								11		*
AlP, AlAs, AlSb													*
GaP, GaAs, GaSb					5		7	8					
InP, InAs, InSb							7						*
ZnS, ZnSe, ZnTe					5		7			10			
CdS, CdSe, CdTe				4	5		7			10			
HgS, HgSe, HgTe					5		7						
GeTe, SnTe									9				
PbTe, PbSe, PbS									9				
Mg_2X, X = Si, Ge, Sn, Pb													*
CuCl, CuBr, CuI, AgI							7						
ZnO, MgO, KCl, AgCl, NaI													*
Al, Ar, Kr, Xe													*

*Work still in progress. Results will be published in due course.

REFERENCES

1. F. Herman, R. L. Kortum, C. D. Kuglin, and R. A. Short, "New Studies of the Band Structure of Silicon, Germanium, and Grey Tin," Quantum Theory of Atoms, Molecules, and the Solid State, P. P. Löwdin, ed. (Academic Press, New York, 1966), pp. 381-428

2. F. Herman, R. L. Kortum, C. D. Kuglin, and R. A. Short, "New Studies of the Band Structure of the Diamond-Type Crystals," J. Phys. Soc. Japan 21 (Supplement) (1966) [Kyoto Semiconductor Conference Issue], pp. 7-14

3. F. Herman, R. L. Kortum, and C. D. Kuglin, "Energy Band Structure of Duamond, Cubic Silicon, and Germanium," International J. Quantum Chemistry, 1S (1966), pp. 533-566

4. J. L. Shay, W. E. Spicer, and F. Herman, "Photoemission Study of the Electronic Structure of CdTe," Phys. Rev. Letters 18 (1967), pp. 649-654

5. F. Herman, R. L. Kortum, C. D. Kuglin, and J. L. Shay, "Energy Band Structure and Optical Spectrum of Several II-VI Compounds," II-VI Semiconducting Compounds, 1967 International Conference, D. G. Thomas, ed. (W. A. Benjamin, Inc., New York, 1967), pp. 503-551; 1428-1440

6. F. Herman, R. L. Kortum, C. D. Kuglin, and J. P. Van Dyke, "New Studies of the Energy Band Structure of Tetrahedrally-Bonded Semiconductors and Semimetals," Energy Bands in Metals and Alloys, L. H. Bennett and J. T. Waber, eds. (Gordon and Breach, New York, 1968), pp. 19-42

7. F. Herman, R. L. Kortum, C. D. Kuglin, J. P. Van Dyke, and S. Skillman, "Electronic Structure of Tetrahedrally-Bonded Semiconductors: Empirically Adjusted OPW Energy Band Calculations," Methods in Computational Physics, B. Alder, S. Fernbach, M. Rotenberg, eds. (Academic Press, New York, 1968), Vol. 8, pp. 193-249

8. F. Herman and W. E. Spicer, "Spectral Analysis of Photoemissive Yields in GaAs and Related Crystals," Phys. Rev. 174 (1968). pp. 906-908

9. F. Herman, R. L. Kortum, I. B. Ortenburger, and J. P. Van Dyke, "Relativistic Band Structure of GeTe, SnTe, PbTe, PbSe, and PbS," Proceedings of the International Conference on IV-VI Semiconducting Compounds, Paris, July 1968 (special issue of Journal de Physique), in press (publication date: January 1969), pp. 61-76

10. D. J. Stukel, R. N. Euwema, T. C. Collins, F. Herman, and R. L. Kortum, "Self-Consistent OPW and Empirically-Refined OPW Band Models for Cubic ZnS, ZnSe, CdS, and CdSe," Phys. Rev. (in press)

11. F. Herman, J. P. Van Dyke, and R. L. Kortum, "Electronic Structure and Optical Spectrum of Silicon Carbide," Proceedings of the International Conference on Silicon Carbide, Pennsylvania State University, October 1968, Materials Research Bulletin, January 1969 (special issue), pp. 1-12

12. F. Herman and J. P. Van Dyke, "New Interpretation of the Electronic Structure and Optical Spectrum of Amorphous Germanium," Phys. Rev. Letters 21 (1968) pp. 1575-1578

INTERBAND MAGNETO-OPTICS IN SMALL BAND GAP SEMICONDUCTORS AND SEMIMETALS

C. R. Pidgeon

Francis Bitter National Magnet Laboratory*

Massachusetts Institute of Technology, Cambridge, Mass.,USA

INTRODUCTION

In the past few years interband and intraband magneto-optical experiments have been successful in determining information about the electronic energy band structure of a variety of semiconductors and semimetals. Representative general reviews of this work have been given by Lax,[1] Dresselhaus and Dresselhaus,[2] and Smith.[3] The two classes of experiments are complementary and both may be necessary for a complete picture:

(i) Intraband free carrier experiments (such as cyclotron resonance or free carrier Faraday rotation) in general give direct information about the effective mass or curvature of a single populated band.

(ii) Interband magneto-optical experiments give energy gaps directly, and reduced masses(and g-values) for valence to conduction band transitions over a wide energy range, at different symmetry points in the Brillouin zone. In particular the interband magnetoreflection technique is necessary for opaque materials such as metals and semimetals, where transmission experiments are not possible. It has the advantage that it can be used to probe both full and empty bands, inaccessible by Fermi studies of the de Haas van Alphen type.

In these lectures I wish to discuss the interpretation of interband magneto-optical experiments in small band gap semiconductors and semimetals with strongly coupled and degenerate energy bands. Specific reference will be made to recent reflection and electroreflection measurements on InSb, HgTe and Grey Tin.

---

*Supported by the U.S. Air Force Office of Scientific Research.

We consider first the simplest possible situation of a semiconductor with a single parabolic conduction band. The ϵ-$\vec{k}$ dispersion relation for this band is of the form:

$$\epsilon = \epsilon_{co} + \frac{\hbar^2 k^2}{2m_c} \quad , \tag{1}$$

where m_c is the effective mass of the electron,

$$\frac{1}{m_c} = \frac{1}{\hbar^2}\frac{d^2E}{dk^2} \quad . \tag{2}$$

In the presence of an external magnetic field, H, this band is coalesced into one dimensional sub-bands (Landau levels):

$$\epsilon = \epsilon_{co} + \frac{\hbar^2 k_z^2}{2m_c} + (n + \tfrac{1}{2})\,\hbar\omega_c \pm \tfrac{1}{2} g_c \beta H \quad , \tag{3}$$

where H is in the z-direction, n is the Landau orbital quantum number, and g_c is the effective g-value. Here $\omega_c = eH/m_c c$ is the cyclotron frequency.

We consider an idealized semiconductor with simple parabolic bands. Direct valence to conduction band transitions, associated with optical excitation, may now occur between Landau sub-bands. If one measures magnetoabsorption or magnetoreflection, one obtains characteristic oscillatory spectra associated with these transitions.

The wave functions associated with the Landau states are proportional to harmonic oscillator functions with quantum number n, so that the selection rule for allowed transitions is: $\Delta n = 0$. A plot of the photon energies of these transitions (i.e., the magnetoreflection maxima or transmission minima) as a function of H gives straight line plots which extrapolate to an energy corresponding to the energy gap, ϵ_g, at zero field. The gradients of these lines give the reduced mass and g-factor for this pair of bands.

In order to understand the meaning of these quantities, and in particular to treat the more complex situation of interacting degenerate energy bands, it is necessary to go into the formalism of effective mass theory. We consider first the situation in the absence of a magnetic field, and derive expressions for the effective mass by two methods-- $\vec{k}\cdot\vec{p}$ perturbation theory[4] and the canonical transformation method of Luttinger and Kohn.[5] A further review of these methods has been given by McLean.[6]

$\vec{k}\cdot\vec{p}$ PERTURBATION THEORY

We consider the Bloch electron moving in the periodic crystal lattice potential $V(\vec{r})$ described by the Schrödinger equation:

$$\left(\frac{p^2}{2m} + V(\vec{r})\right)\psi_{j,k}(\vec{r}) = \epsilon_j \psi_{j,k}(\vec{r}) \tag{4}$$

for the jth band, where

$$\psi_{j,k}(\vec{r}) = u_{j,k}(\vec{r}) e^{i\vec{k}\cdot\vec{r}} , \tag{5}$$

and $\vec{p} \equiv \hbar/i\nabla$. $u_{j,k}(\vec{r})$ is a function with the periodicity of the lattice. Substitution of Eq. (5) in Eq. (4), and operating, leads to

$$\left\{ \frac{p^2}{2m} + \frac{\hbar}{m}\vec{k}\cdot\vec{p} + \frac{\hbar^2 k^2}{2m} + V(\vec{r}) \right\} u_{j,k}(\vec{r}) = \epsilon_j u_{j,k}(\vec{r}). \tag{6}$$

This is the equation for the cell periodic functions and is of the form:

$$(\mathcal{H}_o + \mathcal{H}_{k \cdot p})\, u_{j,k} = \epsilon_j u_{j,k} \; . \tag{7}$$

We expand in terms of the complete set of band edge functions, $u_{i,o}(\vec{r})$, and treat $\mathcal{H}_{\vec{k}\cdot\vec{p}}$ by perturbation theory near $\vec{k} = 0$:

$$u_{j,k} = \sum_i A_i(\vec{k}) u_{i,o} \; , \tag{8}$$

where i runs over all bands.

Expanding to order k^2 we get the $\epsilon - \vec{k}$ dispersion relation for a simple parabolic band j. This defines the effective mass and the effective mass approximation:

$$\epsilon_j(k) = \epsilon_j(0) + k_\alpha D_{jj}^{\alpha\beta} k_\beta \; , \tag{9}$$

where

$$D_{jj}^{\alpha\beta} \equiv \frac{\hbar^2}{2m_o}\delta_{\alpha\beta} + \frac{\hbar^2}{m_o^2}\sum_{i \neq j} \frac{p_{ji}^{\alpha}(0) p_{ij}^{\beta}(0)}{\epsilon_j - \epsilon_i} \tag{10}$$

is the reciprocal effective mass tensor, and α, β run over x, y, and z. $p_{ji} \equiv \langle u_{j,o} | p | u_{i,o} \rangle$, and the linear term in the above expansion, $\frac{\hbar}{m} k \cdot p_{jj}(0)$, is equal to zero for materials with inversion symmetry, since

$$p_{jj}(0) = \frac{\partial \epsilon_j}{\partial k} = 0$$

for an extremum.

We have neglected terms in the wave function of the order of $[\mathcal{H}_{\vec{k}\cdot\vec{p}}/(\varepsilon_j - \varepsilon_i)]^2$ in cutting off the expansion to order k^2. This is equivalent to requiring k to be much smaller than the width of the Brillouin zone, or the reciprocal lattice spacing, i.e., $k \ll 1/a_L$.

An alternative method for getting the same result, which is particularly useful for treating the effect of an external perturbation (e.g., a static magnetic field) is the canonical transformation method of Luttinger and Kohn (LK).[5]

EFFECTIVE MASS BY A CANONICAL TRANSFORMATION

Substituting the expansion of $u_{j,k}$ (Eq. (8)) into Eq. (6) and making use of the orthonormality of the band edge functions, we have the following equation for the A's:

$$\left(\varepsilon_j(0) + \frac{\hbar^2 k^2}{2m}\right)A_j(\vec{k}) + \sum_i \frac{\hbar}{m}\vec{k}\cdot p_{ji}(0)A_i(\vec{k}) = \varepsilon_j A_j(\vec{k}) \quad . \quad (11)$$

In matrix form this may be written

$$\mathcal{H}A = \varepsilon A \quad . \quad (12)$$

We now make a canonical transformation of the basis function, $u_{j,o}$, to remove the interband coupling terms to first order, i.e., A = TB. Details of this are given in LK, but the effect is to diagonalize the Hamiltonian matrix to order k^2, giving the effective mass equation in momentum space,

$$\overline{\mathcal{H}}B(\vec{k}) = T^{-1}\mathcal{H}TB(\vec{k}) = \varepsilon B(\vec{k}) \quad . \quad (13)$$

$\overline{\mathcal{H}}$ is now diagonal to order k^2, and leads to the same $\varepsilon - \vec{k}$ relation as that obtained by perturbation theory:

$$\overline{\mathcal{H}} = \varepsilon_j(k) = \varepsilon_j(0) + k_\alpha D_{jj}^{\alpha\beta} k_\beta \quad . \quad (14)$$

To get to the more usual formulation in real space we define a function,

$$f_j(\vec{r}) \equiv \int_{\text{zone}} e^{i\vec{k}\cdot\vec{r}} B_j(\vec{k}) d^3k \, , \quad (15)$$

the integration as usual being only over the first Brillouin zone. Inverting this Fourier transformation and substituting for $B_j(\vec{k})$ in Eq. (13) leads after some manipulation to the effective mass equation in real space, providing we limit $f_j(\vec{r})$ to be a slowly varying function with respect to the lattice spacing. This limit is equivalent to restricting $\vec{k}$-values to be much smaller than the size of the first Brillouin zone (i.e., $k \ll 1/a_L$). We then have the effective mass equation for the jth band, to order k^2,

$$\left[\epsilon_j(0) + \left(\frac{1}{i}\nabla_\alpha\right) D_{jj}^{\alpha\beta}\left(\frac{1}{i}\nabla_\beta\right)\right] f_j(\vec{r}) = \epsilon_j f_j(\vec{r}) \quad . \tag{16}$$

TWO BAND MODEL

The effective mass approximation defined by Eq. (16) includes the interactions of the jth band with all other bands to order k^2, i. e., describing a simple parabolic band. Of course in the actual small band gap materials we shall be dealing with, we have strongly interacting non-parabolic bands. The simplest example of this is the case where one has two closely spaced bands and can neglect interactions with all other far away bands. This so-called two band model has been remarkably successful in explaining experimental results in materials like bismuth. If we just let i run over two bands (1 and 2) in the A equation (11), the energies are given by the solution of the 2 x 2 determinantal equation

$$\begin{vmatrix} \epsilon_g + \frac{\hbar^2 k^2}{2m} - \epsilon & \frac{\hbar}{m} k \cdot p_{12} \\ \frac{\hbar}{m} k \cdot p_{21} & \frac{\hbar^2 k^2}{2m} - \epsilon \end{vmatrix} = 0 \ , \tag{17}$$

whence

$$\epsilon = \frac{1}{2}\epsilon_g \pm \frac{1}{2}\sqrt{\epsilon_g^2 + 4\frac{\hbar^2}{m^2}k^2 p_{12}^{\ 2}} \tag{18}$$

This is the familiar expression for two "mirror" non-parabolic bands.

EXTERNAL MAGNETIC FIELD

In the presence of an external perturbation (such as a magnetic field) the potential in Eq. (4) is no longer periodic with the periodicity of the lattice--$\vec{k}$ is no longer a good quantum number -- so it is necessary to make a more general expansion,

$$\psi_{j,k} = \sum_i \int A_i u_{i,0}(\vec{r}) e^{i\vec{k}\cdot\vec{r}} d^3k \ . \tag{19}$$

N.B. LK show that these modified Bloch functions form a complete set.

Using the same approach as previously, LK obtain an effective mass equation that has the same form as the zero field equation except that the operator, $\frac{1}{i}\nabla_\alpha$, goes over to $\left(\frac{1}{i}\nabla + \frac{e\vec{A}}{c}\right)_\alpha$, where $\vec{A}$ is the vector potential of the external magnetic field. The effective mass equation becomes

$$\left[D_{jj}^{\alpha\beta}\left(\frac{1}{i}\nabla + \frac{e\vec{A}}{c}\right)_\alpha\left(\frac{1}{i}\nabla + \frac{eA}{c}\right)_\beta + \epsilon_j(0) + \text{spin term}\right]f_j(\vec{r}) + \epsilon_j f_j(\vec{r}) \ . \quad (20)$$

The principal difference here is that the operators no longer commute. The symmetric part of the Hamiltonian leads to the same zero field effective mass; the anti-symmetric part gives rise to an effective g-factor.

The solutions to Eq. (20) are well known from the analogous problem of the free electron in a magnetic field. The energy bands are coalesced into one dimensional sub-bands (Landau levels), given by Eq. (3), where $\vec{H}$ is in the z direction, and $1/m_c$ and g_c are given by the symmetric and anti-symmetric part of the Hamiltonian respectively. The envelope function, $f(\vec{r})$, is given by

$$f_j(\vec{r}) \alpha\, e^{i(k_x x + k_z z)}\, \Phi_n(y - y_0), \quad (21)$$

where Φ_n is a harmonic oscillator function of quantum number n, and k_x labels the different degenerate functions. $y_0 = (\hbar c/eH)\, k_x$ is the centre of the electronic orbits and the degeneracy of k_x is limited by the restriction that this must lie within the crystal dimensions.

The total zeroth order wave function for the electron in the jth band is then given by the product of the envelope function and the band edge cell periodic function,

$$\psi_j = f_j u_{j,o}(\vec{r}) \ . \quad (22)$$

TRANSITION PROBABILITIES

We are now in a position to evaluate selection rules and transition probabilities for the absorption of electromagnetic radiation in a semiconductor with simple parabolic bands in the presence of an external magnetic field.[7] The electromagnetic perturbation can be described by a time-varying electric field whose space variation can be neglected since the wavelength is much longer than the electronic wavelengths involved. The perturbing term in the Hamiltonian is then

$$\mathcal{H}' = \frac{eE_o}{m}\left(\vec{p} + \frac{e\vec{A}}{c}\right) \times \frac{1}{2i\omega}\left[\vec{\epsilon}\, e^{i\omega t} - \vec{\epsilon}^{\,*} e^{-i\omega t}\right] \ , \quad (23)$$

where ω is the frequency of the radiation, E_o the magnitude of the electric field and $\vec{\epsilon}$ is a unit vector parallel to the electric field. By the usual methods of semiclassical radiation theory the optical transition probability for an electron to be raised from state ψ_i (initial) to ψ_f (final) is proportional to the square of:

$$M = \frac{eE_o}{m\omega} \langle \psi_i \,|\, (\vec{p} + \frac{e\vec{A}}{c}) \cdot \vec{\epsilon} \,|\, \psi_f \rangle \ . \tag{24}$$

Substituting for the wave functions from Eq. (22) we have:

$$M = \frac{eE_o}{m\omega} \int \left[f_i^* u_{i,o}^* (\vec{p} + \frac{e\vec{A}}{c}) \cdot \vec{\epsilon} f_f u_{f,o} \right] d\vec{r} \ . \tag{25}$$

Differentiation of the product $(f_f u_{f,o})$ gives:

$$M = \frac{eE_o}{m\omega} \int \left[f_i^* f_f u_{i,o}^* (\vec{p} + \frac{e\vec{A}}{c}) \cdot \vec{\epsilon} u_{f,o} + u_{i,o}^* u_{f,o} f_i^* (\vec{p} + \frac{e\vec{A}}{c}) \cdot \vec{\epsilon} f_f \right] d\vec{r} \ . \tag{26}$$

Since the envelope function, $f(\vec{r})$, and $\vec{A}$ are slowly varying over a unit cell compared to the cell periodic functions, $u_{i,o}$ and $u_{f,o}$, we can break this expression up into an integral over the unit cell involving $u(\vec{r})$ and an integral over the whole crystal involving $f(\vec{r})$. Then,

$$M = \frac{eE_o}{Vm\omega} \Big[\underbrace{(\vec{p} \cdot \vec{\epsilon})_{if} \int f_i^* f_f d\vec{r}}_{\substack{\text{interband} \\ (\Delta n = 0)}} + \underbrace{\int f_i^* (\vec{p} + \frac{e\vec{A}}{c}) \cdot \vec{\epsilon} f_f d\vec{r}}_{\substack{\text{cyclotron resonance} \\ (\Delta n = \pm 1)}} \Big] , \tag{27}$$

where V is the volume of the crystal. The first term represents the interband contribution, since the matrix element of $(\vec{p} \cdot \vec{\epsilon})$ between the band edge functions is non-zero for bands of different parity, but zero for bands of the same parity. The contribution from the vector potential to this matrix element is neglected. The second term gives rise to transitions within the same band (cyclotron resonance), since the band edge functions are orthogonal and $\int u_{i,o}^* u_{f,o} d\vec{r} = \delta_{i,f}$. The selection rules for allowed interband transitions is $\Delta n = 0$, since the harmonic oscillator functions are orthogonal. The polarization selection rules are determined by the term $(\vec{p} \cdot \vec{\epsilon})_{if}$, where $\vec{\epsilon} = \frac{1}{\sqrt{2}}(\epsilon_x \pm i\epsilon_y)$ or ϵ_z, for right and left circularly polarized light and plane polarized light ($\vec{E} \parallel \vec{H}$), respectively.

DEGENERATE PARABOLIC BANDS--GERMANIUM

It is now necessary to consider the more realistic situation of the degenerate p-like valence band of germanium-like semiconductors. LK use the same procedure as that discussed in the previous section, but with the modification that they treat the set of degenerate bands, j', together. Matrix elements of type $p_{jj'} \to 0$ by symmetry, so that this treatment results in a set of coupled equations to order k^2 analogous to Eq. (20):

$$\sum_{j'} \left[D_{jj'}^{\alpha\beta} \left(\frac{1}{i} \nabla + \frac{e\vec{A}}{c} \right)_\alpha \left(\frac{1}{i} \nabla + \frac{e\vec{A}}{c} \right)_\beta + \epsilon_{j'}(0) \delta_{jj'} + \text{spin terms} \right] f_{j'}(\vec{r}) = \epsilon f_j(\vec{r}), \tag{28}$$

where j' runs over the set of three degenerate bands (six with spin). Thus we have a 6 x 6 effective mass Hamiltonian matrix with the inverse effective mass coefficients given by:

$$D_{jj'}^{\alpha\beta} = \frac{\hbar^2}{2m_0} \delta_{\alpha\beta} \delta_{jj'} + \frac{\hbar^2}{m_0^2} \sum_{i \neq j} \frac{p_{ji}^{\alpha}(0)\, p_{ij'}^{\beta}(0)}{\epsilon_0 - \epsilon_i} . \tag{29}$$

i runs over all higher bands, and ϵ_0 is the mean energy of the degenerate set (assumed the same with respect to the higher bands). The total zeroth order wave function for the electron is now the six-component function:

$$\psi = \sum_{j'} f_{j'}(\vec{r}) u_{j',0}(\vec{r}) . \tag{30}$$

LK show that for the case of Ge, at the energies associated with low field microwave CR experiments, it is a good approximation to neglect the spin-orbit split-off valence band. We then have a 4 x 4 effective mass Hamiltonian with respect to the band edge basis functions, $u_{j,0}$, of the light and heavy hole valence band. The correct functions which diagonalize the energies at $\vec{k} = 0$ in the so-called (J, M_J) representation are:

$$\begin{aligned}
u_{3,0} &= \left| \tfrac{1}{\sqrt{2}} (x + iy) \uparrow \right\rangle & &(\tfrac{3}{2}, \tfrac{3}{2}) \\
u_{4,0} &= \left| \tfrac{i}{\sqrt{2}} (x - iy) \downarrow \right\rangle & &(\tfrac{3}{2}, -\tfrac{3}{2}) \\
u_{5,0} &= \left| \tfrac{1}{\sqrt{6}} [(x - iy)\uparrow + 2z\downarrow] \right\rangle & &(\tfrac{3}{2}, -\tfrac{1}{2}) \\
u_{6,0} &= \left| \tfrac{i}{\sqrt{6}} [(x + iy)\downarrow - 2z\uparrow] \right\rangle . & &(\tfrac{3}{2}, \tfrac{1}{2})
\end{aligned} \tag{31}$$

At finite values of magnetic field this leads to a 4 x 4 Hamiltonian matrix in terms of five fundamental band parameters, for materials of diamond symmetry. Three of these constants are the basic effective mass parameters, the other two coming in as a result of the magnetic field. This matrix must be diagonalized for the eigenvalues and eigenvectors of the system. In general it is only possible to carry this out exactly in terms of infinite series of harmonic oscillator functions, but Luttinger[8] has shown that for $\vec{H}$ in the $(1\bar{1}0)$ plane it is possible to find an exact solution with most of the anisotropy included; the remainder may then be treated by perturbation theory. He writes the effective mass Hamiltonian,

$$D = D_0 + D_1 \ , \tag{32}$$

where most of the anisotropy is included in D_0 which can be diagonalized exactly. D_1 is dependent on the second order warping and will be considered in detail in a later section; for the moment we will neglect it. We have also made the approximation that $k_z = 0$. For interband optical transitions one obtains peaks where there are singularities in the joint density of states, so this is a good approximation. D_0 decouples into two 2×2 matrices,

$$D_0 = \left\| \begin{matrix} D_a & 0 \\ 0 & D_b \end{matrix} \right\| \begin{matrix} \tfrac{3}{2} \\ -\tfrac{1}{2} \\ \tfrac{1}{2} \\ -\tfrac{3}{2} \end{matrix} \ , \tag{33}$$

with two-component solutions in terms of single harmonic oscillator functions:

$$f_a^{\pm} = \begin{pmatrix} a_3^{\pm} & \Phi_{n-2} \\ a_5^{\pm} & \Phi_n \end{pmatrix} \ , \tag{34}$$

$$f_b^{\pm} = \begin{pmatrix} a_6^{\pm} & \Phi_{n-2} \\ a_4^{\pm} & \Phi_n \end{pmatrix} \ . \tag{35}$$

Thus, we have four "ladders" of magnetic energy levels corresponding to the light and heavy hole solutions (+ and - respectively) of the a and b matrices. The subscripts on the eigenvectors, a, refer to the associated band edge basis functions of Eq. (31). $a_{3,n}^{\pm}$ and $a_{6,n}^{\pm}$ equal zero for $n = 0$ and 1. In general each level is described by a two-component wave function, from Eqs. (30), (34) and (35), with associated quantum numbers n and (n - 2). By convention we take the highest of these, i.e., n, to label the level. This mixing of the light and heavy hole levels gives rise to the so-called "quantum effects".[8] The spacing of levels at low quantum numbers is not that of the classical light or heavy hole; it is only at high quantum numbers that they tend to the classical limit. In addition, from Eq. (26) we see that it is now possible to have allowed transitions obeying selection rules $\Delta n = 0$ and -2.

INCLUSION OF NON-PARABOLICITY AND DEGENERACY

Finally we must consider the situation in small band gap materials like InSb where, in addition to the complication of a degenerate p-like valence band, one has strongly interacting non-parabolic bands. For the moment we will neglect the effects of inversion asymmetry (present in InSb); these are in any case extremely small, as will be shown in a later section.

The theory of the magnetic energy levels of the valence and conduction bands has been carried out by Pidgeon and Brown[9] using a modification of the LK method.[5] Following Kane's[10] treatment of the zero-field case, the s-like conduction band is treated together with the triply degenerate p-like valence band. The interaction of higher bands with this system is included in the effective mass equation to order k^2, and the resulting 8×8 equation diagonalized in an analagous way to Eq. (28). Thus, the effects of non-parabolic conduction and light hole bands and the "quantum effects" of the light and heavy hole valence bands are included. The principal new feature of the effective mass equation results from the fact that matrix elements of type $p_{jj'}$ do not equal zero for s and p bands. Thus, we obtain an 8×8 equation of form:

$$\sum_{j'}\left[D_{jj'}^{\alpha\beta}\left(\frac{1}{i}\nabla+\frac{e\vec{A}}{c}\right)_{\alpha}\left(\frac{1}{i}\nabla+\frac{e\vec{A}}{c}\right)_{\beta}+p_{jj'}^{\alpha}\left(\frac{1}{i}\nabla+\frac{e\vec{A}}{c}\right)_{\alpha}+\epsilon_{j'}\delta_{jj'}+\text{spin term}\right]f_{j'}(\vec{r})$$
$$= \epsilon f_j(\vec{r}) \quad . \tag{36}$$

The term, $p_{jj'}^{\alpha}\left(\frac{1}{i}\nabla+\frac{e\vec{A}}{c}\right)_{\alpha}$, represents the interaction between s and p-like bands--included here exactly. The term connected with $D_{jj'}^{\alpha\beta}$ now represents only the effect of the higher bands on this system.

Equation (36) can again be broken into two parts, $(D_0 + D_1)\, f = \epsilon f$, where most of the anisotropy is included in D_0 which can be solved exactly (k_z being put equal to zero). As in the previous section we neglect D_1 here; D_0 then decouples into two 4×4 equations which may be diagonalized exactly for the eigenvalues and eigenvectors of the system. For ease of notation we retain the labelling of the valence band levels given in the decoupled scheme of the previous section. The allowed valence to conduction band transitions obey the selection rules $\Delta n = 0$, -2 (N. B. these are not affected by the non-parabolicity of the bands).

INTERBAND MAGNETOREFLECTION IN GREY TIN

We expect the formalism set up in the previous section to give an adequate model for grey tin, which has the "inverted" band scheme described by Groves and Paul.[11] This model is illustrated in Fig. 1 for the Γ_6^-, Γ_7^+ and Γ_8^+ bands, labelled with the double group representations for the diamond lattice. The s-like Γ_7^- band is now submerged between the p-like Γ_8^+ and Γ_7^+ bands, which are split by the spin-orbit splitting, Δ. The Fermi level of the material used here is at the degeneracy point of the doubly degenerate Γ_8^+ band. Thus, the conduction and valence bands are separated by a zero thermal energy gap. As in the case of InSb, the $(\Gamma_7^- - \Gamma_8^+)$ gap is small, giving a large departure from parabolicity of the bands.

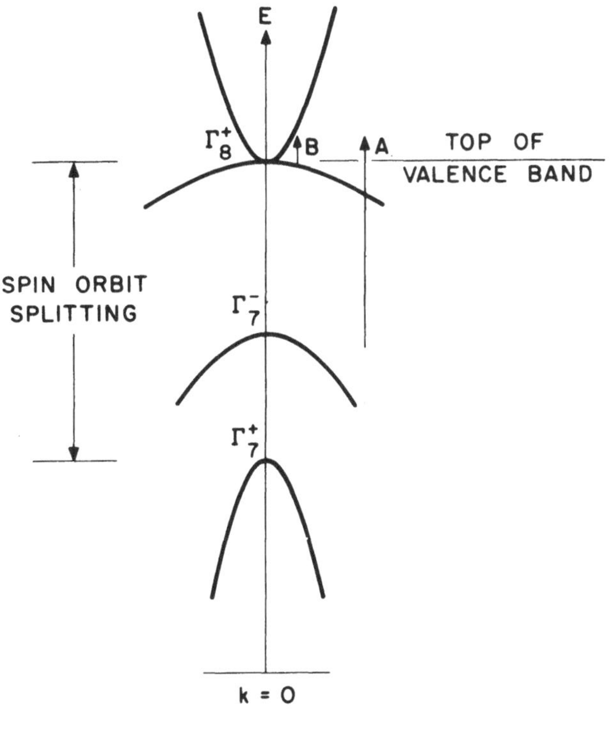

Fig. 1. Grey tin band structure showing highest valence bands and lowest conduction band about $\vec{k} = 0$.

We see from Fig. 1 that direct valence to conduction band transitions across the zero thermal gap are forbidden, at $\vec{k} = 0$, since both bands have p-like symmetry. However, because of the small $(\Gamma_7^- - \Gamma_8^+)$

energy separation, there is a strong admixture of s-like character into the p-like conduction band. This means that at finite values of magnetic field (or of $\vec{k}$), strong forbidden transitions become observable. These are labelled as B in Fig. 1. In addition we expect to see a set of allowed transitions from the submerged s-like band to the p-like conduction band (set A in Fig. 1).

Measurements of magnetoreflection have been carried out by the author in collaboration with Groves of Lincoln Laboratory and Ewald and Wagner of Northwestern University.[12] As predicted, two sets of transitions were observed corresponding to A and B in Fig. 1.

Conventional methods were used to obtain the magnetoreflection spectra, with the radiation reflected from a ⟨100⟩ natural facet of the crystal. The required spectral sensitivity was obtained with a cooled InSb photo-voltaic detector (77°K) for the high energy set of transitions and a helium cooled Cu-doped Ge photoconductive detector for the low energy set. Measurements were made either with the sample submerged in superfluid helium at 1.5°K or on the "cold finger" of a liquid helium dewar at about 20°K (N.B. no observable shift of the spectra with temperature was found in this range). All measurements were carried out in the Faraday configuration ($\vec{E} \perp \vec{H}$), using left and right circularly polarized light (σ_L and σ_R) obtained with a CsI Fresnel rhomb in conjunction with a AgCl pile of plates linear polarizer. The data were

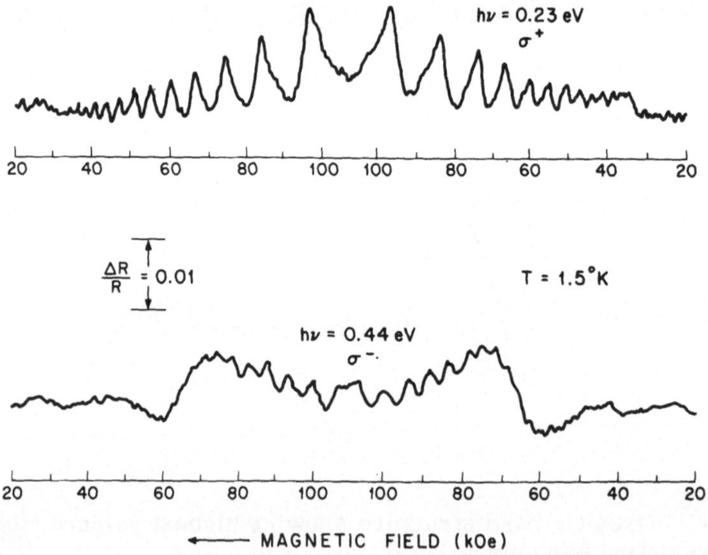

Fig. 2. Experimental traces of magnetoreflection made by sweeping the magnetic field at fixed photon energies (after Ref. 12). Top trace shows peaks due to $\Gamma_8^+ \rightarrow \Gamma_8^+$ transitions, and bottom trace shows small peaks from $\Gamma_8^+ \rightarrow \Gamma_8^+$ transitions superimposed on a large $\Gamma_7^- \rightarrow \Gamma_8^+$ peak.

taken for radiation of different fixed photon energies, by sweeping the magnetic field up and down between zero and 100 kOe.

Typical recorder traces of the detected signal are shown in Fig. 2, for photon energies corresponding to the low energy set of transitions, B ($h\nu$ = 0.23 eV) and high energy set of transitions, A ($h\nu$ = 0.44 eV). It is seen that both allowed and forbidden transitions are of similar strength, but that the former are much broader. This is presumably associated with the greater scattering expected (shorter collision time) for transitions from the submerged band. The low energy set can be seen to high energies, superimposed upon the high energy set.

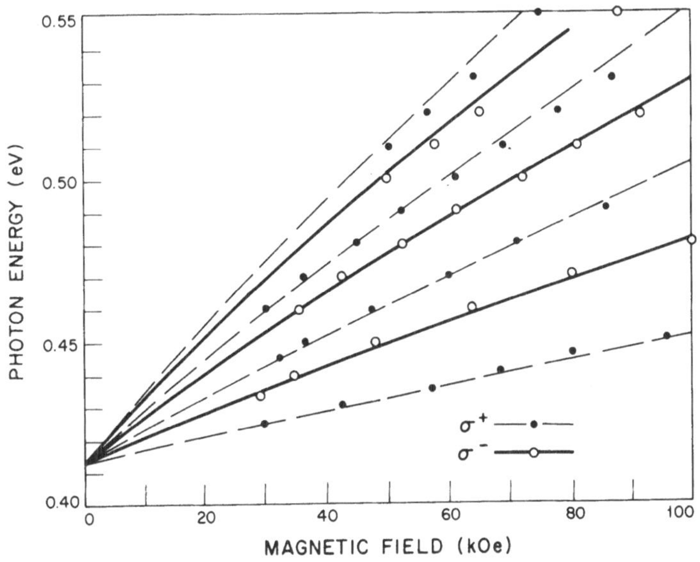

Fig. 3. The open and closed circles give the transition energies in the σ^- and σ^+ polarizations for $\Gamma_7^- \to \Gamma_8^+$ transitions (after Ref. 12). The solid and dashed curves are the theoretical energies for the two polarizations (N.B., $\sigma^- \equiv \sigma_R$ and $\sigma^+ \equiv \sigma_L$, here).

Plots of the photon energies versus the magnetic field values of the reflection maxima are shown in Figs. 3 and 4, for the $\Gamma_7^- \to \Gamma_8^+$ and $\Gamma_8^+ \to \Gamma_8^+$ transitions. From the extrapolation to zero magnetic field we obtain the energy separations $|\Gamma_7^- - \Gamma_8^+| = 0.413$ eV, and $|\Gamma_8^+ - \Gamma_8^+| = 0$ as expected. The solid and dashed lines represent the best fitted theoretical curves, obtained by computation from the 8 × 8 effective mass Hamiltonian of Eq. (36).

The magnetoreflection results for grey tin are very similar to those obtained previously for HgTe by Groves et al.,[13] but are more interesting for a comparison with theory for several reasons. First, as mentioned previously, they are characteristic of a single high symmetry direction

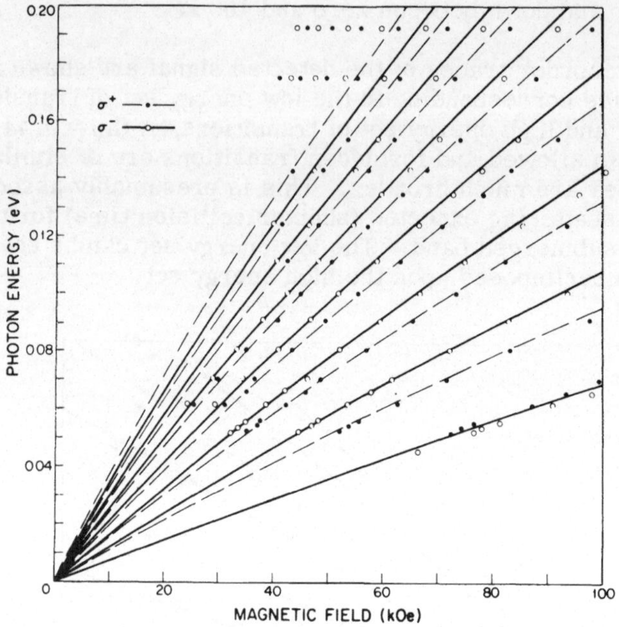

Fig. 4. $\Gamma_8^+ \rightarrow \Gamma_8^+$ transition energies with experimental points and theoretical curves shown with the same convention as in Fig. 3 (after Ref. 12). These transitions have been seen up to 0.55 eV but are only plotted to 0.2 eV here.

($\vec{H} \parallel \langle 100 \rangle$, where the second order warping correction, D_1, is unimportant--as will be shown in a later section); such results could not be achieved in HgTe because of a materials limitation. Secondly, the warping of the Γ_8^+ bands, a parameter which is needed in the theory, has been measured for grey tin.[14] Thirdly, the complications which may be introduced by the inversion asymmetry of the zinc-blende HgTe are absent for grey tin, which has inversion symmetry.

The band edge effective masses corresponding to the theoretical curves of Figs. 3 and 4, in units of free electron mass, are: $m_c(\Gamma_8^+) = 0.028$, $m_v(\Gamma_8^+) = 0.195$, $m(\Gamma_7^-) = 0.0157$ and $m(\Gamma_7^+) = 0.051$.

WARPING AND INVERSION ASYMMETRY IN ZINC-BLENDE CRYSTALS

In the formalism of Eq. (36) and in the exact solution of this equation in terms of single harmonic oscillator functions, two approximations were made: (i) inversion symmetry was assumed: (ii) the second order warping term D_1 was neglected. Usually these approximations cause negligible error, but if two levels, which interact through the omitted terms in the Hamiltonian, are nearly degenerate in energy, considerably admixing of the wave functions can take place. This in turn can cause observable "extra" transitions. Such transitions, associated with the linear-$\vec{k}$ and warping terms in the valence band, have been observed in InSb, by the author in collaboration with S. Groves of Lincoln Laboratory.[15]

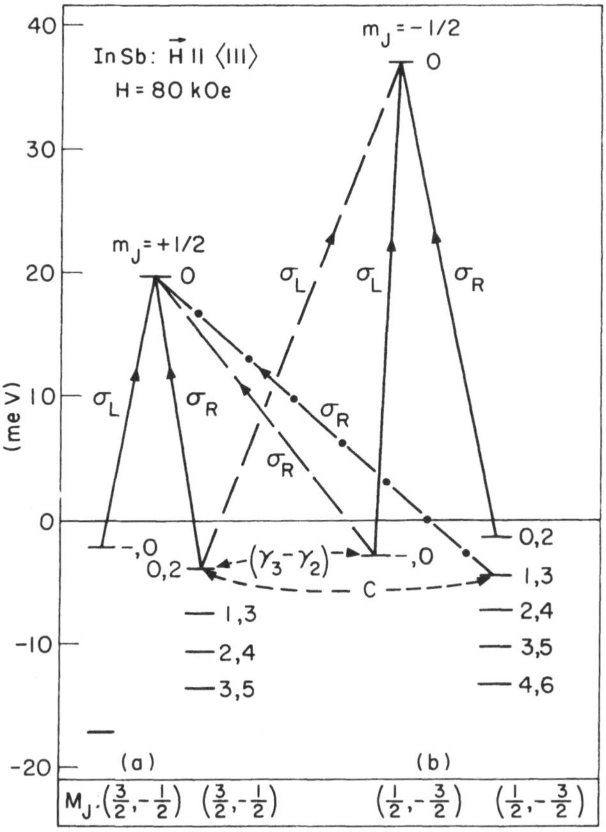

Fig. 5. Lowest magnetic energy levels for the valence and conduction bands of InSb (after Ref. 15). The numbers by the levels in the a and b valence ladders are the Landau quantum numbers, n, for the two-component states. The corresponding total angular momentum quantum numbers, M_J, are given below. Allowed and extra transitions for the $\sigma_L(\Delta M_J = +1)$ and $\sigma_R(\Delta M_J = -1)$ spectra are shown: solid line with arrow, allowed; dashed line with arrow, warping-induced; dot-dashed line with arrow, linear-$\vec{k}$ induced.

The transitions we are concerned with arise from mixing of levels at the top of the valence band. In considering their origin, it is sufficient to work with the valence band solutions alone (i.e., decoupled from the conduction band) from Eq. (33). The absolute energies of the levels are computed from the coupled scheme of Eq. (36). Energy levels calculated by this method are shown in Fig. 5 for the top of the valence band and bottom of the conduction band. The labelling of the levels is in accord with the approximate solution, D_0 , for the light and heavy hole valence band system of Eq. (33), (34) and (35). Each two-component level has two Landau quantum numbers associated with it (n and n - 2), and two total angular momentum quantum numbers, M_J, corresponding to the p-state ($J = \frac{3}{2}$). The lowest two light hole levels have only one Landau quantum number, n, since $a^{\pm}_{3,n}$ and $a^{\pm}_{6,n}$ equal zero for n = 0 and 1.

The solid lines in Fig. 5 show the interband optical transitions which can take place in the Faraday configuration (σ_L and σ_R): electrons can be excited from the two a-set valence band ladders into the $M_J = \frac{1}{2}$ conduction level , or from the two b-set ladders into the $M_J = -\frac{1}{2}$ conduction level. We are concerned here with extra transitions to the lowest conduction band levels caused by the linear-$\vec{k}$ and warping interactions represented by C and $(\gamma_3 - \gamma_2)$ respectively.

The explicit form of D_1 is given in Ref. 8, Eq. (87); in particular we are interested in the term r_3 which is proportional to the warping parameter $(\gamma_3 - \gamma_2)$. In fact we find that this does cause a strong interaction between a pair of nearly degenerate levels of the a and b-ladders for $\vec{H} \parallel \langle 111 \rangle$, but not for $\vec{H} \parallel \langle 100 \rangle$ and $\langle 110 \rangle$. This interaction is labelled $(\gamma_3 - \gamma_2)$ in Fig. 5. The resulting admixing of wave function causes an extra transition to each of the two lowest conduction levels shown by the dashed lines (and observed experimentally, as discussed in the next section).

With the source of these transitions recognized it is desirable to go to the eigenvalue problem for $\vec{H} \parallel \langle 111 \rangle$ where the warping can be treated exactly.[8] The solution to the effective mass equation, $D\langle 111 \rangle f = \epsilon f$, is

$$f_n^s = \begin{pmatrix} A_{3,n}^s & \Phi_n \\ A_{5,n}^s & \Phi_{n+2} \\ A_{6,n}^s & \Phi_{n-2} \\ A_{4,n}^s & \Phi_n \end{pmatrix} \quad \text{for } n \geq 2, 3 \ldots \quad (37)$$

where s runs over the four ladders. Using the band parameters of Ref. 9, we have diagonalized $D\langle 111 \rangle$ numerically, making fine adjustments on the higher band parameters to achieve a good fit to both the relative positions and strengths of the allowed and warping-induced transitions found experimentally (next section).

The antisymmetric potential in zinc-blende materials gives rise to an odd part to the Hamiltonia, D^-, which was neglected in Eq. (36)-- this is zero for materials with inversion symmetry. The explicit form of D^- has been given for $\vec{H} \parallel \langle 001 \rangle$.[16] For $\vec{H}$ in any direction in the $(1\bar{1}0)$ crystal plane we have made the coordinate transformations given in Ref. 8. The details of this are given elsewhere.[15] It is found again that for $\vec{H} \parallel \langle 111 \rangle$ a strong interaction is possible between two nearly degenerate states, shown by C in Fig. 5. This gives rise to an additional σ_R transition to the $M_J = \frac{1}{2}$, n = 0 conduction band level (shown as a dot-dash line). The wave functions of the interacting states (the lowest heavy hole level in the a-set and the second heavy hole level in the b-set) are known, so that, since no other interactions with these levels are present, we may solve this 2 x 2 problem exactly. Hence, from the measurement of the relative strength and energy separation of allowed and linear-$\vec{k}$ induced transition, we may determine the linear-$\vec{k}$ band parameter C directly. Linear-$\vec{k}$ induced transitions may also occur for $\vec{H} \parallel \langle 100 \rangle$ and $\langle 110 \rangle$, but these are expected to be about 5 times weaker and have not been observed in the present work.

MAGNETOREFLECTION IN InSb

Magnetoreflection measurements were made on pure n-type InSb ($N \sim 10^{14}\,cm^{-3}$) at $1.5^\circ K$,[16] in the Faraday configuration. Results are shown in Fig. 6 for two directions of $\vec{H}$, using left and right circularly polarized light.

Because of the large conduction band spin splitting, the transitions to the lowest spin up and spin down conduction band levels are widely separated, permitting unambigous assignments of transitions. We know from earlier work on magneto-absorption[9] that the set of lines between 254 and 260 meV and that between 273 and 277 meV correspond to transitions from levels at the top of the valence band to the first spin up and spin down conduction levels respectively. Thus, for $\vec{H} \parallel \langle 100 \rangle$ we see the pairs of allowed transitions (σ_L and σ_R) to these two levels shown by the solid lines of Fig. 5. We see also that as predicted for $\vec{H} \parallel \langle 111 \rangle$, the lowest σ_R line separates into three components--warping induced (w), allowed and linear-$\vec{k}$ induced (k) transitions--and the second σ_L line into two components--allowed and warping induced (w) transitions. It is seen that the warping induced transitions have strengths comparable with the allowed transitions , but that the linear-$\vec{k}$ transition is about 20 times weaker.

As discussed in the previous section, $D\langle 111 \rangle$ is first diagonalized numerically to give a good fit to both the relative positions and strengths of the allowed and warping induced transitions. This gives a direct measure of $(\gamma_3 - \gamma_2)$. Knowing, then, the relevant wave functions (with the full warping included exactly for $\vec{H} \parallel \langle 111 \rangle$) the 2x2 linear-$\vec{k}$ interaction matrix is diagonalized exactly, and the size of the parameter C determined directly from the relative strengths and energies of the linear-$\vec{k}$ induced and allowed transitions.

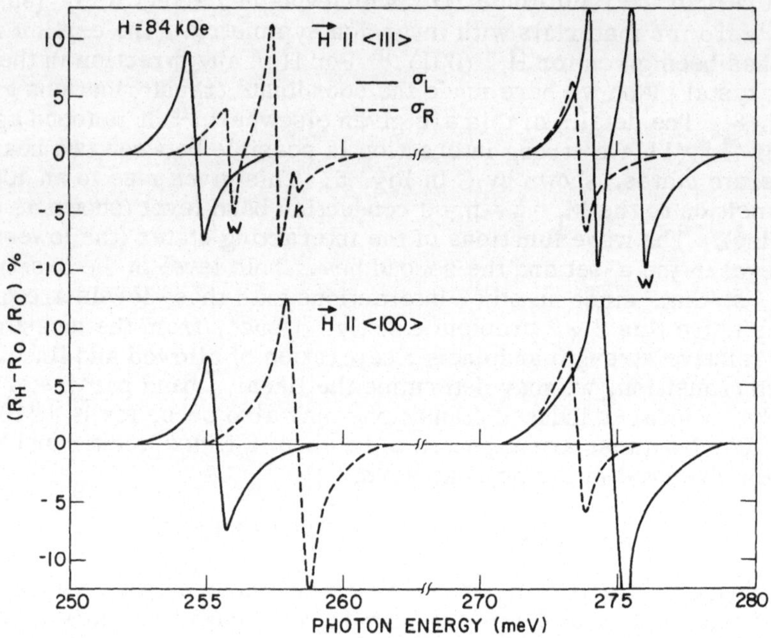

Fig. 6. Magnetoreflection spectra for $\vec{H} \parallel \langle 111 \rangle$ and $\vec{H} \parallel \langle 100 \rangle$ in the Faraday configuration, with H = 84 kOe and T = 1.5°K (after Ref. 15). Transitions to the $M_J = \frac{1}{2}$, n = 0 conduction band level occur to the left hand side of the break in the energy scale and those to the $M_J = -\frac{1}{2}$, n = 0 conduction band level occur to the right hand side. w labels warping-induced transitions and k labels the linear-$\vec{k}$ induced transition.

Results similar to those shown in Fig. 6 were first obtained with the low temperature electroreflectance technique.[17] This method is similar to the transparent electrode arrangement of Seraphin[18] and co-workers, except that it utilizes a thin film dry package (etched sample surface/Kodak photoresist spacer layer/Ni film transparent electrode). The details are described elsewhere.[19] The spectral fine structure, associated with the transitions of Fig. 5, was perhaps more easily observed with this method, but because of the complications of interpreting the electroreflectance line shapes we prefer to go to the simple magneto-reflection spectra for quantitative analysis.

WARPING AND LINEAR-$\vec{k}$ BAND PARAMETERS IN InSb

Following Johnson[20] and others, we identify the sharp symmetric lines of Fig. 5 with exciton ground states. We make the assumption that the binding energies for all transitions to the same n = 0 conduction band level are equal; so the differences in energy between transitions are given in terms of the Landau level theory. This is reasonable since all the valence levels involved have about the same mass($\sim$ 20 times greater than the conduction mass).

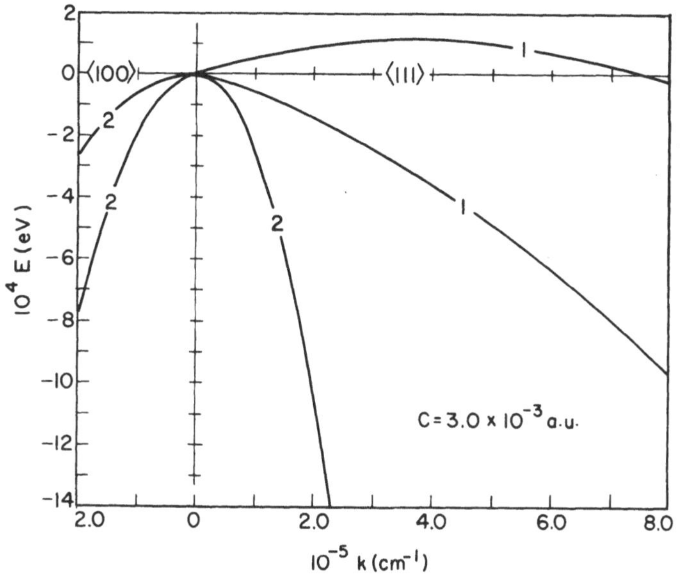

Fig. 7 Valence band dispersion relations for $\vec{k}$ along ⟨100⟩ and ⟨111⟩ directions which result from the parameters determined here (after Ref. 15). The numbers indicate the degeneracy of the band. In all other directions both light and heavy mass bands are split. The maximum splitting occurs for the ⟨111⟩ heavy mass band.

From the foregoing comparison of experiment and theory the following values were obtained for the warping and linear-$\vec{k}$ valence band parameters: $(\nu_3 - \nu_2) = 1.0 \pm 15\%$, $C = 3.0 \times 10^{-3}$ atomic units $\pm$ 30%. The valence band ε - $\vec{k}$ dispersion relations for $\vec{k}$ along ⟨100⟩ and ⟨111⟩ directions, for the parameters determined above, are shown in Fig. 7. The maximum upbending of the heavy hole band, resulting from the linear-$\vec{k}$ term, is about 1 meV. The result for $(\nu_3 - \nu_2)$ is in good agreement with the cyclotron resonance work of Bagguley et al.[21] The value for C is about three times smaller than that given in a cyclotron resonance experiment by Robinson,[22] but is in good agreement with the earlier theoretical estimate of Kane.[10]

At low magnetic fields the linear-$\vec{k}$ terms will have a large perturbing effect on the magnetic energy levels, and the perturbation approach discussed here will break down. In this case a general energy level computation, such as that given recently by Bell and Rogers,[23] becomes necessary. However, the small size of the linear-$\vec{k}$ transition observed experimentally justifies our approach. We have seen this transition in the region from 30 to 100 kOe, where the small pertrubation method is found to be valid.

CONCLUSION

In conclusion, the coupled band scheme developed in Ref. 9 provides a good picture of zero band gap cubic semiconductors of the grey tin type. Measurements of magnetoreflection in grey tin and HgTe have been interpreted quantitatively in terms of this model.

We have found experimentally that the inversion asymmetry terms in the valence band of InSb do not represent a large perturbation on the quasi-Ge scheme of Ref. 9 in the high field limit (i.e., for fields greater than about 20 kOe). However, the second order warping interaction may become extremely strong for directions of magnetic field other than $\vec{H} \parallel \langle 100 \rangle$ and $\langle 110 \rangle$, and must be treated explicitly, as discussed in this paper. Thus, with these provisos, the coupled band scheme of Ref. 9 also provides a good model for describing experiments in small band gap zinc-blende semiconductors of the InSb type.

ACKNOWLEDGEMENTS

I am grateful to Dr. S. Groves for helpful discussions concerning most of the subject matter of this paper.

REFERENCES

1. B. Lax, in Semiconductors, ed. R. A. Smith (Academic Press, Inc., New York, 1963) p. 240.

2. G. Dresselhaus and M. S. Dresselhaus, in The Optical Properties of Solids, ed. J. Tauc (Academic Press, Inc., New York, 1966) p. 198.

3. S. D. Smith, Handbuch der Physik, XXV/2a, 234 (1967).

4. A general review of $\vec{k}\cdot\vec{p}$ perturbation theory has been given by E. O. Kane, Semiconductors and Semimetals I, ed. Willardson and Beer (Academic Press, Inc., New York, 1967) p. 75.

5. J. M. Luttinger and W. Kohn, Phys. Rev. 97, 869 (1955).

6. T. P. McLean, in Semiconductors, ed. R. A. Smith (Academic Press, Inc., New York, 1963) p. 479.

7. The treatment here follows that given by: L. M. Roth, B. Lax and S. Zwerdling, Phys. Rev. 114, 90 (1959).

8. J. M. Luttinger, Phys. Rev. 102, 1030 (1956).

9. C. R. Pidgeon and R. N. Brown, Phys. Rev. 146, 575 (1966).

10. E. O. Kane, J. Phys. Chem. Solids 1, 249 (1957).

11. S. H. Groves and W. Paul, Phys. Rev. Letters 11, 505 (1963).

12. S. H. Groves, C. R. Pidgeon, A. W. Ewald and R. J. Wagner, to be published in Proc. Int. Conf. on Physics of Semiconductors, Moscow (1968).

13. S.H. Groves, R. N. Brown and C. R. Pidgeon, Phys. Rev. 161, 779 (1967).

14. B. L. Booth and A. W. Ewald, to be published (Phys. Rev.).

15. C. R. Pidgeon and S. H. Groves, Phys. Rev. Letters 20, 1003 (1968); and a more extensive account to be published (Phys. Rev.).

16. G. Dresselhaus, Phys. Rev. 100, 580 (1955).

17. C. R. Pidgeon and S. H. Groves, to be published in Proc. Int. Conf. on Physics of Semiconductors, Moscow (1968).

18. B. O. Seraphin and R. B. Hess, Phys. Rev. Letters 14, 138 (1965); B. O. Seraphin, Phys. Rev. 140, A1716 (1965).

19. C. R. Pidgeon, S. H. Groves and J. Feinleib, Solid State Comm. 5, 677 (1967).

20. E. J. Johnson, Phys. Rev. Letters 19, 352 (1967).

21. D. M. S. Bagguley, M. L. A. Robinson and R. A. Stradling, Phys. Letters 6, 143 (1963).

22. M. L. A. Robinson, Phys. Rev. Letters 17, 963 (1966).

23. R. L. Bell and K. T. Rogers [Phys. Rev. 152, 746 (1966)] have numerically solved the full Hamiltonian, truncated to order 240 x 240, for one assumed set of band parameters, with $\vec{H} \parallel \langle 100 \rangle$.

ANISOTROPIC MAGNETO-OPTICAL EFFECTS IN SEMICONDUCTORS WITH CUBIC SYMMETRY

B. Donovan

Westfield College, University of London

1. Introduction

Investigation of magneto-optical phenomena is now well established as a valuable means of obtaining information on the energy band structure of semiconductors. In particular, the changes in the polarization of the radiation which occur in longitudinal or transverse magnetic fields, for both the transmitted and reflected beams, offer a wide range of experimental possibilities. These have been extensively studied with infra-red radiation and, to a lesser degree, in the microwave region.

In single crystals these effects are, in general, not isotropic, i.e. they depend upon the orientations of the magnetic field and the initial polarization with respect to the crystal axes. We consider here some typical examples of this anisotropy, as exhibited by the free carrier Faraday and Voigt effects in cubic semiconductors. The essential quantity in this analysis is the high frequency magneto-conductivity tensor $\mathbf{S}(\omega, \vec{H})$, which relates the components of the current density and electric field:

$$J_i = S_{ik}(\omega, \vec{H}) E_k \quad . \tag{1}$$

Several important results may be established by symmetry arguments alone, i.e. by examining the invariance of $\mathbf{S}$ under suitable transformations appropriate to the cubic symmetry of the crystal.

2. Symmetry of the Magneto-conductivity Tensor

The system of axes (x, y, z) is chosen so that the magnetic field $\vec{H}$ is parallel to the z-direction. The cubic axes will be denoted by 1, 2, 3.

The simplest situation is obtained when $\vec{H}$ is parallel to [001] or to [111]. These are respectively 4- and 3-fold symmetry axes and, by applying a rotation about the z-axis, it is easily shown that in both cases $\underset{\sim}{S}$ has the form

$$\underset{\sim}{S} = \begin{pmatrix} S_{xx} & S_{xy} & 0 \\ -S_{xy} & S_{xx} & 0 \\ 0 & 0 & S_{zz} \end{pmatrix} \tag{2}$$

The distinguishing feature of (2) is that it applies irrespective of the orientation of the x and y axes, which is not true for any other direction of $\vec{H}$ in the crystal.

If $\vec{H}$ is restricted to lie in a given plane, we first apply a transformation which rotates the z-axis by π about an axis perpendicular to this plane. Then the "back transformation" required to restore $\vec{H}$ to its original direction is simply a reversal of the field, hence we use the Onsager relation

$$S_{ik}(\omega, \vec{H}) = S_{ki}(\omega, -\vec{H}) \; . \tag{3}$$

Suppose, for example, that $\vec{H}$ lies in the (100) plane at an angle ϕ with the 3-axis [001], the x-axis remaining parallel to [100]. Then $\underset{\sim}{S}$ has six independent components and reduces to the form

$$\underset{\sim}{S} = \begin{pmatrix} S_{xx} & S_{xy} & S_{xz} \\ -S_{xy} & S_{yy} & S_{yz} \\ -S_{xz} & S_{yz} & S_{zz} \end{pmatrix} \tag{4}$$

In the special case $\phi = \pi/4$ ($\vec{H}$ parallel to [011], y parallel to [01$\bar{1}$]) S_{xz} and S_{yz} are both zero. When $\phi = 0$ ($\vec{H}$ parallel to [001]) (4) reduces to (2).

For a further example we take $\vec{H}$ in the (1$\bar{1}$0) plane at an angle ϕ with the 3-axis. The x and y axes must, of course, be specified and we consider a symmetrical arrangement in the crystal, with the x and y axes inclined at an angle $\cos^{-1}\{\frac{1}{2}(1+\cos\phi)\}$ with respect to the 1 and 2 axes respectively. Then $\underset{\sim}{S}$ again has six components, with the form

$$\underset{\sim}{S} = \begin{pmatrix} S_{xx} & S_{xy} & S_{xz} \\ S_{yx} & S_{xx} & S_{yz} \\ S_{yz} & S_{xz} & S_{zz} \end{pmatrix} \tag{5}$$

In the special case $\phi = \pi/2$ ($\vec{H}$ parallel to [110]) S_{xz} and S_{yz} are both zero.

If we consider $\vec{H}$ in the $(1\bar{1}0)$ plane but keep the x-axis parallel to $[1\bar{1}0]$, then the general form of $\underset{\sim}{S}$ is given by (4). This vital significance of the x and y axes is particularly relevant when considering the disposition of the electric field components.

The results in this section are, of course, independent of the method of calculation of $\underset{\sim}{S}$ which in general must be carried out numerically for a particular model. Only in the case of ellipsoidal energy surfaces is it possible to obtain closed expressions valid for arbitrary $\vec{H}$. The procedure is essentially a high frequency generalization of standard magnetoresistance theory and the S_{ik} are in general complex quantities.

3. Complex Propagation Constants

We consider initially plane polarized radiation with angular frequency ω propagating in a direction either parallel to, or perpendicular to, the magnetic field. In each case two independent modes exist in the crystal and the propagation constants may be derived in terms of the components of the conductivity tensor $\underset{\sim}{S}$ and the dielectric constant ($\epsilon_{ik} = \epsilon\delta_{ik}$ for cubic symmetry). It is convenient to introduce the abbreviation (using Gaussian units):

$$\Phi_{ik} = S_{ik} + \frac{i\omega}{4\pi}\epsilon_{ik} . \tag{6}$$

(i) Faraday configuration: propagation parallel to $\vec{H}$, i.e. along z.

We regard the electric vector $\vec{E}$ as confined to the xy plane, although this is not rigorously correct in all circumstances. Then Maxwell's equations lead to [1]

$$\frac{\partial^2 E_x}{\partial z^2} = AE_x + BE_y \quad , \qquad \frac{\partial^2 E_y}{\partial z^2} = CE_x + DE_y \quad , \tag{7}$$

where A, B, C, D are the xx, xy, yx and yy components respectively of $\gamma\Phi$, with $\gamma = 4\pi i\omega/c^2$. These equations signify, in general, two elliptically polarized modes and the field components are constructed from solutions of the form $\exp(i\omega t - \mu z)$. The two propagation constants ($\mu = \alpha + i\beta$) are given by

$$2\mu_\pm^2 = A + D \pm [(A-D)^2 + 4BC]^{1/2}, \tag{8}$$

where μ_+ refers to the component which rotates in a clockwise sense to an observer looking along the positive z-direction.

When the tensor $\underset{\sim}{S}$ has the form shown in (2) we have A = D and B = C, as in the isotropic case. Then (8) reduces to

$$\mu_\pm^2 = A \mp iB \tag{9}$$

and the component waves are circularly polarized.

The radiation transmitted through the specimen is elliptically polarised and the ellipticity Δ_F arises, essentially, as a result of the differential attenuation $\bar{\alpha} = \alpha_- - \alpha_+$. The angular displacement of the major axis of the ellipse is the Faraday rotation θ_F and is a consequence of the difference in phase velocities $\bar{\beta} = \beta_- - \beta_+$.

(ii) Voigt configuration: propagation perpendicular to $\vec{H}$ and specifically along x direction.

Here the electric vector of the incident radiation lies in the yz plane but a longitudinal component E_x is produced through the mechanism of the Hall effect. E_x is determined by the condition that the total current in the x-direction must be zero, and a pair of equations is obtained similar to (7) but with x, y, z replaced by y, z, x respectively.[2,3] The coefficients are, however, more complicated; the appropriate A for example is given by

$$A = -\frac{\omega^2}{c^2}\epsilon + \gamma\left(S_{yy} - \frac{S_{yx}S_{xy}}{\Phi_{xx}}\right). \qquad (10)$$

The propagation constants are derived exactly as in (8), the two independent modes being elliptically polarized in general.

An important special case is obtained when $\vec{H}$ is along [100]-, [111] - or [110] -type directions. The form of the tensor $\underset{\sim}{S}$, noted in the previous section, makes B = C = 0 and the propagation constants are given by

$$\mu_{\perp}^2 = A, \quad \mu_{\parallel}^2 = D, \qquad (11)$$

where the suffices denote the orientation of the $\vec{E}$ vector with respect to $\vec{H}$. The two modes are linearly polarized and, since D contains only S_{zz} in this case, the parallel component is unaffected by the magnetic field.

The polarization characteristics of the transmitted radiation may be analysed and studied experimentally, as in the Faraday configuration, but the interpretation is different in the two cases. Since $\bar{\beta} = \beta_{\parallel} - \beta_{\perp} \neq 0$, a phase shift η is developed in distance x and the Voigt ellipticity Δ_V is given by $\eta/2$. The displacement of the major axis of the ellipse from the initial polarization direction is the Voigt rotation Θ_V, and is due to the differential attenuation ($\bar{\alpha} = \alpha_{\parallel} - \alpha_{\perp}$).

The analysis is simplified somewhat when $\omega\tau \gg 1$ and $\omega \gg \omega_c$ (infra-red approximation), in which case the behaviour in the two configurations may be compared as follows:

$$\text{Faraday} \qquad \Theta_F \sim \frac{H}{m^2\omega^2}, \quad \Delta_F \sim \frac{H}{m^2\omega^3};$$

$$\text{Voigt} \qquad \Delta_V \sim \frac{H^2}{m^3\omega^3}, \quad \Theta_V \sim \frac{H^2}{m^3\omega^4}, \qquad (12)$$

where m is the appropriate effective mass.

The significant distinction is that the Voigt effect is proportional to H^2 and is anisotropic in cubic crystals even in the low field limit.[4] In contrast, the Faraday effect is isotropic to order H and investigation of anisotropy is more profitable in the microwave region rather than the infra-red, where very high fields (~ 200 kOe) would normally be required.

4. Anisotropy of Faraday Rotation

The general expressions for Θ_F and Δ_F have been given by Donovan and Webster.[1] Their complexity is due to the elliptical polarization of the component waves; in particular, the real and imaginary parts of the $\mu_{\pm}$ both enter into the expressions for Θ_F and Δ_F, in contrast to the isotropic theory, where Θ_F and Δ_F depend only on $\bar{\beta}$ and $\bar{\alpha}$ respectively.

The following typical results apply to n-type germanium, where the energy surfaces are four ellipsoids oriented along the [111] directions. The conductivity tensor is found for each ellipsoid; it is then necessary to transform to the axes (x,y,z) specified in § 2, and sum over the four ellipsoids.[5,6] The data refer to a non-degenerate sample (T = 300°K): d.c. conductivity = 9×10^{10} e.s.u., carrier density = 1.3×10^{14} cm^{-3}, and $\epsilon = 16$.

The effect of varying the magnetic field orientation is shown in Figure 1, where ϕ is the angle defined in § 2 and the two situations correspond to the tensors (4) and (5). The initial direction of the electric vector is along the x-axis. The

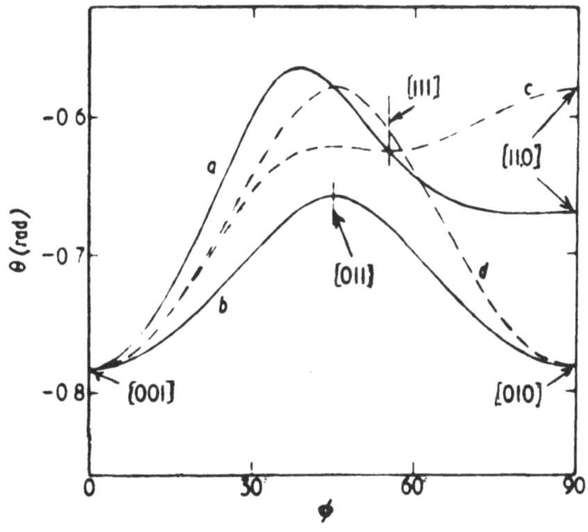

Figure 1. Variation of Faraday rotation with orientation of magnetic field: a, in (1$\bar{1}$0) plane; b, in (100) plane. The broken lines represent $-\frac{1}{2}\beta z$: c, in (1$\bar{1}$0) plane; d, in (100) plane. $\lambda = 3$ cm, $H = 2 \times 10^4$ Oe, $z = 5$ mm.

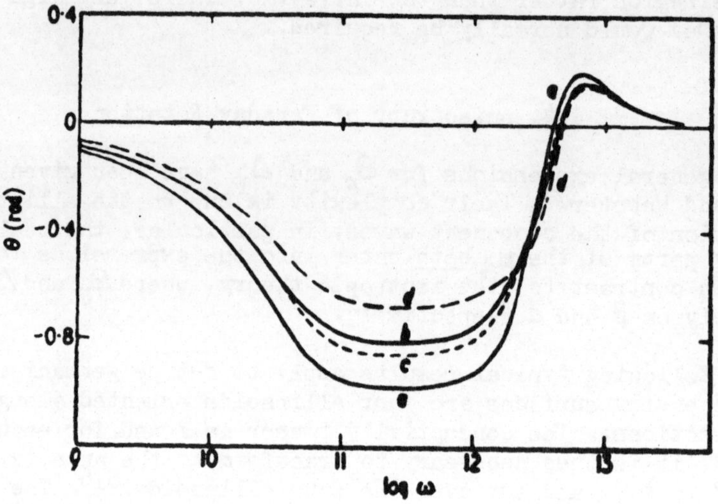

Figure 2. Variation of Faraday rotation with frequency for various orientations of magnetic field: a, along [100]; b, along [111]; c, along [011], with $\phi = 0$; d, along [011], with $\phi = 45°$. $H = 2 \times 10^4$ Oe, z = 5 mm.

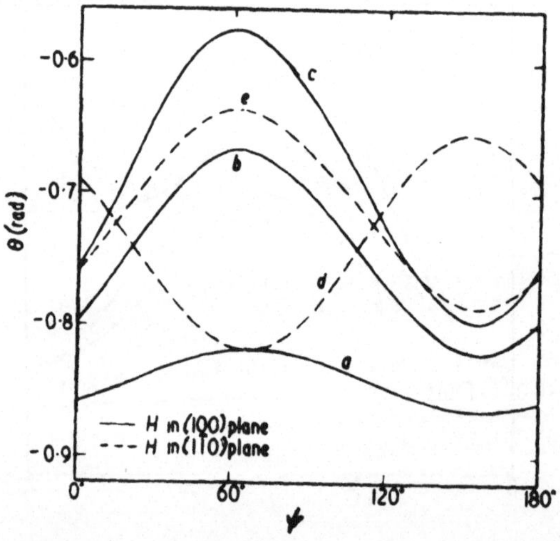

Figure 3. Variation of Faraday rotation with orientation of plane of polarization for various orientations of magnetic field in the planes indicated: a, $\phi = 15°$; b, $\phi = 30°$; c, $\phi = 45°$; d, $\phi = 35°$; e, $\phi = 70°$. $\lambda = 3$ cm, $H = 2 \times 10^4$ Oe, z = 5 mm.

frequency ω is 6.3×10^{10} sec^{-1} (3 cm. radiation) and the behaviour shown is typical of the microwave region. The broken lines (c and d) illustrate the anisotropy in the propagation constants; note that curves a and c intersect at the [111] direction, where the special case (2) applies and Θ_F is given by the "isotropic" formula.

The frequency dependence of Θ_F, for the same values of field strength and sample thickness as Figure 1, is shown in Figure 2, which emphasizes that anisotropic behaviour is most prominent in the microwave region. Curves a and b apply to the two special propagation directions specified by (2), and their separation is due solely to the anisotropy in $\mu_\pm$ since the two component waves are circularly polarized. For all other directions this polarization is elliptical and, as a result, Θ_F is dependent upon the orientation of the initial plane of polarization with respect to the crystal axes.

This azimuthal variation is illustrated in Figure 3, where the angle ψ specifies the orientation of the $\vec{E}$ vector, measured from its initial direction i.e. the x-axis in §2. The broken lines have been displaced through 45° for convenience. The effect is most pronounced for propagation along [110]-type directions and

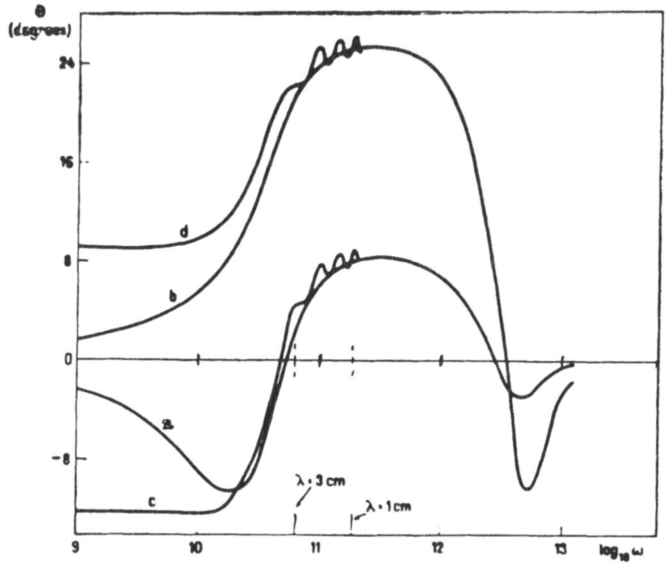

Figure 4. Variation of Voigt rotation with frequency, with $H = 3 \times 10^4$ oersteds, $x = 0.5$ cm, a and c, H along [001] ; b and d, H along [011]; c and d, with multiple reflections included.

here the relative change in Θ_F is of the order of 30% with the field shown. Curves c and d of Figure 2 refer to the orientations $\phi = 0$ and 45° respectively on curve c of Figure 3. These anisotropic effects have been investigated experimentally in n-type germanium by Bouwknegt and Volger,[7] who obtained excellent agreement with the theoretical predictions.[6]

5. Anisotropy of Voigt Rotation

The anisotropic Voigt effect[2,3,8] may be analysed by a procedure analogous to that in the Faraday configuration and the general expressions for Θ_v and Δ_v are rather lengthy. The anisotropic behaviour of Δ_v was first demonstrated experimentally in n-type germanium by Palik,[9] using infra-red radiation. In the following we shall deal with Θ_v, which (cf.(12)) is of greater interest in the microwave region.[10] The numerical results relate to a typical non-degenerate specimen of n-type germanium, with the same parameters as in §4.

We consider specifically the case of propagation along the [100] direction, with $\vec{H}$ therefore confined to the (100) plane. The frequency dependence of Θ_v is shown in Figure 4 and it is evident that the two extreme field orientations give markedly different results. The high frequency sign change occurs at $\omega\tau \sim 1$ for $\omega_c \tau < 1$ and at $\omega \sim \omega_c$ for $\omega_c \tau > 1$, ω_c being the appropriate cyclotron frequency and τ the relaxation time. Curves c and d show the effect of taking into account multiple reflections in the specimen.[11] This is not normally a serious correction in the microwave region.

The variation of Θ_v with H in the microwave region (1 cm. radiation) is shown in Figure 5. An interesting feature is that, unlike the Faraday rotation, Θ_v approaches a limiting value independent of H in very strong fields. This can be shown to follow when the required tensor components are evaluated in the approximation $\omega\tau \ll 1$ and $(\omega_c\tau)^2 \gg 1$.

If the magnetic field is rotated round the propagation direction Θ_v varies as shown in Figure 6, where ϕ is the angle between $\vec{H}$ and the [001] direction and the situation is described by the tensor in (4). The lack of symmetry about the [011] direction should be noted. This is brought about by the component S_{yz}, which vanishes when $\phi = \pi/4$ and is antisymmetric with respect to the [011] direction.[3]

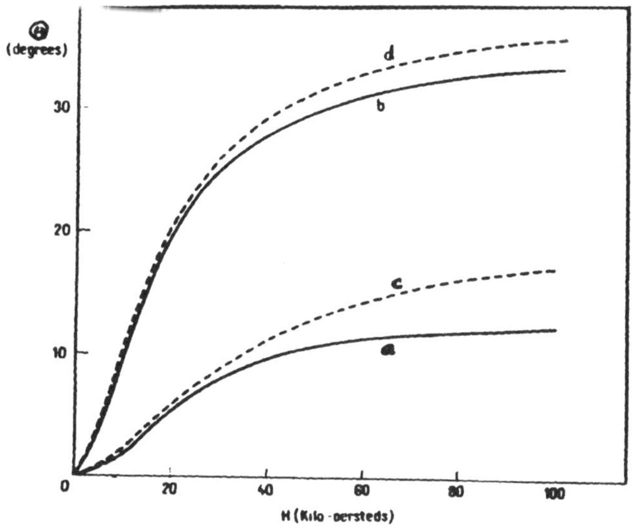

Figure 5. Variation of Voigt rotation with field strength, with $\omega = 1.88 \times 10^{11}$ sec^{-1}, $x = 0.5$ cm, a and c, H along [001] ; b and d, H along [011] ; c and d, with multiple reflections included.

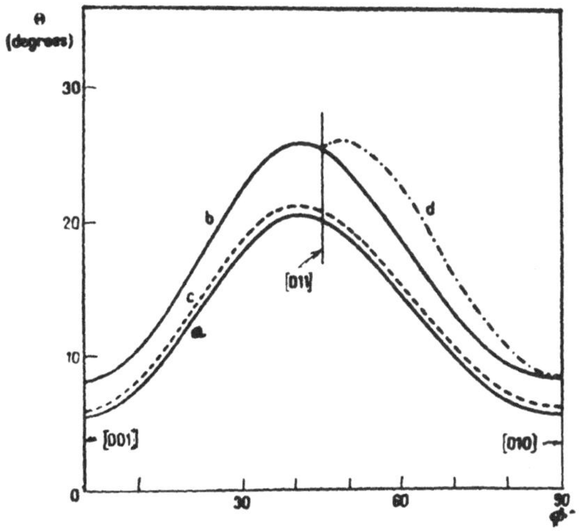

Figure 6. Variation of Voigt rotation with orientation of magnetic field in (100) plane, with $\omega = 1.88 \times 10^{11}$ sec^{-1}, $x = 0.5$ cm, a and c, $H = 2 \times 10^4$ oersteds; b, $H = 3 \times 10^4$ oersteds; c, with multiple reflections included; d, section of b for $0 < \phi < \pi/4$ reflected about [011] direction.

References

1. B. Donovan and J. Webster, Proc. Phys. Soc. 79, 46, 1081 (1962).
2. S. Teitler, J. Phys. Chem. Solids, 24,1487 (1963).
3. J. Webster and B. Donovan, Brit. Journ. Appl.Phys. 16, 25(1965).
4. M. Cardona, Helv. Phys. Acta 34, 796 (1961).
5. B. Abeles and S. Meiboom, Phys. Rev. 95, 31 (1954).
6. B. Donovan and J. Webster, Proc. Phys. Soc. 81, 90 (1963).
7. A. Bouwknegt and J. Volger, Proc. Int. Conf. Semiconductor Physics, Paris, 1964, p. 281.
8. J. Webster and B. Donovan, Physics Letters, 2, 330 (1962).
9. E.D. Palik, J. Phys. Chem. Solids, 25, 767 (1964).
10. B. Donovan and J. Webster, Proc. Int. Conf. Semiconductor Physics, Paris, 1964, p. 275.
11. B. Donovan and T. Medcalf, Brit. Journ. Appl. Phys. 15, 1139(1964).

CALCULATION OF MAGNETO-CONDUCTIVITY TENSOR FOR SEMICONDUCTORS WITH WARPED ENERGY SURFACES

B. Donovan and Rosemary Herbert

Westfield College, University of London

The energy surfaces for the holes near the top of the valence band in Ge, Si and diamond consist of two sets of warped spheres, which are degenerate at $\vec{k} = 0$. Owing to the complicated form of these surfaces a lengthy numerical calculation is, in general, required to solve the Boltzmann equation and obtain the magneto-conductivity tensor. In weak magnetic fields an expansion in powers of H is possible, and was used by Lax and Mavroides[1,2] to discuss d.c. galvanomagnetic effects in p-type Ge and Si. A formal method was given by McClure,[3] based on a Fourier expansion of the carrier velocity in a plane normal to $\vec{H}$. This can be applied to warped spheres[4,5] but in general a large number of Fourier components is required. An approximate solution of the high frequency problem was used by Zeiger, Lax and Dexter[6] (ZLD) in the analysis of cyclotron resonance, and involved a Fourier expansion of the perturbed distribution function. A similar approach is used in the present work, which is aimed at deriving the high frequency magneto-conductivity tensor components, for arbitrary field strengths, in a form suitable for computer programming.

Following ZLD, the relation between energy ϵ and momentum $\vec{p}$ is expressed in the form

$$\epsilon = -\frac{p^2}{2m^*}(1+g), \tag{1}$$

where m^* is an average effective mass and g is a small angular term. In terms of spherical polar coordinates (p, θ', ϕ'), referred to a cubic axis, we have

$$g(\theta', \phi') = K\left[\sin^4\theta'(\cos^4\phi' + \sin^4\phi') + \cos^4\theta' - \frac{2}{3}\right], \tag{2}$$

and K and $m^{\times}$ can be obtained for each type of hole from cyclotron resonance data. For p-type Ge, the values of K are roughly 1.08 and 0.15 for the heavy and light holes respectively.

We shall seek a solution of the Boltzmann equation, taking the electric field $\vec{E} \propto e^{i\omega t}$ and assuming the relaxation time τ to be independent of energy. If the distribution function is written as $f = f_0 - \Phi \, \partial f_0 / \partial \epsilon$ the equation to be solved is

$$-e\tau \vec{E}\cdot\vec{v} + \frac{e\tau}{c}(\vec{v}\times\vec{H})\cdot \mathrm{grad}_p \Phi - (1+i\omega\tau)\Phi = 0. \quad (3)$$

The two types of hole give separate solutions and the total conductivity tensor $\underset{\sim}{S}(\omega,\vec{H})$ is the sum of the two contributions.

We now transform to coordinates (p,θ,ϕ), referred to $\vec{H}$ as the z-axis, and write

$$\Phi = \frac{e\tau}{m^{\times}} p \, \chi(\theta,\phi) . \quad (4)$$

For $\vec{H}$ in a given plane, the function g is expanded as a series in $\cos n\phi$ and $\sin n\phi$, and the coefficients $g_{\pm n}$ are functions of θ and θ_H, the angle between $\vec{H}$ and the [001] direction (see ZLD, Appendix). Introducing the unit vector $\vec{u}(=(m^{\times}/p)\,\vec{v})$ parallel to the velocity, and using (1) and (4), equation (3) may be written as

$$\vec{E}\cdot\vec{u} + \omega_0\tau\left(1+g+\frac{\cot\theta}{2}\cdot\frac{\partial g}{\partial\theta}\right)\frac{\partial\chi}{\partial\phi}$$
$$-\frac{\omega_0\tau}{2}\left(\cot\theta\frac{\partial g}{\partial\phi}\right)\frac{\partial\chi}{\partial\theta} - \left(\frac{\omega_0\tau}{2}\cdot\frac{\partial g}{\partial\phi} - 1 - i\omega\tau\right)\chi = 0 \quad (5)$$

where $\omega_0 = eH/m^{\times}c$, with the same $m^{\times}$ as in (1) and (4). Multiplying through by $\sin^2\theta$ to remove the singularities and putting $\vec{u}\sin^2\theta = \vec{w}$ then leads to an equation of the form

$$\vec{E}\cdot\vec{w} + A\frac{\partial\chi}{\partial\phi} - B\frac{\partial\chi}{\partial\theta} - C\chi = 0. \quad (6)$$

We solve this by expanding χ in a double Fourier series

$$\chi = \sum_{nq} \chi_{nq} e^{in\phi} e^{iq\theta} \tag{7}$$

and likewise for A, B, C and $\vec{w}$. The coefficient B_{oo} is zero and all the C_{nq} are pure imaginaries except C_{oo}. The coefficients in (7) are then obtained from the equation

$$D(n)\chi_{nq} = \vec{E}\cdot\vec{w}_{nq} + \sum_{n'q'}{}' \left(in' A_{\substack{n-n'\\q-q'}} - iq' B_{\substack{n-n'\\q-q'}} - C_{\substack{n-n'\\q-q'}} \right) \chi_{n'q'} \tag{8}$$

where $D(n) = C_{oo} - in A_{oo}$

and the prime on the summation denotes exclusion of the term $n = n'$, $q = q'$. Equation (8) is solved by iteration and leads finally to χ_{nq} in the form

$$\chi_{nq} = \sum_{\alpha} V_{nq}^{\alpha} E^{\alpha}, \quad (\alpha = x, y, z) . \tag{9}$$

If the initial value of χ_{nq} is obtained from the first term in (8) we have $n = 0, \pm 1 \ldots\ldots\ldots \pm 5$ and $q = \pm 1, \pm 3 \ldots\ldots \pm 7$ (odd values only); hence $|n-n'| \leq 4$ and $|q-q'| \leq 6$. It is, however, preferable to include all the coefficients independent of the warping and to use the initial value of χ_{nq} given by the spherical approximation (K = 0). Preliminary computer calculations indicate that reasonably self-consistent solutions are obtainable with about 4 or 5 iterations. Although the number of n and q values increases with each iteration it is not necessary to retain all the components in every stage.

The current density (for one type of hole) is given by

$$\vec{J} = \frac{e}{4\pi^3\hbar^3} \int \vec{v}\, \Phi \frac{\partial f_o}{\partial \epsilon}\, d\vec{p} \tag{10}$$

and if we substitute from (4) and introduce the carrier density N_o as in ZLD, we have

$$\vec{J} = a \iint \chi(\theta,\phi)\, \vec{P}(\theta,\phi) \sin\theta\, d\theta\, d\phi \tag{11}$$

where $a = \dfrac{3N_0 e^2 \tau}{m^{\times}}$

and the vector $\vec{P}$ is defined by

$$\vec{P} = \vec{u}\,(1+g)^{-5/2}\left[\iint (1+g)^{-3/2} \sin\theta\, d\theta\, d\phi\right]^{-1} . \quad (12)$$

In the notation of (9) the components of the conductivity tensor are given by

$$S^{\alpha\beta} = a \iint P^{\alpha} V^{\beta} \sin\theta\, d\theta\, d\phi . \quad (13)$$

Expanding the components of $\vec{P}$ and $\vec{V}$ as in (7) and carrying out the angular integrations, we have finally

$$S^{\alpha\beta} = 4\pi a \sum_{n} \sum_{qq'} \left[(q+q')^2 - 1\right]^{-1} P^{\alpha}_{nq} V^{\beta}_{-nq'} . \quad (14)$$

Note that n runs over positive and negative values and the terms for $\pm n$ can be combined as in the treatment of McClure[3], where the quantity $1 + i(\omega - n\omega_0)\tau$ is just D(n) in equation (8). Thus the magnetic field appears in the numerator of the off-diagonal components but cancels out in the diagonal terms.

References

1. B. Lax and J.G. Mavroides, Phys. Rev. 100, 1650 (1955).
2. J.G. Mavroides and B. Lax, Phys. Rev. 107, 1530 (1957).
3. J.W. McClure, Phys. Rev. 101, 1642 (1956).
4. A.C. Beer and R.K. Willardson, Phys. Rev. 110, 1286 (1958).
5. J. Kolodziejczak and S. Zukotynski, Acta Physica Polonica 23, 783 (1963).
6. H.J. Zeiger, B. Lax and R.N. Dexter, Phys. Rev. 105, 495 (1957).

MEANING OF AN ANOMALY IN THE X-RAY SCATTERING OF ZnSe*

P. M. Raccah

Lincoln Laboratory, Massachusetts Institute of Technology

Lexington, Massachusetts 02173

ABSTRACT

Accurate relative X-ray intensity measurements were made on ZnSe. Comparison of the results with the values calculated from the theoretical atomic scattering reveals systematic discrepancies. The discrepancies are different for the three families of reflections allowed by the structure, and they occur for all observable peaks. An explanation is suggested based on electron transfer.

INTRODUCTION

Numerous works[1-3] have been published on the observed discrepancy between measured form factors and calculated scattering functions. This has so far been attributed to the Hartree-Fock atomic or ionic wave functions. It has been pointed out,[4] however, that in comparing theory and experiment, use is made of the usual X-ray scattering formulas in which the overlap terms are neglected. More explicitly, a model is adopted in which the crystal wave function is approximated by a single Slater determinant made up of nonorthogonal orbitals, but the charge density is calculated as if they were orthogonal orbitals. Such an approach may prove to be particularly inadequate in the case of metals and broad-band semiconductors in which electrons are quite delocalized.

In view of this situation, it seemed interesting to investigate a broad-band semiconductor. Choice was made of ZnSe because of its simple structure and of the high purity of the available materials. It was found

*This work was sponsored by the U. S. Air Force.

that a large discrepancy exists between the observed and calculated intensities, even extending to very large angles.

EXPERIMENTAL

Two different samples were used: one on a phosphor-pure commercial product, the other prepared by firing stoichiometric quantities of spectroscopically pure Zn and Se in an evacuated silica tube. The mass spectrometer indicated both to be of suitable purity.

Relative intensity measurements were made on two Norelco diffractometers using CuKα radiation and two different counting techniques. One method was by step scanning at 0.01° intervals, the other consisted of scanning several times across each peak to accumulate the total count. Diffracted-beam monochromators were used in both cases. Reproducibility of the results was checked from sample to sample, run to run, and instrument to instrument.

RESULTS AND DISCUSSION

Since this study is a comparison of experimental results with theory, it seemed convenient to present our data as a plot (Fig. 1)

$$100(I_{obs}/I_{calc}) \text{ versus } (\sin\theta/\lambda) \quad ,$$

where I_{obs} = the observed relative intensity, I_{calc} = the relative calculated intensity using the form factors resulting from Hartree-Fock calculations, θ = the Bragg angle, and λ is the wavelength used.

In the calculation of the intensities, the scaling factor s, the Zn atomic temperature factor B_{Zn}, and the Se atomic temperature factor B_{Se} were considered as parameters and were used to minimize the function

$$R = \sum_{hkl} \frac{|I_{calc}(hkl) - I_{obs}(hkl)|}{\sum_{hkl} I_{obs}(hkl)} \quad ,$$

with

$$I_{calc}(hkl) = sm(hkl)\, P(\theta) |F_c|^2 \quad ,$$

$$|F_c| = |f_{Zn} + f_{Se} \exp[\pi i/2(h+k+1)]| \quad ,$$

$$f_{Zn} = [f_{Zn}^{0}(\theta) + \Delta f_{Zn'} + i\Delta f_{Zn''}] \exp(-B_{Zn} x^2) \quad ,$$

and

$$f_{Se} = [f_{Se}^{0}(\theta) + \Delta f_{Se'} + i\Delta f_{Se''}] \exp(-B_{Se} x^2) \quad ,$$

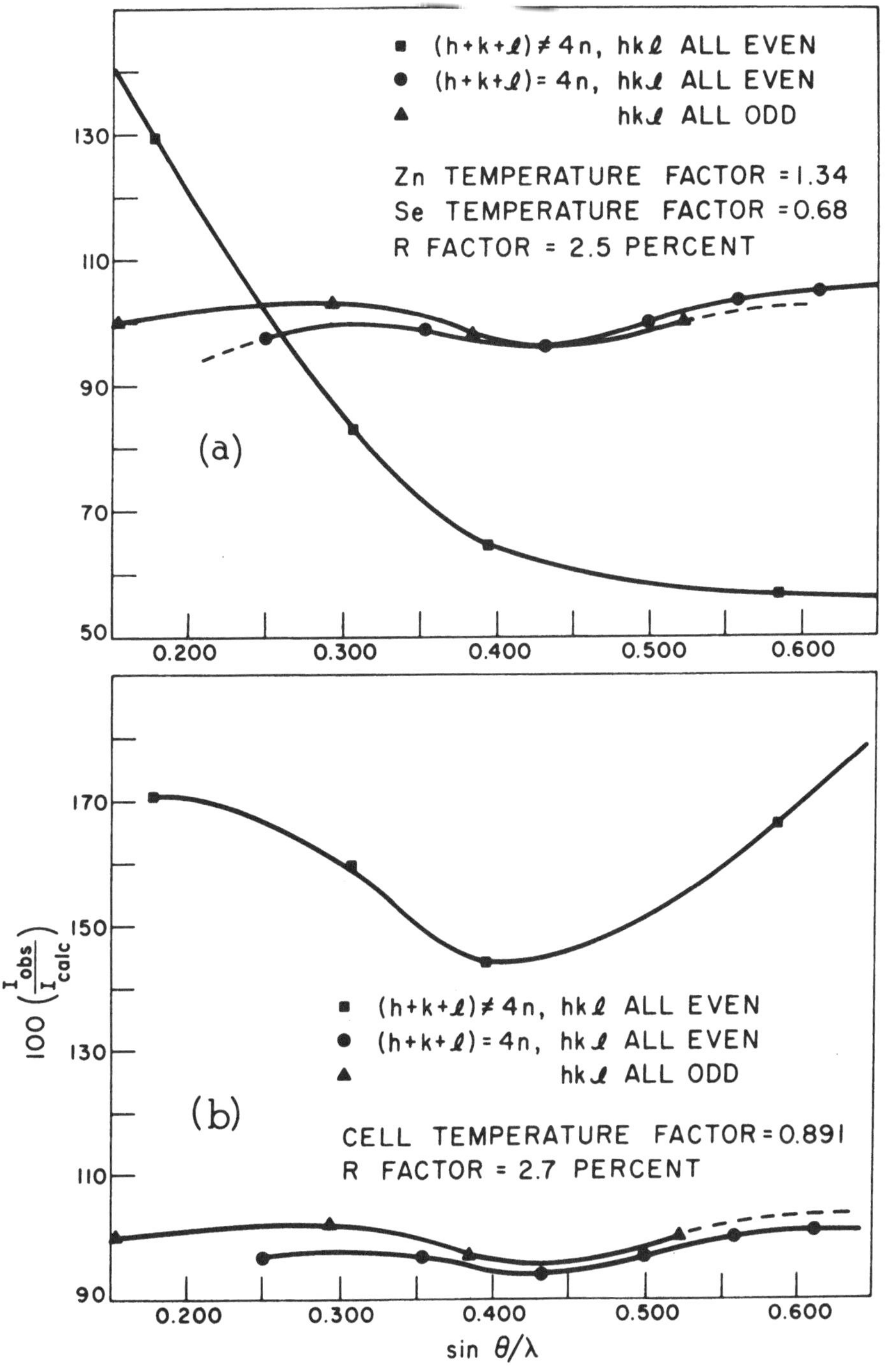
(h+k+ℓ)≠4n, hkℓ ALL EVEN
(h+k+ℓ)=4n, hkℓ ALL EVEN
hkℓ ALL ODD
Zn TEMPERATURE FACTOR = 1.34
Se TEMPERATURE FACTOR = 0.68
R FACTOR = 2.5 PERCENT
(a)
(h+k+ℓ)≠4n, hkℓ ALL EVEN
(h+k+ℓ)=4n, hkℓ ALL EVEN
hkℓ ALL ODD
CELL TEMPERATURE FACTOR = 0.891
R FACTOR = 2.7 PERCENT
(b)
100 (I_obs/I_calc)
sin θ/λ

TABLE I. COMPARISON OF OBSERVED X-RAY INTENSITIES AND STRUCTURE FACTORS FOR ZnSe WITH VALUES CALCULATED USING OPTIMAL TEMPERATURE FACTORS $B_{Zn} = 1.34$, $B_{Se} = 0.68$

h k l	2θ Values	Observed intensities	Calculated intensities	Ratio $100(I_{obs}/I_{calc})$	Observed structure factor	Calculated structure factor
1 1 1	27.25	195 983.000	195 982.993	100.00	145.904	145.904
2 0 0	31.57	1567.000	1210.016	129.50	17.514	15.390
2 2 0	45.23	135 351.000	138 672.031	97.60	166.534	168.564
3 1 1	53.58	83 934.000	81 575.800	102.89	110.096	108.538
2 2 2	58.16	455.000	547.279	83.18	14.710	16.133
4 0 0	65.84	20 821.000	21 067.804	98.83	133.591	134.381
3 3 1	72.63	31 064.000	31 726.008	97.91	88.726	89.667
4 2 0	74.83	808.000	1252.392	64.54	14.648	18.237
4 2 2	83.46	38 516.000	39 999.834	96.29	108.867	110.944
4 4 0	100.38	12 451.000	12 451.000	100.00	93.121	92.121
5 3 1	106.93	23 065.000	19 885.780	103.38	80.839	79.507
6 2 2	128.58	725.000	1277.788	56.73	14.150	18.786
4 4 4	140.50	7534.000	7189.945	104.78	69.881	68.267

TABLE II. COMPARISON OF OBSERVED X-RAY INTENSITIES AND STRUCTURE FACTORS FOR ZnSe WITH VALUES CALCULATED ASSUMING TEMPERATURE FACTORS $B_{Zn}=B_{Se}$=CELL TEMPERATURE FACTOR

h k l	2θ Values	Observed intensities	Calculated intensities	Ratio $100(I_{obs}/I_{calc})$	Observed structure factor	Calculated structure factor
1 1 1	27.25	195 983.000	195 982.996	100.00	146.155	146.155
2 0 0	31.57	1567.000	917.906	170.07	17.544	13.428
2 2 0	45.23	135 351.000	139 842.742	96.79	166.820	169.566
3 1 1	53.58	83 934.000	82 161.898	102.16	110.285	109.115
2 2 2	56.16	455.000	284.867	159.71	14.735	11.659
4 0 0	65.84	20 821.000	21 448.042	97.08	133.821	135.821
3 3 1	72.63	31 064.000	32 029.525	96.99	88.879	90.249
4 2 0	74.83	808.000	561.008	144.03	14.673	12.227
4 2 2	83.46	38 516.000	40 995.584	93.95	109.054	112.510
4 4 0	100.38	12 451.000	12 818.165	97.14	93.282	94.647
5 3 1	106.93	23 065.000	23 064.997	100.00	63.218	63.218
6 2 0	118.41	20 558.000	20 546.933	100.00	80.978	80.956
6 2 2	128.58	725.000	436.518	166.28	14.175	10.999
4 4 4	140.50	7534.000	7444.675	101.19	70.001	69.585

where hkl are the Miller indices, m(hkl) = the multiplicity, $P(\theta)$ = Lorenz-polarization factor including the monochromator correction $|F_c|$ = calculated structure factor, f = form factor including the thermal vibration term and the correction for anomalous dispersion, $x = (\sin\theta/\lambda)$, $f^0(\theta)$ = tabulated[5] form factors, $\Delta f' + i\Delta f''$ = tabulated[5] real and imaginary parts of the correction for anomalous dispersion. This refinement yielded a value of 2.5% for the R factor and the following values for the atomic temperature factors

$$B_{Zn} = 1.34 \text{ and } B_{Se} = 0.68 \ .$$

Table I presents a comparison of the calculated and observed intensities as well as the calculated and observed structure factors (uncorrected for thermal vibration). Our computer program also allowed us to impose the restriction

$$B_{Zn} = B_{Se} = \text{cell temperature factor} \ .$$

This second calculation yielded a value of 2.7% for the R factor and the value B = 0.891 for the over-all thermal vibration correction. Table II presents the results of this fitting. It appears that the introduction of atomic temperature factors does not improve significantly the fitting even when B_{Zn} and B_{Se} are very different. The agreement coefficient R is unaffected by a discrepancy, even large, on the small intensities. In addition, it does not differentiate a systematic arrangement of the points into subsets. To us its small values, in the two cases considered, do not give a complete representation of the results, and a plot (I_{obs}/I_{calc} versus $(\sin\theta/\lambda)$ reveals more details.

The results in Table II ($B_{Zn} = B_{Se}$) are presented in Fig. 1(b). The points are split into three families [Fig. 1(b)] corresponding to

hkl all even, $(h+k+1) \neq 4n$,
hkl all even, $(h+K+1) \neq 4n$,
hkl all odd .

Also the small lines, which correspond to

$$F = f_{Zn} = f_{Se} \ ,$$

i.e., to a difference in the scattering powers of the atom, are in very large error (average of 50%) with respect to the calculated values, whereas the fit is very satisfactory for the large and medium lines corresponding to

$$F = f_{Zn} + f_{Se} \ ,$$

and

$$F = f_{Zn} \pm if_{Se} \quad .$$

This discrepancy is considerably larger than what has been reported in the work on metals.[1-3] In these, the low-angle lines only were investigated, the agreement with the theory being supposed excellent at high angle. One can see that this is not the case, for the result presented, and that the disagreement covers the whole reflection circle. From a comparison of Figs. 1(a) and 1(b), it can be seen that the introduction of atomic temperature factors does not improve the situation in the sense that neither does it produce a merging of the curves nor does it reduce the discrepancy on the small lines. The effect observed can thus be explained only by (1) a systematic error such as to generate this peculiar effect, or (2) a theoretical error, which could be either an inadequacy of the Hartree-Fock theory or of the theoretical model as discussed above, or both.

POSSIBLE EXPERIMENTAL CAUSES

We list and discuss briefly below why the usual causes of error do not account for the families observed.

Preferred orientation. This possibility cannot be retained since the points divide into the classes of reflections indicated and not into cleavage and noncleavage lines.

Extinction. Samples with different history gave the same result, and extinction, to account for the separation, would have to reduce all of the strong lines by 50% over the whole 2θ, a requirement beyond what can be expected.

Half-wavelength contribution. Since a monochromator was used and since all of the weak peaks have strong peaks at one-half of this d value, this was considered a definite possibility. However, samples run without the monochromator or at voltages too low to excite the half-wavelength radiation gave essentially the same results.

Finally the three following checks were made.

External consistency. Using the same equipment, we measured the relative intensity of MgO and were able to reproduce closely the Burley[6] results.

Experimental arrangement. Since the Norelco is a vertical diffractometer, these results were reproduced on a Siemens diffractometer, which is horizontal, with excellent accuracy.

TABLE III. COMPARISON OF OBSERVED X-RAY INTENSITIES AND STRUCTURE FACTORS FOR ZnSe WITH VALUES CALCULATED ASSUMING TEMPERATURE FACTORS $B_{Zn} = 0.9879$, $B_{Se} = 0.7361$, $\alpha = 0.71$, $\beta_{Zn} = 11.45$, $\beta_{Se} = 14.92$

h k l	2θ Values	Observed intensities	Calculated intensities	Observed structure factor	Calculated structure factor
1 1 1	27.25	195 983.000	189 607.875	148.773	146.333
2 0 0	31.57	1567.000	1567.024	17.858	17.859
2 2 0	45.23	135 351.000	135 224.750	169.808	169.729
3 1 1	53.58	83 934.000	79 969.812	112.261	109.578
2 2 2	56.16	455.000	454.974	14.999	14.999
4 0 0	65.84	20 821.000	20 849.766	136.218	136.312
3 3 1	72.63	31 064.000	31 354.023	90.471	90.892
4 2 0	74.83	808.000	854.022	14.936	15.356
4 2 2	83.46	38 516.000	40 131.664	111.007	113.312
4 4 0	100.38	12 451.000	12 644.336	94.952	95.687
5 3 1	106.93	23 065.000	23 064.766	64.351	64.350
6 2 0	118.41	20 558.000	20 424.063	82.429	82.160
6 2 2	128.58	725.000	725.001	14.428	14.428
4 4 4	140.50	7534.000	7458.637	71.255	70.898

Direct verification. Jennings,[7] using one of our samples, made absolute intensity measurements, using Mo Kα radiation, on the (220) and the (200) lines. His findings were in close agreement with ours.

POSSIBLE EXPLANATION

It has been seen that when B_{Zn} and B_{Se} are allowed to vary independently, their optimum values differ by approximately a factor two. Such a result is in itself inacceptable, since the masses of these two atoms are close, and its meaning might be that the radiuses and densities of Zn and Se in the solid are grossly different from the free atom. This would imply a redistribution of the outer electrons and the structure factor would have to be rewritten:

$$|F_c| = [f'_{Zn}(x) + \phi_{Zn}(x)]\exp(-B_{Zn}x^2) + (i)^{h+k+1}$$
$$[f'_{Se}(x) - \phi_{Se}(x)]\exp(-B_{Se}x^2)$$

with $f'(\theta) = f^0(\theta) + \Delta f' + i\Delta f''$.

In order to check this hypothesis one has to assume a form for $\phi_{Zn}(x)$ and $\phi_{Se}(x)$ with the requirement that $\phi_{Zn}(0) = \phi_{Se}(0)$ so that the total number of electrons remains the same. We have assumed various reasonable functions in real space and used their Fourier transform for $\phi_{Zn}(x)$ and $\phi_{Se}(x)$. The results reported in Table III are for the Gaussian case where $\phi_{Se}(x)$ and $\phi_{Zn}(x)$ turn out to be themselves Gaussians:

$$\phi_{Se}(x) = \alpha \exp(-\beta_{Se}x^2)$$
$$\phi_{Zn}(x) = \alpha \exp(-\beta_{Zn}x^2) \ .$$

The values determined for α, the number of electrons redistributed, B_{Se} and B_{Zn} did not differ much for the various real space functions we have used: $a \exp(-br^2)$, $ar^2\exp(-br^2)$, $a\, r\exp(-b|r|)$, etc. In all cases we found

$$\alpha = 0.7 \pm 0.1$$
$$B_{Zn} = 0.98 \pm 0.02$$
$$B_{Se} = 0.74 \pm 0.01$$

One will notice that B_{Zn} and B_{Se} are now acceptably close, also the discrepancies affecting the difference lines has vanished while the fit to the other lines has not worsened.

This very simple model appears to represent adequately the data. It implies that in ZnSe the atoms are not neutral like in a metallic alloy and on the average 0.7 ± 0.1 electrons are spending more time close to the Se nucleus than to the Zn nucleus.

REFERENCES

1. B. W. Batterman, D. R. Chipman, and J. J. DeMarco, Phys. Rev. 122, 68 (1961).
2. L. D. Jennings, D. R. Chipman, and J. J. DeMarco, Phys. Rev. 135, A1612 (1964).
3. M. J. Cooper, Phil. Mag. 7, 2059 (1962).
4. T. A. Kaplan and W. H. Kleiner (private communication).
5. International Tables for X-ray Crystallography, edited by K. Lonsdale (The Kynoch Press, Birmingham, England, 1962), Vol. III.
6. G. Burley, J. Phys. Chem. Solids 26, 1605 (1965).
7. L. D. Jennings, Army Materials Research Agency.

IMPURITIES IN SEMICONDUCTORS - I

D. H. Parkinson

Royal Radar Establishment, Malvern, England

Introduction

1. Impurity centres dominate the electrical properties of semiconductors because they introduce extra energy levels into the forbidden electronic energy band gap. In spite of being studied intensively for a long time, even in Germanium and Silicon, our knowledge of their behaviour is most imperfect. In these papers attention will be focussed primarily on these relatively simple materials.

Besides a knowledge of the position of the impurity levels on the energy scale, information is needed, for example, on their scattering cross sections for current carriers, their absorption cross section for radiation, or whether they act as traps for current carriers or as recombination centres.

The position of the impurity energy levels is apparently the simplest parameter and will be discussed here.

Shallow Impurity Levels and the "Hydrogen Model"

2. The most well known techniques for finding the position of the impurity levels are those in which the Hall coefficient and electrical conductivity are examined as functions of temperature. Such techniques have been thoroughly discussed by Putley[1] in his book which has been republished recently. A second method consists of examining the photo response as a function of wavelength[2]. These methods lead to the 'ionisation energy' of the impurity centre. With them it was shown that the group III elements as impurities in

Germanium and Silicon produced shallow acceptor levels and that the group V elements produced shallow donor levels. The results of such early investigations are shown in Table I.

TABLE I

GROUP III AND V IMPURITIES IN Si AND Ge

IMPURITY	D OR A	ENERGY IN e.v. FROM NEAREST BAND	
		Si	Ge
P	D	.044	.0120
As	D	.049	.0127
Sb	D	.039	.0097
Bi	D	.069	
B	A	.045	.0104
Al	A	.057	.0102
Ga	A	.065	.0108
In	A	.16	.0112

For both types of impurity in Germanium the ionisation energies E_o are all about .01 e.v. but have some dependence on the nature of the impurity atom itself. In Silicon there is a much greater variation in E_o but a rounded figure of about .05 e.v. can be taken.

3. It is usual to discuss these impurities in terms of the 'hydrogen model' which has been reviewed by W. Kohn[3]. For the present purposes some features of this model and the methods used in its development are picked out. Consider first a donor atom in Germanium or Silicon, such as phosphorus, imagine that it is ionised by the removal of one electron. If it is now inserted substitutionally into the lattice the remaining four electrons can satisfy the four tetrahedral homopolar bonds. But the Phosphorus atom is charged (+ e) and will polarise the lattice. At a considerable distance the mean potential will be e/Kr where K is the

static dielectric constant. Immediately in the vicinity of the impurity atom the potential well will be that of the atom itself.

If now the electron which was removed from the P atom is inserted, it will be a matter of energetics how it will behave. Let us suppose that it remains mostly under the influence of the potential e/Kr. As the valence band is full it can be supposed that it goes into states represented by wave functions from the conduction band. The Shroedinger equation could then be written as

$$\left(- \frac{\hbar^2}{2m^*} \nabla^2 - \frac{e}{Kr} \right) F(r) = EF(r) \qquad \ldots\ldots (1)$$

m^* is the effective mass from the conduction band. To simplify the model, for the moment it will be assumed that the material has a zone centred conduction band.

Equation (1) is simply that of the hydrogen atom but for the insertion of m^* and K. The resulting energy levels are those of the hydrogen atom but modified by m^*/K^2. We find at once the ionisation energies are not too unreasonable and also that there should be a series of excited levels associated with each impurity.

Such levels have been found first by Burstein[4] and his co-workers in Si using infra-red spectroscopy. It is, therefore, worth developing the crudely stated hydrogen model somewhat further.

Thus the Shroedinger equation for an impurity atom in an otherwise unperturbed lattice could be written as

$$\left(- \frac{\hbar^2}{2m} + V(r) - \frac{e^2}{Kr} \right) \psi = E\psi \qquad \ldots\ldots (2)$$

Provided e^2/Kr is small it can be assumed that the solutions of equation (2) can be written as a series in terms of the original Bloch functions of the unperturbed lattice,

$$\text{i.e.} \quad \psi' = \sum_k A_{nk} \cdot U_{nk}(r) \cdot e^{ik \cdot r}$$

In attempting to evaluate the coefficients A_{nk}, Kohn has shown that with small e^2/Kr the important terms pick out wave functions from the bottom of the conduction band. Further the wave equation is indeed hydrogen like but the complete solutions are of the form

$F(r).\phi(r)$ where $\phi(r)$ are Bloch functions and $F(r)$ is a slowly varying envelope function modulating the rapidly varying Bloch functions. In this analysis one also finds justification for replacing m by m^*.

The functions $F(r)$ are the important parts of the solutions and give them their hydrogen like quality. Thus in calculating the matrix elements for transitions from the ground to excited states the $\phi(r)$ are unimportant, the $F(r)$ govern the selection rules.

With Silicon and Germanium we know from cyclotron resonance and other measurements that the conduction bands are more complex than we have so far considered (equn. (1)). In Germanium there are four equivalent minima at the zone edge in the (111) directions. In Silicon there are six equivalent minima part way across the zone in the (100) direction. In a full analysis for these materials equation (2) must be replaced by a set of equations one for each minimum and the solutions are in the form

$$\sum_j \alpha_j F_j(r) \cdot \phi_j(r)$$

The ground states are therefore j. fold degenerate unless account is taken of the tetrahedral symmetry of the lattice, in which case this degeneracy is partly lifted. An analysis such as this can be carried through equally well for acceptor states.

In table II are shown the calculated and observed energy differences between the ground state and excited states for donors in Silicon

TABLE II
COMPARISON OF "HYDROGEN MODEL" WITH EXPERIMENT
DONOR IN Si

	THEORETICAL ENERGIES	ENERGY DIFFERENCES			
		THEORY	P	As	Sb
1s	- 2.9				
		1.8	3.45	4.21	3.18
2p, m = 0	- 1.09				
		0.50	0.50	0.53	0.47
2p, m = ± 1	- 0.59				
		0.30	0.31	0.32	0.34
3p, m = ± 1	- 0.29				

The calculated relative positions of the excited states are in reasonable agreement with experiment but the ionisation energies are badly wrong and the model does not, of course distinguish between different impurity atoms. These results are no better than the basic assumptions permit. Nevertheless this model gives some results which must be noted. First the wave functions are very widely spread covering many hundreds of atoms. In Germanium the equivalent Bohr radius of the ground state is ~ 30 atom spacings and in Silicon ~ 20 atom spacings. This implies that interactions between adjacent impurity sites can be expected at quite low concentrations (e.g. 10^{16} cms^{-3} impurity atoms) and that there can be interactions between an impurity centre and many atoms of the parent lattice (e.g. due to electron or nuclear spins). One can also see that the ionisation energy is sensistive to the potential well near the impurity atom.

The amplitude of the wave function $|\psi_0|$ at the impurity nucleus can be estimated from the hyperfine structure shown in spin resonance experiments[5]. Thus with phosphorus (nuclear spim $\frac{1}{2}$) in silicon the splitting of the resonance line yields $|\psi_0|^2$. Using this information the model can be improved but in a somewhat empirical way.

Impurity Conduction and Hopping Processes

4. Overlapping of the wave functions of shallow impurity centres occurs easily and it is natural to enquire whether electrons can pass from impurity centre to impurity centre so contributing to conduction processes. If such effects occur they will be essentially low temperature phenomena. Two possible processes can be distinguished.

(a) "metallic-like"conduction: for this to occur overlap must be large and according to Mott[6] such conduction sets in when the concentration (hence overlap) becomes high and is greater than some critical value (~ 10^{18} to 10^{19} cms^{-3} impurity centres). The impurity centres are in random array and such conduction forms part of the study of disordered structures. This type of conduction will not be discussed further here.

(b) Hopping processes: it can be imagined for example, that electrons can be excited at an impurity centre by phonons and can then tunnel to the adjacent centres without entering the conduction band. For such a process to take place there must be some impurity centres unoccupied by electrons, thus there must be a degree of compensation. The problem then for donors is that of the transport of electrons in a random lattice from one positively charged donor to another in the field of fixed negatively charged acceptors.

If donors are considered, the two simplest cases are, first, where $N_D >> N_D - N_n$ so that there are a small number of electrons in donor states most of which are empty, and second, where $N_D >> N_A$ so that there are a small number of vacancies for electrons in donor states. For general purposes however we require an analysis of the situation for any degree of compensation ($0 < K = N_A/N_D < 1$). Miller and Abrahams[7] have attempted such an analysis. They have calculated the probability per unit time ω_{ij} that an electron jumps to a neighbouring site as a function of the site separation, energy being conserved by the emission or adsorption of a phonon. To calculate the consequent conduction in the specimen, they have shown that the equation governing the rate of change of the probability of an electron occupying a given site and the net current flow in an electric field are equivalent to Kirchoifs laws for charge flow in a three-dimensional random resistance network. Each link in the network corresponds to two impurity centres while the link impedance is inversely proportional to ω_{ij} for the same two centres. The network resistivity can then be computed assuming that it arises from suitably chosen non-intersecting chains of impedances taken in parallel.

There have been a considerable number of observations of impurity conduction but the most significant for our purposes have been those of Fritzche and Cuevas[8] who published a series of measurements on p-type Germanium with acceptor (Ga) concentration ranging from 8×10^{14} to 1.3×10^{18} cm^{-3} and compensation (As and Se) ratio K fixed at 0.4. These impurities were introduced by slow neutron bombardment and the ratios of their concentrations were fixed by the capture cross sections for neutrons and the consequent decay schemes for the Germanium isotopes. The total amount of impurities was varied by varying the total exposure to neutrons. Studies of the resistivity and Hall coefficient as a function of temperature at low temperature ($\sim 4^{\circ}K$) enabled the activation energy ϵ for impurity conduction to be estimated as a

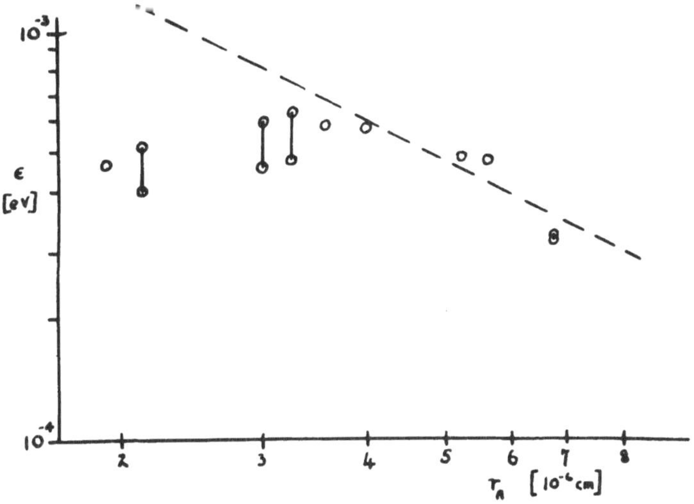

Figure 1 The activation energy ϵ of impurity conduction. Experimental points compared with Miller and Abrahams[7] theory. (Figure after Mott and Twose[6]).

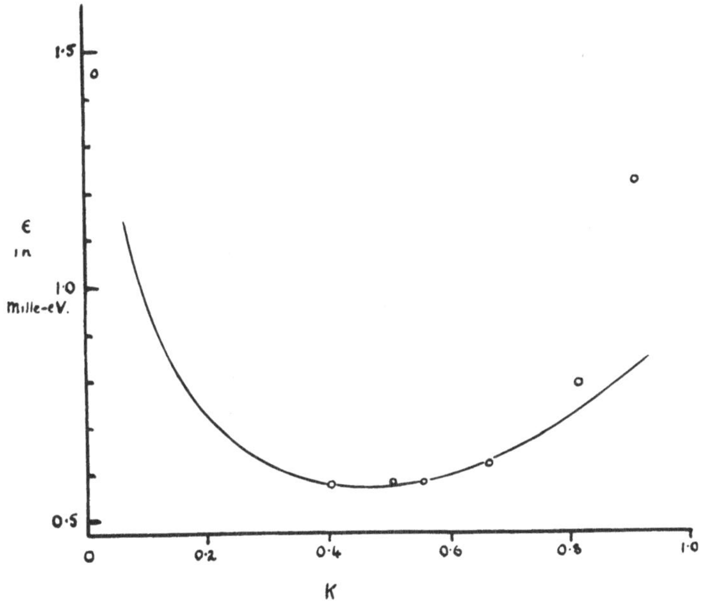

Figure 2 The activation energy ϵ of impurity conduction as a function of compensation ratio K. Experimental points compared with Miller and Abrahams[7] theory.

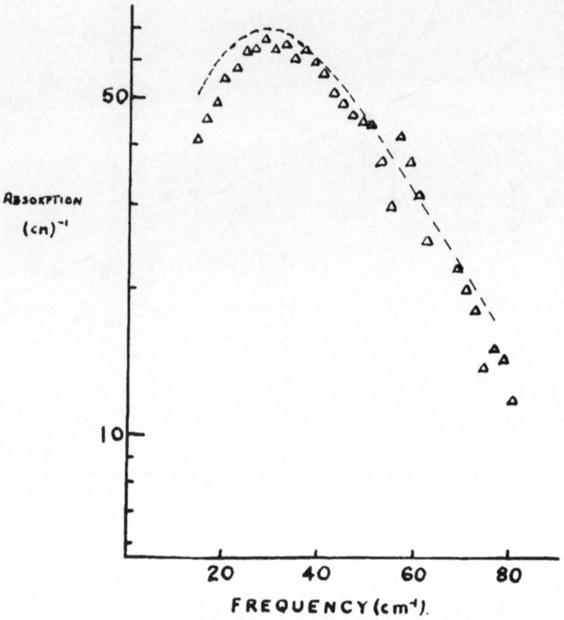

Figure 3 Far I.R. absorption due to photon-induced hopping, in compensated n-type Silicon. Triangles are experimental points, dashed line follows theory of Blinowski and Mycielski[9]. After Neuringer and Milward[10].

function of impurity separation and also as function of K. Figures 1 and 2 show these results. The theory is reasonably good for low impurity concentrations, the region to which it should apply.

Perhaps the most convincing evidence for hopping processes between impurity centres in semiconductors comes from studies of photon induced hopping. Blinowski and Mycielski[9] have developed a theory using techniques similar to those of Miller and Abrahams. These authors consider two possibilities, the first is simple photon induced hopping and the second, an "indirect" process, photon plus phonon induced hopping. Both processes should give rise to absorption bands in the very far infra-red around the 300 to 500 μ range but obviously the simple photon induced hopping should be much more probable than the three body process.

Milward and Neuringer[10,11] using a Fourier transform interferometer have observed absorption bands which can be ascribed to these effects. Figure 3 shows a set of their results. The agreement between theory and experiment in this instance is reasonably good.

References

1. Putley, E.H., "Hall Effect and Semiconductor Physics" Dover Publications Inc., New York, 1968.

2. Lifshitz, T.M., Likhtman, N.P., and Sidarov, V.I., 1968. JETP Letters 7, 84.

3. Kohn, W., "Solid State Physics", Vol. 5., Academic Press, New York, 1957.

4. Burstein, E., Picus, G.S., Henvis, B., and Wallis, R., 1956, J. Phys. Chem. Solids, 1, 65.

5. Feher, G., Wilson, D.K., and Gere, E.O., 1959, Phys. Rev. Letters, 3, 25.

6. Mott, N.F. and Twose, W.D., 1961, Advances in Phys., 10, 107.

7. Miller, A. and Abrahams, E., 1960, Phys. Rev., 120, 745.

8. Fritzche and Cuevas, 1960, Phys. Rev., 119, 1238.

9. Blinowski, J., and Mycielski, J., 1964, Phys. Rev., 136, A266.

10. Milward, R.C., and Neuringer, L.J., 1965, Phys. Rev. Letters, 15, 604.

11. Neuringer, L.J., Milward, R.C., and Aggarwal, R.L., 1966, Proc. 8th Int. Conf. Semiconductors, Kyoto, p583.

IMPURITIES IN SEMICONDUCTORS - II

D. H. Parkinson

Royal Radar Establishment

Malvern, England

1. In this paper attention is turned to the 'deep' impurity levels in semiconductors concentrating in the first place on Germanium and Silicon.

It is useful to discuss impurities in terms of the homopolar bonding model which has already been referred to for shallow impurities. Thus, if an element from Group II of the periodic table enters the Germanium or Silicon lattice substitutionally, it can be imagined that the two valence electrons can satisfy two of the tetrahedral bonds leaving two unsatisfied. Such an impurity should then be a double acceptor, the first electron entering the lower level and only when this is occupied will there be an upper one for the second to enter. It is feasible for such a substitutional impurity to show donor characteristics also in some circumstances so that an electron is excited from it leaving three unsatisfied bonds at impurity. Such a donor level can be expected to be far from the conduction band.

It is now evident that it is important to know whether the impurity is substitutional or interstitial, for if the Group II atom appears in an interstitial site it can only be a double donor. Group VI impurities can be treated by a similar simple line of argument. Likewise the elements of Group I, Copper, Silver and Gold, on substitution in the diamond lattice of Group IV should appear as triple acceptors, for there will be three unsatisfied bonds at such impurity sites. It is feasible that they can also be singly ionisable donor centres only. Similar reasoning could be used for Group VII impurities.

2. Before examining the experimental situation for deep impurities in Germanium and Silicon, reference must be made to another method of obtaining information on the electronic nature of the impurity sites: that is by studies of solubilities as functions of carrier concentration. Shockley and Moll[1] have examined the interaction between impurities which have deep levels (e.g. Copper) and shallow donors and acceptors. They have shown, at a given temperature, that it depends only on the ratio of free hole (or electron) concentration to the intrinsic carrier concentration and is, therefore, independent of the particular donor or acceptors used. It can also be shown that the solubility can be used to give the multiplicity, r, of the energy levels associated with the deep impurity. Thus if the solubility of substitutional copper, C^s, in extrinsic n-type Germanium is taken as an example, and one can assume that the acceptor levels are below the middle of the energy gap, then for non-degenerate material the solubility, C^s, should be given by $C^s = C^s{}_i\ (n/n_i)^r$.

Where n is the free carrier concentration and the subscript i implied intrinsic material. For Copper in Germanium r should be 3.

Hall and Racette[2] have made extensive investigations of the solubility of Copper in a number of host lattices using radioactive trace techniques (Cu^{64}, half life 12.8 h). Their results indicate that in both Germanium and Silicon substitutional Copper should be a triple acceptor while interstitial Copper should be a single donor. The form of their results in p-type Silicon are shown in Figure 1., the minimum in the curve showing a very small ratio of substitutional to interstitial solubility in intrinsic Silicon. Curves of the form

$$C = C^i{}_i\ (n_i/n) + C^s{}_i\ (n/n_i)^3$$

can be drawn through the experimental points, the superscript i referring to interstitial impurities. Our interest is in the index 3.

3. In figures 2 and 3 the present information on most deep impurity levels in Germanium and Silicon is summarised. The position of the energy levels is measured from the nearest band edge. Most of this information is derived from Hall effect measurements using counter-doping techniques to reveal the deep levels (see Putley[3]). Figure 2 refers to Germanium[4]. There it can be seen that the Group II materials Zn, Cd and Hg, do indeed create double acceptor levels. The Group I materials Cu, Ag and Au, all create triple acceptor level but Gold also shows a single donor level near to the valence band edge. This behaviour is what we expect from the simple homopolar bonding model. The group six elements are

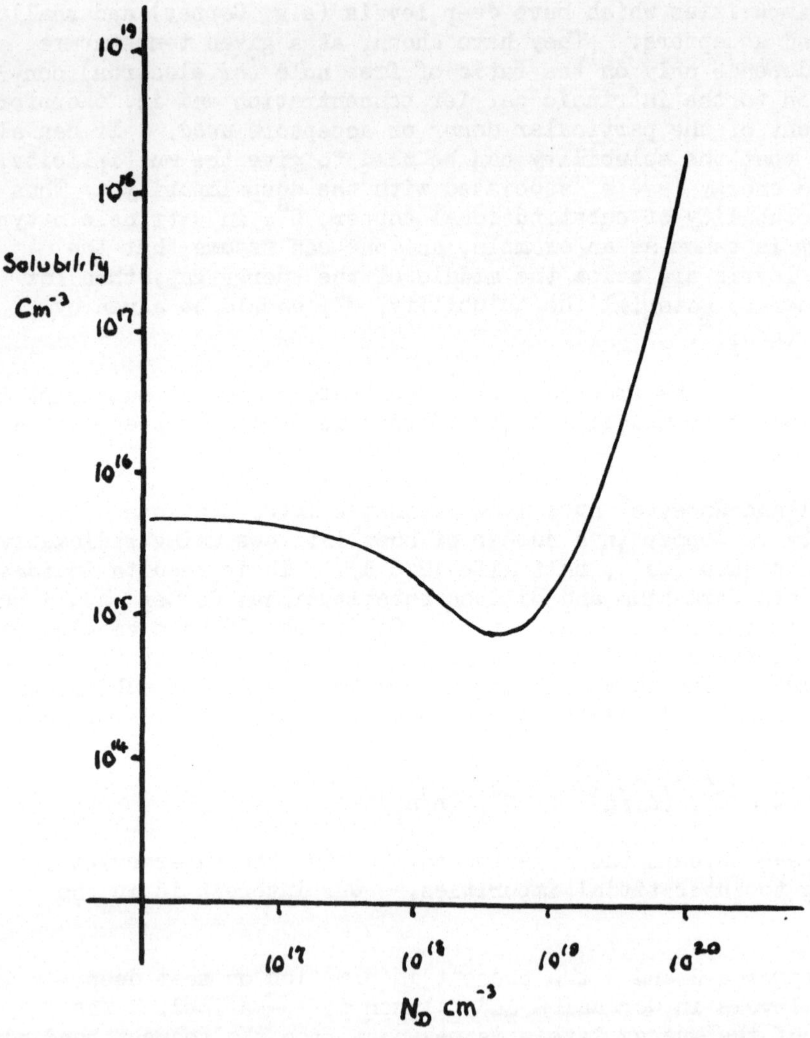

Figure 1. Form of results of Hall and Racette for solubility of Copper as function of concentration of donor concentration.

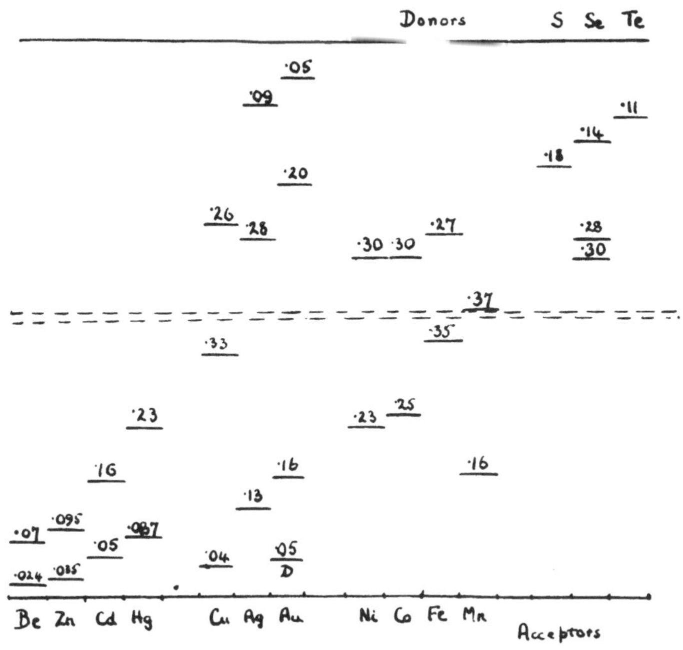

Figure 2. Deep impurity levels in Germanium.

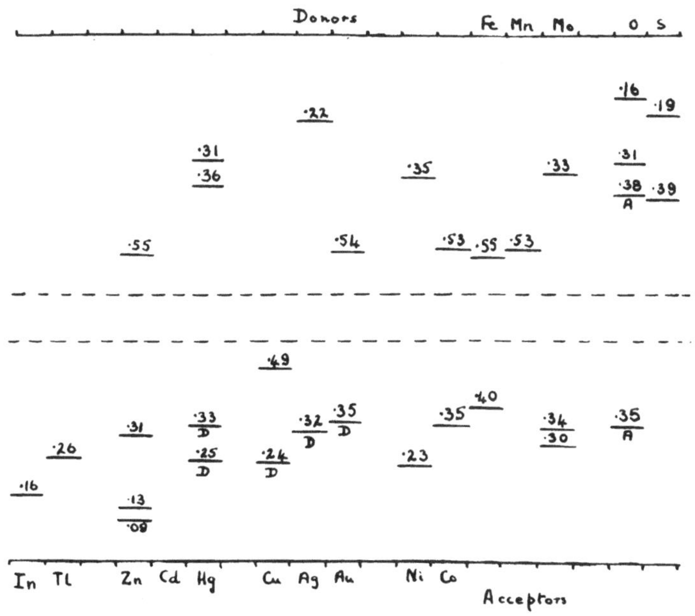

Figure 3. Deep impurity levels in Silicon.

donors as expected but are not clearly established as double donors. The transition metals Mn, Fe, Co Ni, are apparently all double acceptors while cobalt also shows a donor level.

4. In studying Figure 3 it must be recalled that Silicon is the most important technical electronic material at the moment and is likely to remain so for some years. The single acceptor levels of In and Tl are shown, Tl being the deepest of all the III, V impurities. Zn, Cd and Hg have the acceptor levels shown attributed to them. The shallow levels at .09 and .13 ev with Zn only appear in material which is already fairly heavily doped with Boron. They were first reported in the 1950's but recently work points to them being associated with Zn-B pairs in Silicon and not characteristic of Zn as a substitutional atom.

The levels associated with Cu, Ag and Au, are a deep acceptor and donor in each case. A result for Copper which is at variance with the indirect conclusions drawn by Hall and Racette. These authors have also shown that at an elevated temperature interstitial Copper is far more soluble (10^4) than substitutional Copper and probably precipitates out rapidly in a specimen on cooling. They suggest that the acceptor level at .49 ev and the donor level at .24 are characteristic of the precipitates rather than of substitutional Copper. This seems reasonable and if so, we know nothing about the position of the energy levels of substitutional Copper in Silicon. Similar reasoning cannot be used with the present evidence[5] for the levels associated with Au and Ag. The behaviour of Mn, Fe, Ni and Co in Silicon are in contrast to their behaviour in Germanium. Mn and Fe are now donors while Co and Ni remain double acceptors. It is not clear whether the observations on Mn and Fe refer to substitutional atoms or to interstitial, the weight of evidence leans to the former.

The differences in behaviour between the same substitutional atoms in Si and Ge are due to the differences in the host lattices, namely, in the lattice spacing and hence the potential wells. The significant differences must be in those regions near the boundaries of the 'atomic Polyhedra'. It is difficult to define satisfactory polyhedra except in the simplest lattices, however there must be a "unit cell" about each atom such that there is a 'ridge' of potential in crossing the boundaries from cell to cell[6]. The potential will vary in height round the perimeter of such a cell. This variation must be sensitive to the nature of the impurity atom in a given lattice, as well as to the lattice surrounding a given impurity, and must also govern the behaviour of the wave functions of the electrons associated with an impurity. Unfortunately there is no satisfactory analysis of this problem at present.

5. Returning to Figure 3, both oxygen and sulphur act as donors in silicon as might be expected. Oxygen, however, is present in large quantities in most silicon specimens. "Oxygen free" silicon grown by the floating zone technique probably contains less than 10^{16} atoms cm^{-3}. Most oxygen is dissolved interstitially and usually forms the electrically inactive group Si_2O, the presence of which is revealed by an absorption band at 9 μ (amongst the two and three phonon lattice absorption bands). There is also evidence for an SiO_4 complex with which the donor level at .16 ev below the conduction band is probably associated. The two acceptor levels shown are also probably associated with complexes.

To add further to the difficulties with Silicon recent attempts[7] to carry out accurate analyses of the carbon content point to concentrations above 18 atoms cms^{-3} in all specimens. In view of the possibility of ion pairing and complex formation between impurities themselves and between impurities and vacancies such results are disturbing.

6. If the homopolar bonding model is applied to the III-V compounds the possible behaviour of impurity atoms increases in complexity for there are now two sites to choose from. Fortunately most groups of impurities show preferential substitution on one site. Thus on present evidence Group II elements such as Zn and Cd are single acceptors substituting on a Group III site. The ionisation energies vary widely from as low as .0075 e.v. for Zn and Cd in indium antimonide to .05 e.v. for Cd in indium phosphide the 'hydrogen' model indicates $\sim$ .05 e.v. Likewise the Group VI impurities substitute on a Group V site as single donors.

The elements of Group IV are interesting for there seem to be three choices:

a) they can occupy only one sub-lattice so acting either as single donors (on Group III sites) or as single acceptors only (on Group V sites).

b) they could be sited in pairs on adjacent sites one on each sub-lattice so being overall electrically neutral.

c) they can be distributed statistically between the two sub-lattices, the concentration in a particular sub-lattice depending on the nature and concentration of the impurity and external conditions.

In fact in most III-V compounds a given Group IV impurity substitutes preferentially on one sub-lattice usually that with the largest "ionic radius" but there are exceptions to this rule, e.g. Sn (radius 1.4Å) substitutes on Ga (radius 1.26Å) sites in GaAs.

With Si in GaAs it has been shown[8] that the carrier (n-type) concentration at first increases linearly with Si atom concentration. At near 10^{19} atoms cms^{-3} the carrier concentration levels off as the Si concentration increases. This can be explained if at high concentrations Si atoms increasingly occupy As sites so acting as acceptors and partially compensating the Si atoms on Ga sites.

Impurities from Group I introduced substitutionally usually act as acceptors but in some instances the situation is complicated by the formation of complexes with vacancies and other defects. Indeed such complexes can also be formed with Group II, IV and VI impurities.

In this discussion it is not possible to cover the II, VI compounds, the student can work out the variety of possibilities for substitutional and interstitial atoms if the homopolar bonding model is used. However, in these materials variations from that behaviour are more frequent.

In summary one can say that the simple homopolar bonding model is extremely useful. It can be applied in Si and Ge with caution, in the III and V compounds with caution squared, and in the II-VI compounds with caution cubed.

It is to be noted that going from Ge and Si to the III-V compounds and then to the II-VI compounds is also the direction in which the ionicity of the atoms of the classes of materials increases.

REFERENCES

1. Shockley, W. and Moll, J. L., 1960, Phys. Rev., 119, 1480

2. Hall, N. J., and Racette, J. H., 1964, Jour. Appl. Phys., 35, 379.

3. Putley, E. H., 1968, "Hall Effect and Semiconductor Physics", Dover Publications Inc., New York.

4. Putley, E. H., 1964, Physica Status Solidi, 6, 571.

5. Schibli, E., and Milnes, A. G., 1967, Mats. Sci. and Eng, 2, 173

6. Bullis, W. M., 1966, Solid State Electronics, 9, 143.

7. Ziman, J., 1967, Proc. Phys. Soc., 91, 701.

8. Schink, N., 1965, Solid State Electronics, 8, 767.

9. Whelan, J. M., Struthers, J. D., Ditzenberger, J. A., 1967, Proc. International Conference on Semiconductors, Prague, p.943.

EMISSION FROM EXCITED TERMINAL STATES OF BOUND EXCITON COMPLEXES

D. C. Reynolds

Aerospace Research Laboratories, Wright-Patterson

Air Force Base, Ohio

ABSTRACT

Emission from the excited terminal states of bound exciton-donor complexes has been observed in CdS and CdSe crystals. Studying these optical transitions allows one to determine the donor ionization energies, the electron effective masses as well as the electron g-values in these materials. A good theoretical fit to the experimental data was obtained, using the effective mass approximation. Emission from the excited terminal states of bound exciton-acceptor complexes has not yet been observed in these materials. There is no basic reason why such transitions should not occur. Studying transitions of this type would allow one to obtain fundamental information concerning the acceptor impurities in these materials.

I. INTRODUCTION

Measurements of magnetic field splittings of donor and acceptor impurities in semiconductors allows one to study the ionization energies of these impurities as well as the effective masses of the carriers. Studies of this type have been made on a number of different impurities in Si and Ge.[(1),(2),(3)] These investigations were made by studying infrared absorption in a magnetic field.

The highest quality crystals obtainable in II-VI materials are platelet type crystals. These crystals are very thin (0.5μ-50μ), as a result they are not suitable for infrared absorption studies. A detailed description of the bound exciton complexes in one of these materials (CdS) has been presented by Thomas and Hopfield.[(4)] They pointed out that an exciton bound to a neutral donor (or acceptor), upon decaying, could leave the donor (or acceptor) in an excited state. They observed transitions in CdS that were characterized by large magnetic field splittings and negative diamagnetic shifts which they tentatively identified with transitions of this type. Investigations of this type offer another approach to the study of donor or acceptor impurities in a particular host lattice.

It is the purpose of this paper to report on bound exciton transitions in CdS in which the impurity is left in an excited state. The transitions have been associated with donor type complexes in which the initial state is that of an exciton bound to a neutral donor and the final state is that of an excited state of the electron on the donor.

In dealing with this type of complex in a material having the wurtzite symmetry it is possible to examine directly the magnetic field splitting of the electron on the donor. It was previously shown[(4)] that for the wurtzite symmetry the electron g-value is isotropic, (g_e = isotropic). It was further shown that the hole g-value is completely anisotropic ($g_h = g_{h\parallel} \cos\theta$), where θ is the angle between the "c" axis of the crystal and the magnetic field direction. If one does an emission experiment in the orientation C ⊥ H, then in the upper state (see Fig. 2) the electron spins are paired to give a bonding state and one is left with a degenerate hole spin. For this orientation the hole g-value is zero so there will be no magnetic field effect on this state. In the lower state, since the electron g-value is isotropic, the electron will split according to the state it occupies. In the ground state one will observe the usual electron spin splittings. In an excited state one will observe orbital splitting as well. In this configuration one is observing transitions from a singlet upper state to a magnetically split lower state. This provides a direct observation of the ground and excited states of the electron on the donor.

From zero field measurements we have been able to determine donor binding energies and magnetic field measurements have permitted the determination of the electron effective mass. Measurements in other orientations ($0° < \theta < 90°$) have demonstrated the additional multiplicity due to hole splitting.

In this material we have not identified excited state complexes that could be associated with an exciton bound to an acceptor site. Similar data has been obtained from CdSe crystals.

II. EXPERIMENTAL

The crystals used in these experiments were of the platelet type and were grown from the vapor phase.[(5)] It has been observed that the platelet type crystals are the highest quality crystals available and are therefore the most desirable for the measurements.

The platelet samples ranged in thickness from 0.5 to 50μ and were glued on one end (relatively strain free) to a sample holder which was in turn placed in the tip of a glass helium dewar. The mounting was arranged so that the samples were immersed in liquid He. Provision was made for pumping on the liquid He and the temperature was measured by means of vapor-pressure thermometry, using an oil manometer. All of the experiments were conducted at approximately 1.2°K. The dewar tip was inserted in the air gap of a conventional dc electromagnet, the pole tips of which were separated by 5/16 in. The maximum field strength of this magnet was 45,000G. A 500-W Hg lamp (Osram high pressure), equipped with a blue filter, was used for fluorescence excitation. Spectral analysis of the crystal emission was made with a Bausch and Lomb 2-m grating spectrograph. The spectrograph employed a large, high resolution, diffraction grating and produced a reciprocal dispersion of approximately 2A/mm in first order. With this grating, the spectrographic aperture was about f.16. All of the spectra were photographically recorded on Kodak type 103 a-F (CdS) and 1N (CdSe) spectroscopic plates.

III. RESULTS AND DISCUSSION

A. Identification of Complexes

A number of bound exciton complexes have been observed in CdS,[(4),(6)] and many of them have been identified with neutral donors or acceptors.[(4)] In the present experiment five lines are observed in zero field at 2.5214 eV, 2.5220 eV, 2.5235 eV, 2.5254 eV and 2.5262 eV shown in Table 1. The magnetic field splitting of these lines identified them as excited states of bound exciton complexes. Only one of the ground state bound exciton complexes in CdS has been positively identified as an exciton bound to a neutral acceptor complex. This is the I_1 (4888.5Å) line. It has been shown[(7)] that the I_1 transition can be removed from CdS by doping with Cd. It has further been shown that the lines associated with excited state transitions are still present when the I_1 line is practically eliminated by Cd doping. From this it is concluded that the excited states are not associated with the I_1 complex.

In order to relate the energy states of a particular series it is necessary to calculate the "hydrogen like" ground and excited states of a neutral donor. The method chosen is analogous to the one used by Wheeler and Dimmock[(8)] where they derived the energies of an exciton in an anisotropic crystal. The effective mass of the electron in the neutral donor "hydrogenic" model is assumed to be isotropic; however, the dielectric constant in the z direction is different from that in the x and y directions. This gives the following Hamiltonian:

$$H = H_0 + H_\alpha \tag{1}$$

TABLE I

Bound Exciton Emission Lines in CdS T = 1.2°K

Line	Exciton Complex	g_e (electron) g-value isotropic	g_h (hole) g-value C H
I_1(4888.47Å)	Neutral Acceptor	-1.76	-1.03
I_2(4867.17Å)	Neutral Donor	-1.76	-1.76
I_3(4861.66Å)	Ionized Donor	-1.74	
I_5(4869.14Å)	Neutral Donor	-1.75	-1.76
4907.15Å	Neutral Donor with terminal	-1.76	
4908.7Å	excited state	-1.76	
4912.4Å		-1.76	
4915.32Å		-1.76	-1.76
4916.5Å		-1.76	-1.76

where

$$H_o = -\frac{\hbar}{2m_e^*} \nabla^2 - \frac{e^2}{\varepsilon\eta^{1/2}} \left(\frac{1}{r}\right) \qquad (2)$$

and

$$H_\alpha = -\frac{e^2}{\varepsilon\eta^{\frac{1}{2}}} \left\{ (X^2 + Y^2 + \eta^{-1} Z^2)^{-1/2} - \frac{1}{r} \right\} \qquad (3)$$

In the above equations ε is the dielectric constant transverse to the z direction, and $\varepsilon\eta$ is the dielectric constant parallel to the z axis. H_o is the hydrogenic operator which is the unperturbed part of the Hamiltonian, and H_α, which can be written as

$$H_\alpha = -\frac{e^2}{\varepsilon\eta^{\frac{1}{2}} r} \left(1/2\alpha \, \mathrm{Cos}^2\, \theta + \frac{3}{8}\, \alpha^2\, \mathrm{Cos}^4\, \theta + \ldots .\right) \qquad (4)$$

is assumed to be accurately calculated by first order perturbation theory. In equation (4)

$$\alpha = 1 - \eta^{-1} \qquad (5)$$

and in CdS α has the value of 0.095, so only terms to α^2 are considered.

If one now has an external constant magnetic field in the x direction, two more terms are added to the Hamiltonian of equation (1), i.e.

$$H_2 = \frac{\beta_o H_x}{m_e^*} \hat{L}_x \qquad (6)$$

and

$$H_3 = \frac{1}{8} \frac{e^2}{m_e^* c^2} H_x^2 (y^2 + z^2) \qquad (7)$$

The term in Eq. (6) is linear in magnetic field and is responsible for the linear Zeeman splitting, Eq. (7) gives the diamagnetic or quadratic Zeeman effect. Again, these terms are assumed to be accurately calculated by first order perturbation theory. Adding the spin momentum interacting with the magnetic field ($\mu^* \cdot H$) to the Hamiltonian given in Eq. (1) and treating as a first order perturbation one obtains the g-values for the electron spin splittings. The resulting energies obtained from the above Hamiltonian, for the n=2 and n=3 states of the donor complexes, as a function of the magnitude of the magnetic field are given as solid and dashed lines in Fig. 1. It is apparent that there are two overlapping sets of

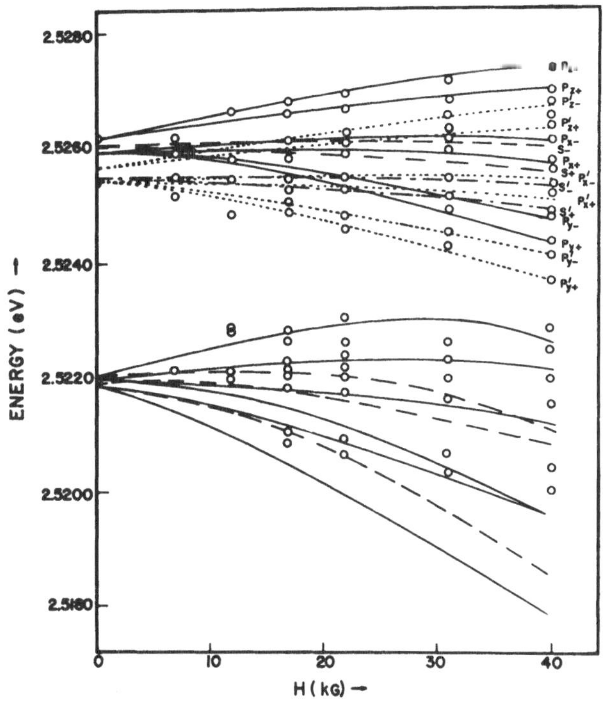

Fig. 1 Photon energy of the n=2 and n=3 states of the I_{2c} and I_5 neutral donor bound exciton complexes as a function of magnetic field for the orientation $C \perp H$.

n=2 states derived from separate donors. The calculated energy of the n=2 state of the donor was matched to the experimentally observed n=2 state. From the calculated binding energy of the donor (0.026 eV) one can calculate the ground state donor energy. Applying this binding energy to the solid and dashed lines of Fig. 1 gives a good fit to the I_5 and I_{2c} (4870.2A)[9] levels respectively.

In the calculated energy levels of Fig. (1) an electron effective mass $m_e^* = 0.18\ m_o$ was used. This gives a good theoretical fit to the experimental data.

The calculated energies for the n=3 states are shown in Fig. 1. In this calculation the spin was not added for the sake of clarity, likewise only the n=3 states of the I_5 donor are included. The fit between theory and experiment is not as good for the n=3 states as it is for the n=2 states. This probably results from the large radius of the donor causing mixing of states of different principle quantum number, through the diamagnetic term, for larger quantum numbers. It is apparent however that the experimental energies of the n=3 states do come in the region where one calculates them to be, and that they do fit the hydrogenic series for the calculated donor binding energies.

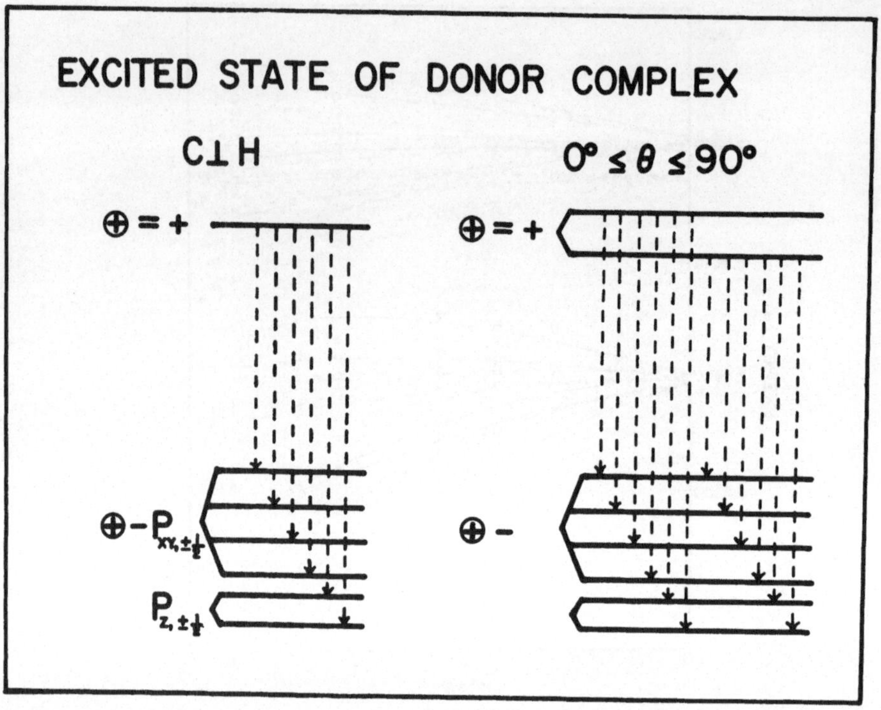

Fig. 2 Model for the decay of a bound donor complex in which the decay of the exciton leaves the terminal state in an excited configuration, in this case a p-state.

The model used to describe the magnetic field splittings is shown in Fig. 2. It is evident from the model that the lowest energy transitions result from the highest excited states. The model also demonstrates how the positive diamagnetic shift resulting from Eq. (7) is translated into a negative diamagnetic shift. Negative diamagnetic shifts are observed for the transitions in Fig. 1. It can further be seen from Fig. 2 that as the "c" axis of the crystal is rotated with respect to the magnetic field direction a contribution from the hole spin should be observed. Figure 3 shows the splitting of the zero field lines at 2.5214 eV and at 2.5220 eV as θ is varied from 0° - 90° for a constant magnetic field of 40 KG. For the orientation C ∥ H only three lines are observed for these excited states. In the magnetically field split lines of Fig. 3 the outer pair of lines (high and low energy) split as the sum of the g-values ($g_e + g_h$), and the inner pair of lines split as the difference of the g-values ($g_e - g_h$). If the electron and hole g-values are nearly equal the inner pair of lines will show no splitting and the outer pair of lines are not allowed by spin consideration. This explains the splitting for the C ∥ H orientation in Fig.3. It was previously shown[6] that the electron and hole g-values for the I_2 and I_5 complexes in the ground state were approximately equal.

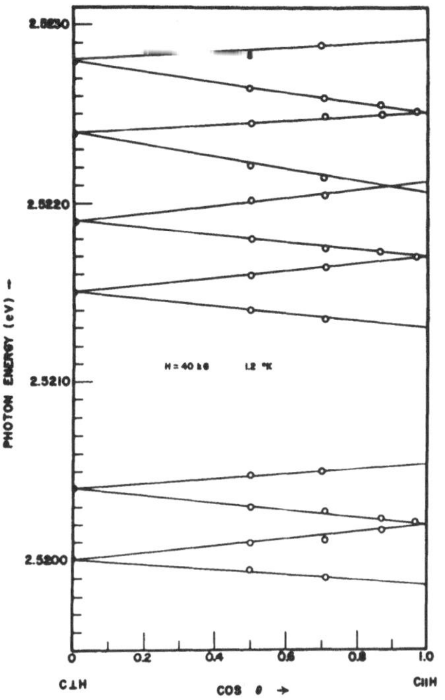

Fig. 3 Magnetic field splitting of the zero field 2.5220 eV line as a function of cos θ for a constant field of H = 40 KG.

A number of bound exciton complexes have also been observed in CdSe[10] however, they have not been identified with donor or acceptor complexes. The magnetic field splitting of the excited states of some of these complexes can positively identify them as excitons bound to neutral donor sites.

In applying the theory to the CdSe data it was found that to obtain the best fit to the experimental results it was necessary to include the mixing of the different principle quantum number states by the diamagnetic term of Eq. 7.

The reported values for the dielectric constants for CdSe are $\varepsilon = 9.53$ and $\varepsilon\eta = 10.65$.[11] These constants give a value of 0.105 for the crystal field perturbation parameter. In the calculation for the orientation C∥H the best fit to the experimental results was obtained using an effective electron mass of 0.13 m_e. The above values give a donor ionization energy of 0.018 eV and an n=2 state binding energy of 0.0045 eV. The results of the calculation for neutral donor complexes are the solid lines of Fig. 4 and Fig. 5. In these figures the photon energy is plotted as a function of the magnetic field. The experimental energies are plotted as the circles. The zero field calculated energy of the n=2 state of the donor was matched to the experimentally observed n=2 state. In both Fig. 4 and Fig. 5 the n=2 states of two separate donors are observed. At high magnetic fields it is noted that the experimental points lie above the calculated energies for the $2p_{-1}$ states. The chief reason

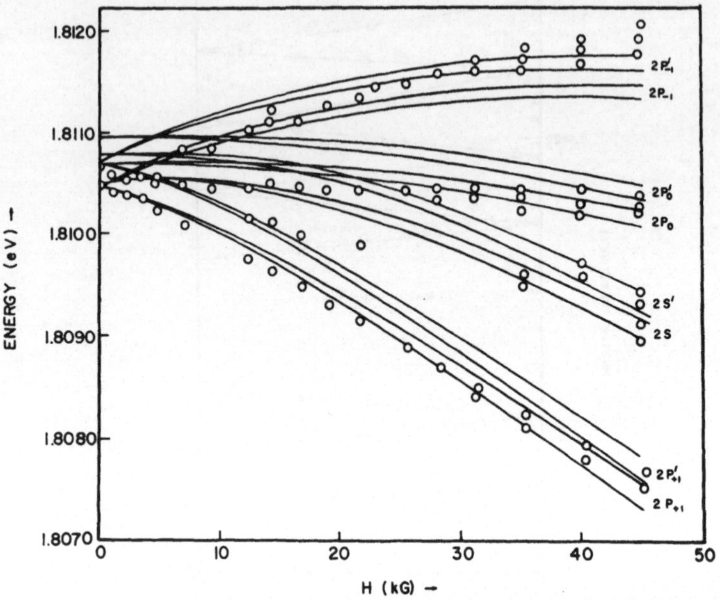

Figure 4 - Photon energy as a function of magnetic field for the I_{8a} and I_{8b} donors for C ∥ H. The solid lines are the theoretically calculated energies, the circles are the experimental energies.

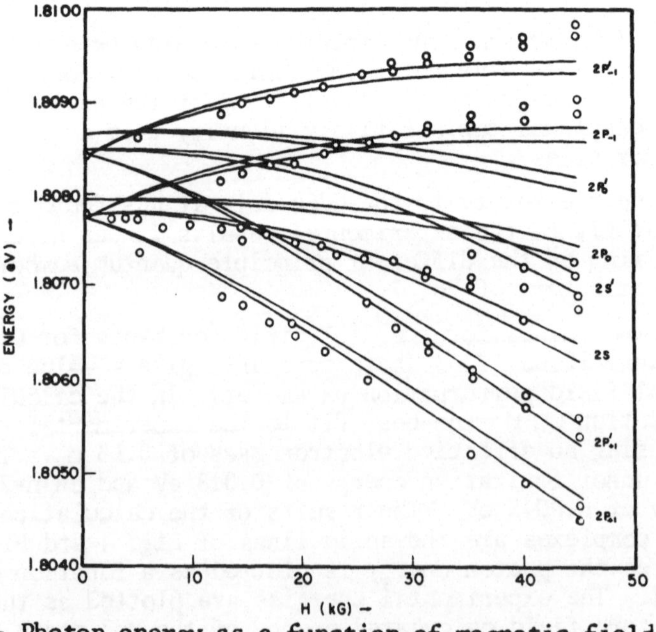

Figure 5 - Photon energy as a function of magnetic field for the I_{9a} and I_{9b} donors for C∥H. The solid lines are the calculated energies, the circles are the experimental energies.

for this difference is that in the calculated energies, only mixing through the n=3 states was considered.

Having fixed the n=2 state of the donor exciton complex, the donor ground state can then be determined from the donor binding energy. Using the donor binding energy of 0.01 eV it is found (see Table 2) that the I_2 and I_3 lines are the ground state donor complexes giving rise to the n=2 states I_{8a} and I_{8b} respectively. Likewise the ground state I_4 and I_5 lines give rise to the n=2 states I_{9a} and I_{9b} respectively. The nomenclature for the ground state energies is taken from reference (10). Some of the n=3 states are observed and identified by their diamagnetic shift, however, they are not resolved well enough to permit analyses of their magnetic field splittings.

Another interesting observation is the apparent quenching of the hole spin. One expects the hole spin to be anisotropic ($g_h = g_{h\parallel} \cos\theta$; see Fig. 2). The contribution of the hole spin should be observed for the orientation C∥H. In the case of CdSe one observes little dependence on the angle θ and also the multiplicity is accounted for at all angles including θ = 0 by the orbital and spin splittings of the n=2 state electrons. These observations lead to assigning the I_2, I_3, I_4 and I_5 lines to neutral donor complexes, whereas they were tentatively assigned to ionized donor complexes in Ref (10) because no hole splitting was observed.

IV. CONCLUSIONS

The study of the terminal states, after decay of bound exciton complexes, is a technique for gaining information about impurity states in semiconductors. It has been shown that after exciton decay the donor electron is left in an excited state in a number of transitions. From the differences in energies of the transitions to various excited states one can determine the donor binding energies. Studying the same excited states in a magnetic field has yielded carrier effective mass information.

In the case of CdS and CdSe only excited terminal states were observed for donor complexes. There is no apriori reason why similar effects should not be observed for acceptor complexes. The technique should be applicable to any material where well resolved bound exciton lines are observed.

TABLE 2

A Summary of Some of the Bound Exciton Lines Observed in Emission in CdSe.

LINE	WAVELENGTH (Å)	ENERGY (eV)	g-VALUE
I_1	6790.9_5	1.8254	0.52
I_2	6792.60	1.8250	0.52
I_3	6794.1_2	1.8246	0.52
I_4	6804.1_8	1.8219	
I_5	6805.8_5	1.8214	0.53
I_6	6821.8_4	1.8172	0.53
I_7	6827.4_3	1.8157	1.96/0.50
I_{8a}	6846.6_2	1.8106	
I_{8b}	6847.3_9	1.8104	
I_{9a}	6855.1_4	1.8083	
I_{9b}	6857.3_5	1.8078	

References

1. H. Y. Fan and P. Fisher, J. Phys. Chem. Solids 8, 270 (1959)
2. W. S. Boyle, J. Phys. Chem. Solids 8, 321 (1959)
3. S. Zwerdling, K. J. Button and B. Lax, Phys. Rev. 119, 875 (1960)
4. D. G. Thomas and J. J. Hopfield, Phys. Rev. 128 2135 (1962)
5. D. C. Reynolds in the Art and Sciences of Growing Crystals, edited by J. J. Gilman (John Wiley & Sons, Inc., New York, 1963), p. 62.
6. D. C. Reynolds and C. W. Litton, Phys. Rev. 132, 1023 (1963)
7. C. W. Litton and D. C. Reynolds in II-VI Semiconducting Compounds, edited by D. G. Thomas (W. A. Benjaming, Inc., New York 1967), p. 694,
8. R. G. Wheeler and J. O. Dimmock, Phys. Rev. 125, 1805 (1962)
9. E. T. Handelman and D. G. Thomas, J. Phys. Chem. Solids 26, 1261 (1965)
10. D. C. Reynolds, C. W. Litton and T. C. Collins, Phys. Rev. 156, 881 (1967)
11. D. Berlincourt, H. Jaffe and L. R. Shiozowa, Phys. Rev. 129, 1009 (1963).

STUDIES OF NONRADIATIVE RECOMBINATION IN INSULATORS AND SEMICONDUCTORS VIA THE PRINCIPLE OF DETAILED BALANCE

D. F. Nelson

University of Southern California

Los Angeles, California

ABSTRACT

Experimental confirmation of detailed balance to an accuracy of 3% is described for the R lines of ruby. The method avoids the problems introduced by the presence of vibronic transitions or nonradiative transitions. Separate measurements demonstrate the actual presence of a small amount of nonradiative de-excitation in ruby whose dependence on temperature suggests a multiphonon process. An apparent failure of detailed balance in bound exciton emission in donor-doped GaP or Si at low temperature can be well explained by an Auger process occurring in a complex consisting of an exciton bound to a neutral donor.

INTRODUCTION

Direct study of nonradiative de-excitation of crystals is difficult because of the need to detect the emitted phonons or energetic charged particles which carry off the excitation energy. If the excitation has a measurable probability of radiative decay, the indirect approach of studying the radiative characteristics can sometimes be used to deduce the properties of the competing nonradiative processes. For instance, if the decay lifetime of the fluorescence of some excitation is shorter than the radiative lifetime that is predicted from the absorption strength of the excitation via the principle of detailed balance, then it can be concluded that there is another decay process competing with the radiative process. Study of this discrepancy versus external influences such as temperature then gives information concerning the competing process.

The principle of detailed balance states that the probability per unit time for a transition upward in energy is equal to that for a transition between the same states downward. This is not true in general but does always hold in the first Born approximation where the transition operator is replaced by the potential.(1) When absorption and emission of light are considered, this means that nonlinear optical processes, such as harmonic generation are neglected. In other words, detailed balance should hold for interactions with low intensity light beams.

The Einstein relations are a consequence of detailed balance. For isotropic media they state that

$$B_{21} = \frac{g_1}{g_2} B_{12} = \frac{v^3}{8\pi h \nu^3} A_{21}, \tag{1}$$

where B_{12} is the probability of absorption per unit time per unexcited atom per unit of radiation density (units of $Jm^{-3}Hz^{-1}$), B_{21} is the probability of stimulated emission per unit time per excited atom per unit of radiation density, and A_{21} is the probability of spontaneous emission per second per excited atom. The quantity A_{21} is thus the inverse of the radiative lifetime τ_R. In Eq. (1) g_1 and g_2 are the statistical weights of the lower and upper energy levels, respectively, v is the velocity of light in the medium, ν its frequency, and h is Planck's constant. The Einstein relations were originally derived, not from quantum mechanics, but from a consideration of thermal equilibrium of an ensemble of two level atoms immersed in a black body distribution of radiation. Here, also, nonlinear optical processes were neglected.

It is a simple consequence(2) of the Einstein relations that the absorption cross section $\sigma(\nu)$ is given by

$$\sigma(\nu) = \frac{v^2 g_2 g(\nu)}{8\pi g_1 \tau_R \nu^2}, \tag{2}$$

where $g(\nu)$ is the line shape function whose integral over all frequencies is unity. If, as is true of almost all absorption mechanisms in the visible spectrum, the frequency width is small compared with the frequency, then Eq. (2) yields

$$\int_0^\infty \sigma(\nu)\, d\nu = \frac{v^2 g_2}{8\pi g_1 \tau_R \nu^2}. \tag{3}$$

This equation allows a prediction of the spontaneous emission lifetime from measurement of the integrated absorption cross section.

It is a simple matter to generalize the above formulae to anisotropic media so as to be applicable to crystals. Equations (1) are replaced by

$$B_{21}(\theta,\varphi,p) = \frac{g_1}{g_2} B_{12}(\theta,\varphi,p) = \frac{v^3(\theta,\varphi,p)}{h\nu^3} A_{21}(\theta,\varphi,p). \qquad (4)$$

These Einstein coefficients apply to a radiation density of a particular polarization state p traveling in a unit solid angle about a direction specified by the spherical angles θ and φ. The spontaneous emission lifetime τ_R is now given by

$$\frac{1}{\tau_R} = \sum_{p=1}^{2} \int_0^{2\pi} \int_0^{\pi} A_{21}(\theta,\varphi,p) \sin\theta \, d\theta \, d\varphi, \qquad (5)$$

and the cross section by

$$\sigma(\nu,\theta,\varphi,p) = \frac{g_2 g(\nu)\, v^2(\nu,\theta,\varphi,p)\, A_{21}(\theta,\varphi,p)}{g_1 \nu^2} \qquad (6)$$

If we neglect the usually small anisotropy of v and once again consider an emission line with a frequency width small compared to its frequency, Eq. (6) leads to

$$\sum_{p=1}^{2} \int_0^{2\pi} \int_0^{\pi} \int_0^{\infty} \sigma(\nu,\theta,\varphi,p)\, d\nu \sin\theta \, d\theta \, d\varphi = \frac{g_2 v^2}{g_1 \tau_R \nu^2} \qquad (7)$$

For a uniaxial crystal, such as ruby, the cross section can be expressed as

$$\sigma(\nu,\theta,\varphi,p = 1) = \sigma_\perp(\nu), \qquad (8)$$

$$\sigma(\nu,\theta,\varphi,p = 2) = \sigma_\perp(\nu) \cos^2\theta + \sigma_\parallel(\nu) \sin^2\theta, \qquad (9)$$

where p = 1 and p = 2 correspond respectively to the ordinary and extraordinary waves and $\perp$ and $\parallel$ refer to the orientation of the electric vector of the light with respect to the optic axis. The small anisotropy of the velocity has also been ignored in writing Eq. (9). Combining Eqs. (7), (8) and (9) we obtain for a uniaxial crystal

$$\int_0^\infty \left[\frac{2}{3}\,\sigma_\perp(\nu) + \frac{1}{3}\,\sigma_{||}(\nu)\right] d\nu = \frac{g_2 v^2}{8\pi g_1 \tau_R \nu^2} \,. \qquad (10)$$

DEMONSTRATION OF DETAILED BALANCE IN THE R LINES OF RUBY[1]

If apparent failures of detailed balance are to be attributed to competing nonradiative decay, it is important to demonstrate that the methods used are capable of confirming detailed balance under the appropriate circumstances. It is interesting to note in this regard that the only precise numerical check of detailed balance in crystals is that performed on the R lines of ruby to be described here.[1] Detailed balance was, however, carefully checked for several resonance lines of monatomic vapors many years ago. The lines studied were the 2288 and 3261 Å lines of Cd,[3] the yellow doublet of Na[4], and the 2563 Å line of Hg[5].

The fluorescent spectrum of ruby consists of two lines (the so-called R_1 and R_2 lines) in the red at 14,419 and 14,448 cm^{-1} respectively, whose excited state populations are in thermal equilibrium under normal observation conditions. At low temperature ($T \lesssim 100$°K) each of the R lines are doublets with a splitting of 0.38 cm^{-1} because of the ground state splitting. There is also a broad weak continuum of emission on the low frequency side of the R lines at low temperature due to vibronic transitions in which a phonon is emitted along with the photon.

The R lines appear in absorption at exactly their position in emission[1]. The vibronic transitions in absorption appear only on the high frequency side of the R lines at low temperature. Since the vibronic transitions in emission and absorption involve different initial and final states, detailed balance need not apply to them. Thus, in checking detailed balance for the R lines the effects of the vibronic transitions must be eliminated from the comparison.

To meet these problems the following procedure was employed[1]. The absorption coefficients of the R lines were measured in the two principal polarizations ($E \perp C$ and $E \parallel C$). The concentrations of the Cr^{3+} ions in the ruby crystals used were measured by optical absorption in the green as calibrated by Cr concentration determinations using a wet chemical method[6], x-ray fluorescence[6], emission spectrography[6], and neutron activation[1]. The absorption coefficients divided by the Cr^{3+} concentration gave the cross sections used in Eq. (10). By taking the statistical weights $g_1 = g_2 = 2$ Eq. (10) then yielded values for τ_{R_1} and τ_{R_2}, the predicted radiative lifetimes of the R_1 and R_2 lines. Because of the thermal equilibrium of the excited state populations these were combined by

$$\tau_R^{-1} = \left[1 + \exp - \Lambda/kT\right]^{-1} \left[\tau_{R_1}^{-1} + \tau_{R_2}^{-1} \exp - \Lambda/kT\right] \qquad (11)$$

to give the radiative lifetime τ_R of the R_1 plus R_2 lines. Here $\exp - \Lambda/kT$ is the Boltzman factor for the energy separation Λ of the initial states of the R_1 and R_2 transitions.

The measured <u>decay</u> lifetime τ_D of the excited state population is given by

$$\tau_D^{-1} = \tau_R^{-1} + \tau_V^{-1} + \tau_N^{-1} \qquad (12)$$

where τ_V is the lifetime of emissive vibronic transitions and τ_N is the lifetime of any nonradiative transitions. The directly measured lifetime is τ_D, while the absorption data predict τ_R. Measurement of τ_D is done with small, low-doped crystals to avoid lifetime lengthening from self-absorption. In order to make a comparison of τ_D and τ_R the lifetime τ_C, defined by

$$\tau_C^{-1} = \tau_V^{-1} + \tau_N^{-1}, \qquad (13)$$

must be separately measured.

It can, in fact, be measured directly in large ruby crystals. The emissive vibronic transitions, particularly at low temperature, cannot be self-absorbed by the crystal. Neither can nonradiative transitions either, of course. However, the R lines, being resonance lines, can be self-absorbed by the crystal. If the crystal is large enough, (dimensions $\gg$ absorption length for either R line), the major portion of the volume of the crystal cannot de-excite by R line emission, but only by vibronic and nonradiative transitions. In this case the observed decay lifetime becomes equal to τ_C. Thus, a comparison of the difference of two measured quantities $\tau_D^{-1}-\tau_C^{-1}$, is made with the measured integrated absorption cross section to check detailed balance. The comparison is shown in Table I. The agreement is seen to be excellent. Since the agreement holds for two temperatures, it must hold for the R lines individually.

TABLE I

Comparison of the emission and absorption probabilities in the R lines of ruby to check the principle of detailed balance. The transition probability τ_R^{-1} is calculated from the emission data by Eqs. (12) and (13) and from the absorption data by Eqs. (10) and (11).

Temperature	Emission			Absorption
T(°K)	τ_D^{-1}(sec^{-1})	τ_C^{-1}(sec^{-1})	τ_R^{-1}(sec^{-1})	τ_R^{-1}(sec^{-1})
77	239 ± 3	63 ± 1	174 ± 3	173 ± 7
20	262 ± 3	63 ± 1	199 ± 3	189 ± 10

NONRADIATIVE DECAY IN RUBY[1]

The above confirmation of detailed balance in the ruby R lines says nothing about the presence or absence of competing nonradiative decay. However, τ_N can be now determined if τ_V can be evaluated. This can be done in the following way. The emission spectrum of ruby is used to measure the radiative efficiencies $\eta\perp, \eta\|$ of the R lines in the two principal polarizations. The radiative efficiency of the R lines is the fraction of the emission that is emitted in the R lines compared to the total emission including vibronic emission. The radiative efficiency, so measured, is assumed to apply to the R lines individually. Equation (10) is then replaced by

$$\int_0^\infty \left[\frac{2}{3}\,\frac{\sigma\perp(\nu)}{\eta\perp} + \frac{1}{3}\,\frac{\sigma\|(\nu)}{\eta\|}\right] d\nu = \frac{g_2 v^2}{8\pi g_1 \tau_R v^2} \,. \tag{14}$$

This form is used for both τ_{R_1} and τ_{R_2} in Eq. (11) to predict the radiative lifetime of the excited population distribution at low temperature arising from vibronic and electronic transitions. At room temperature and above additional terms must be included in Eq. (11) to account for the population and radiative lifetime of the higher $2T_1$ levels. In this manner the lifetime of the excited population distribution due to all radiative mechanisms can be calculated versus temperature.

The results are shown in Fig. 1 along with the experimental results[1]. In measuring the lifetime at low temperature minute samples of very low doping were used to avoid lifetime lengthening from self-absorption. It can be seen that a nonradiative process is present with a lifetime of 0.037 ± 0.016 sec at 20°K, 0.038 ± 0.013 sec at 77°K, 0.029 ± 0.010 sec at 195°K, 0.023 ± 0.008 sec at 273 °K, and 0.0075 ± 0.0012 sec at 373°K. Thus it appears that there is a component of the nonradiative decay at low temperature which is independent of temperature and another component which increases as the temperature increases. This is qualitatively what is expected of a multi-phonon decay process. Recently Tsukerblat and Perlin[7] have calculated the temperature independent contribution of the multi-phonon decay for ruby. They obtain a lifetime of 0.7 sec which is about 20 times longer than was observed. The many approximations that were used in the complex calculation suggest, however, that their result may be only an order of magnitude estimate.

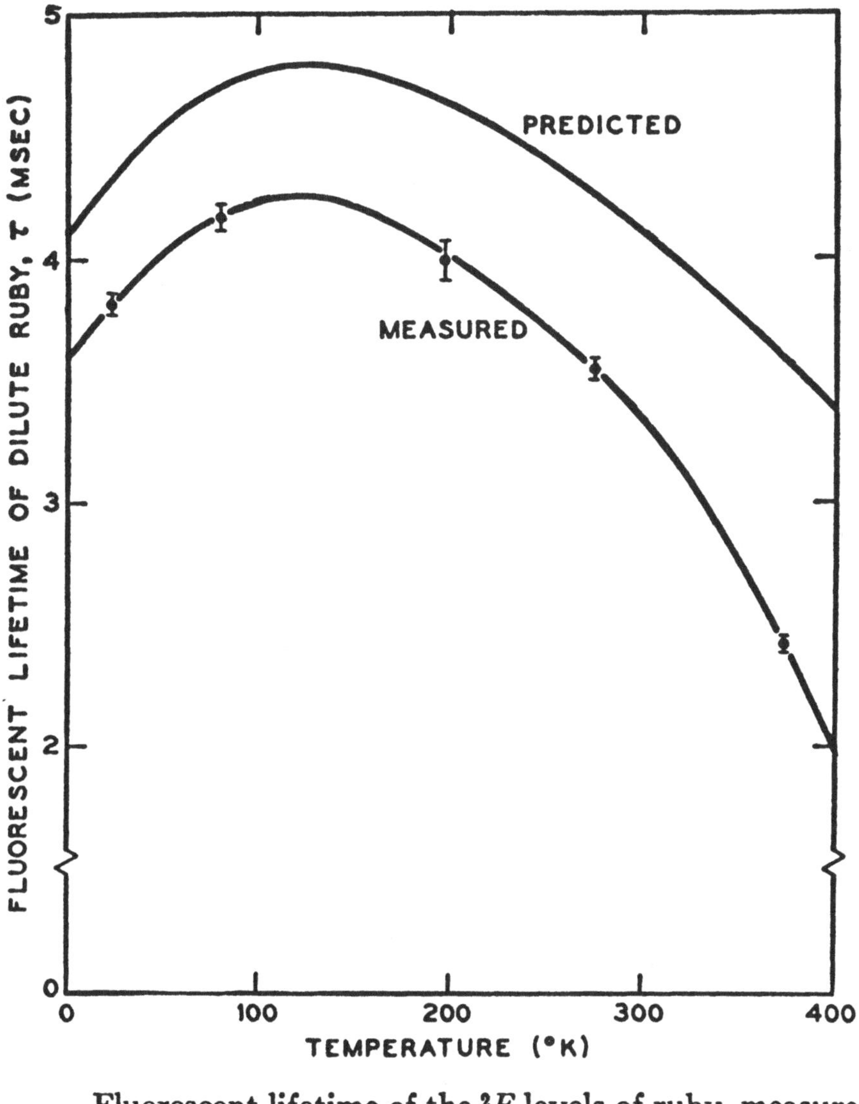

Fluorescent lifetime of the ^2E levels of ruby, measured in optically thin crystals, as a function of temperature.

NONRADIATIVE DECAY IN GALLIUM PHOSPHIDE AND SILICON[8]

In a very pure semiconductor at temperatures of a few degrees Kelvin recombination can proceed by radiative annihilation of free excitons, mobile complexes consisting of a hole and an electron bound by coulomb attraction. If a donor is added to the crystal, the excitons become bound to the donors, which are in their neutral charge state at such temperatures, before recombination occurs. Radiative recombination of such a bound exciton complex is seen as a very sharp emission line.

Such an emission line has been observed in S-doped GaP at low temperature. It can be observed in absorption at the same frequency. Vibronic sidebands of the emission are very small -- no more than 6% of the strength of the resonance line. The line has been identified as due to an exciton bound to a neutral sulfur atom. From absorption measurements on the line on crystals of known S concentration the cross section integrated over the emission frequencies was found to be 7×10^{-21} cm^2 eV[9]. Taking the degeneracy of the neutral S donor state $g_1 = 2$, the degeneracy of the state representing the exciton bound to a neutral S donor $g_2 = 4$, the wavelength $\lambda = 5367$ Å, and the refractive index $n = 3.47$, we find from Eq. (3) that the radiative lifetime of the complex is predicted to be 11 μsec. The decay lifetime of the emission at low temperature (1.6°K) was measured to be 21 ± 4 nsec. At this temperature ionization of the complex is extremely improbable and hence the decay lifetime must be determined by recombination processes within the complex itself.

We conclude that the decay of this bound exciton complex proceeds mainly by a nonradiative process which is some 500 times stronger than the radiative process. This complex consists of a singly positively charged S ion on a P site to which are bound two electrons and one hole. Radiative recombination of one of the electrons with the hole yields an emitted photon. We believe the much more probable recombination mechanism is an Auger process in which one of the electrons and the hole recombine to give their energy to the second electron, which is freed into the conduction band and dissipates its excess kinetic energy through collisions with the lattice.

The theory of the relative probabilities of the Auger and radiative processes has many similarities to the calculation of the internal conversion coefficient of nuclear physics. Thus a rough estimate of the ratio of the probabilities of the Auger process to the radiative process can be obtained from the internal conversion coefficient α, as adapted to the present situation. For electric dipole radiation,

$$\alpha = 4/(a_0\kappa)^3\ ka, \tag{15}$$

where a factor of $\frac{1}{2}$ has been introduced to account for there being only one electron to which energy can be coupled (rather than the two K electrons of the internal conversion problem). Here $a_0 = \hbar^2\varepsilon/m_i{}^*e^2$ is the Bohr radius altered to include the dielectric constant ε of the crystal and the effective mass $m_i{}^*$ of the electron in its initial state, $a = \hbar^2 n^2/m_f{}^*e^2$, where $m_f{}^*$ is the effective mass of the Auger electron, $k \cong [2m_f{}^*\omega/\hbar]^{1/2}$ is the approximate wave number of the Auger electron, and $\kappa = \omega n/c$ is the wave number of the light quantum from the radiative process. Therefore,

$$\alpha = \left(\frac{\sqrt{2}cm_i{}^*}{\varepsilon}\right)^3 \left(\frac{1}{\hbar\omega}\right)^{7/2} \left(\frac{e^2}{\hbar}\right)^4 \left(\frac{m_f{}^{*1/2}}{n^5}\right). \tag{16}$$

For the exciton bound to the neutral S donor in GaP, $\omega = 3.51\times10^{15}$ rad/sec, $\varepsilon = 11.0$, $n = 3.47$, and we estimate $m_i{}^* \approx 0.25m_0$ and $m_f{}^* \approx m_0$. Therefore $\alpha = 960$, which is in reasonable agreement with the absorption-emission discrepancy figure of 500. Since the assumptions in the derivation of Eq. (15) are not wholly applicable to the problem at hand, this agreement must be regarded as somewhat fortuitous. However, it seems clear from this calculation that the Auger process will predominate over the radiative one in this bound-exciton complex in GaP.

Other experimental results on GaP support our interpretation. It is known that luminescence from either donor-acceptor pairs or excitons bound to isoelectronic traps in GaP can have near unit quantum efficiency at low temperatures and low doping levels. This is possible since the Auger process proposed here cannot occur for either of these luminescence mechanisms. On the other hand, the bound exciton luminescence, which must compete with the Auger process, is very inefficient. We have measured its quantum efficiency in photoluminescence at 4.2°K in a number of crystals to be one part in 700 ± 200. This agrees with the ratio of the measured decay time of the bound exciton emission to the calculated radiative lifetime from Eq. (3). Since light of photon energies greater than the band gap energy was used for the quantum efficiency measurements, this agreement also indicates that the Auger process is the dominant recombination process for all excited carriers in these crystals at low temperatures.

Similar measurements were performed on As-doped Si. Since the emission from excitons bound to neutral As atoms at low temperature has a vibronic sideband of considerable strength, a radiative efficiency (0.31) must be included as in Eq. (14) (there is no anisotropy, of course, for a point defect in a Si crystal). With

$\lambda = 1.078\,\mu$, $n = 3.56$, $g_1 = 2$, and $g_2 = 4$, the absorption measurements on As-doped Si at low temperature then yield a predicted radiative lifetime of 750 μsec. The decay lifetime of the resonance line light was measured to be 0.08 μsec at 1.6°K. This yields an absorption-emission discrepancy of 9400 which we interpret as due to the predominance of Auger recombination. The quantum efficiency of the bound exciton complex was measured to be one part in 4000, in substantial agreement with the lifetime ratio. With $\varepsilon = 11.7$, $m_i^* = 0.25\, m_0$ and $m_f^* \approx m_0$, Eq. (16) predicts the ratio to be 7900, again in good agreement with experiment and in support of the Auger-recombination interpretation. Since the quantum efficiency measurement was made with light of photon energies greater than the band gap, we also conclude that the Auger process is the dominant recombination process for all excited carriers in As-doped Si crystals at very low temperatures.

REFERENCES

1. D. F. Nelson and M. D. Sturge, Phys. Rev. 137, A1117 (1965).
2. A. C. G. Mitchell and M. W. Zemansky, Resonance Radiation and Excited Atoms (Cambridge University Press, Cambridge, England, 1934), p. 92.
3. W. Kuhn, Naturwiss. 14, 48 (1926); M. W. Zemansky, A. Physik 72, 587 (1931).
4. R. Ladenberg and E. Thiele, Z. Physik 72, 697 (1931); H. H. Hupfield, Z. Physik 54, 484 (1929).
5. W. Zehden and M. W. Zemansky, Z. Physik 72, 442 (1931); P. H. Garrett, Phys. Rev., 40, 449 (1932).
6. D. M. Dodd, D. L. Wood, and R. L. Barns, J. Appl. Phys., 35, 1183 (1964).
7. B. A. Tsukerblat and Yu E. Perlin, Opt. Spectry. 20, 6 (1966).
8. D. F. Nelson, J. D. Cuthbert, P. J. Dean and D. G. Thomas, Phys. Rev. Letters 17, 1262 (1966).
9. M. Gershenzon, D. G. Thomas and R. E. Dietz in Proceedings of the International Conference on the Physics of Semiconductors Exeter, July 1962. (The Institute of Physics and the Physical Society, London, 1962) p. 752.

THE ELECTRO-OPTIC EFFECT AS A PROBE FOR STUDYING THE OPTICAL WAVEGUIDE PROPERTIES OF P-N JUNCTIONS

D. F. Nelson

Department of Physics, University of Southern California

Los Angeles, California

ABSTRACT

Observations of light transmitted along the plane of reverse-biased GaP p-n junctions shows that the depletion layer is birefringent and produces mode confinement of the light. Intensity and phase measurements of the transmitted light are interpreted in terms of a waveguide model which includes effects due to the linear and quadratic electro-optic effects. The latter is invoked to interpret an anomalous modulation most visible in diodes with their junction field in the [100] direction. The origin of the higher optical dielectric constant in the junction, which is mainly responsible for mode confinement, is not as yet understood.

INTRODUCTION

The natural occurrence of a planar optical waveguide in the p-n junction region of laser diodes is an important phenomenon since it eliminates otherwise large diffraction losses and so allows attainment of laser oscillation at reasonable current levels. The determination of the origin of the waveguide has been a difficult problem. It was felt that the use of visible light in the study of such waveguides could possibly aid in this determination. Hence, a study of the optical waveguide properties of p-n junctions in gallium phosphide, which is transparent for wavelengths longer than those in the green, was undertaken even though GaP cannot be used in laser diodes because it is an indirect bandgap semiconductor.

Early observations[1] of polarization changes of the light propagating through GaP p-n junctions when placed in reverse bias led to the conclusions that they were due to the linear electro-optic (Pockels) effect of the large junction electric field. GaP belongs to point group $\bar{4}3m$ and so lacks a center of symmetry — a necessary condition for the existence of a linear electro-optic effect. It has turned out that the ability of GaP diodes to modulate laser beams passing through the junction waveguide by varying the reverse bias of the junction is sufficiently large to have possible device application.[2,3] This has added to the motivation to understand the waveguide properties of p-n junctions in GaP.

The study of p-n junction waveguides is made particularly difficult by two problems. One is the fact that the optical properties, such as the index of refraction, vary significantly over a distance comparable to the wavelength of light. Thus the regions of different optical properties cannot be probed separately with light; rather, a beam of light necessarily interacts with all the regions at once. Hence, these variations in properties must be determined less directly, that is, by comparing observations with the predictions of various waveguide models of the junction. For this reason many waveguide models have been explored.[4] The second major problem is the existence of a significant fraction (~10%) of the light at the output side of the diode being in continuum modes of the waveguide.[5] This affects significantly the accuracy of measurements on the discrete modes of the waveguide, particularly since the amount and form of the continuum mode light depends critically on the focusing of the light into the waveguide. The origin of this continuum mode light is in the matching of the electromagnetic boundary conditions at the entrance to the waveguide at the input surface of the diode. If diodes over a centimeter long could be fabricated, the continuum modes would spread out and drop to an insignificant intensity at the output surface. Unfortunately, materials problems have prevented fabricating such diodes so far.

MEASUREMENT OF THE LINEAR ELECTRO-OPTIC COEFFICIENT

Since solutions of Maxwell's equations for the phase of the light in the presence of a linear electro-optic effect in a very thin waveguide have significant terms containing the square of the electro-optic coefficient[1,4], it is important to have an accurate value of it for the wavelengths of interest. For this reason measurements were made on bulk GaP crystals to determine it.[6] A value of $r_{41} = -0.97\times10^{-12}$ m/V, believed accurate to about 2%, was found at 6328 Å wavelength. This was measured both from the phase modulation of a high frequency (~70 MHz) electric field as detected by an optical heterodyne arrangement and from the polarization modulation caused by an abrupt electric field pulse as detected by an optical compensator technique. In the latter method it was necessary to account for the piezoelectrically excited acoustic resonances of the crystal.[6]

INTENSITY MEASUREMENTS

It is crucial to the interpretation of the phase measurements on the GaP diodes to establish whether or not discrete modes of a waveguide are present. This can be done by the intensity measurements[7] of the light on the exit surface of the diode shown in Fig. 1. To obtain these data a 6328 A laser beam was focused on the input side of the diode. The diode, whose junction field was in the [111] direction, was placed in full reverse bias (-28V). The image of the output surface was scanned photoelectrically using a narrow slit. The broad low curves of Fig. 1 in each polarization were obtained when the input beam was focused on the surface of the homogeneous crystal on the n-side of the junction. When the position of the focus was moved laterally over to the p-n junction, the peak intensity of the light at the exit surface was increased by a factor of five and the width of the distribution was narrowed by a similar factor. This factor is almost the same for each principal polarization, being only slightly larger for the TM mode (electric field of light parallel to the junction electric field) than the TE mode (electric field of light perpendicular to junction electric field). This indicates that the higher optical dielectric constant inside the p-n junction, that causes mode confinement, is nearly the same for each polarization. This is true in spite of the fact that for the TM mode the linear electro-optic effect increases the dielectric constant while for the TE mode it decreases it. From these facts we conclude that the "built-in" higher optical dielectric constant in the junction is large compared to the electro-optic contribution. The latter, for the TM mode, is calculated to be 4×10^{-4} fractional change in the dielectric constant from the junction electric field at full reverse bias determined from capacitance measurements. Hence the "built-in" fractional dielectric constant change must be large compared to this value. If it is taken as 1.5×10^{-3} (a typical value found from the phase measurements) and a waveguide width comparable to typical junction widths is assumed, then there will exist one discrete mode of each polarization if the ratio of twice the difference of the optical dielectric constants on the two sides of the waveguide to the sum of their deviations from the intrinsic value is less than 0.16[4], that is, if the asymmetry is small. From free carrier effects on the dielectric constant on each side of the junction this ratio should be only 0.01 to 0.02 (which, however, is sufficient to account for the asymmetry of the profiles of Fig. 1). Hence we conclude that discrete modes of the waveguide do exist in the diodes.

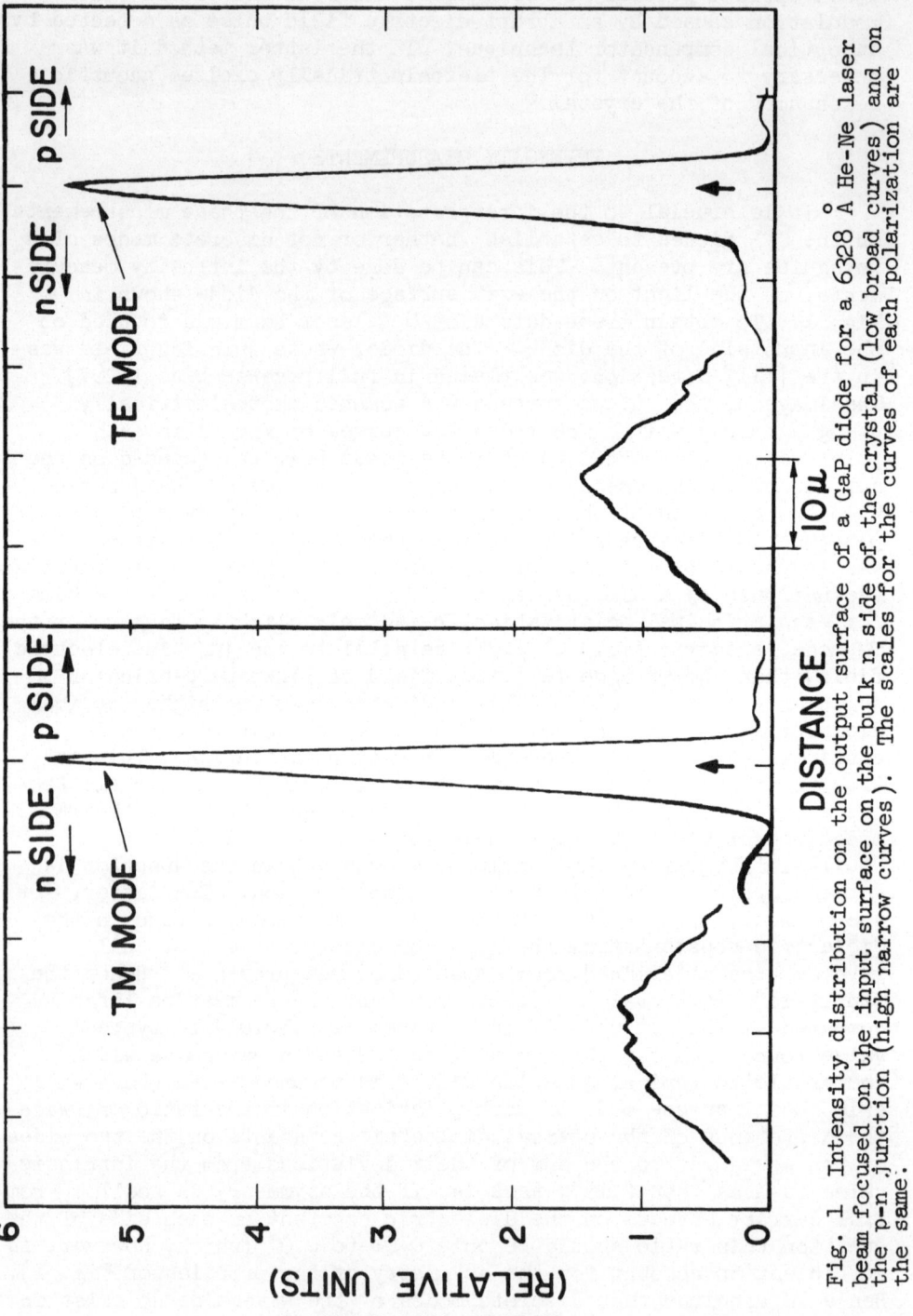

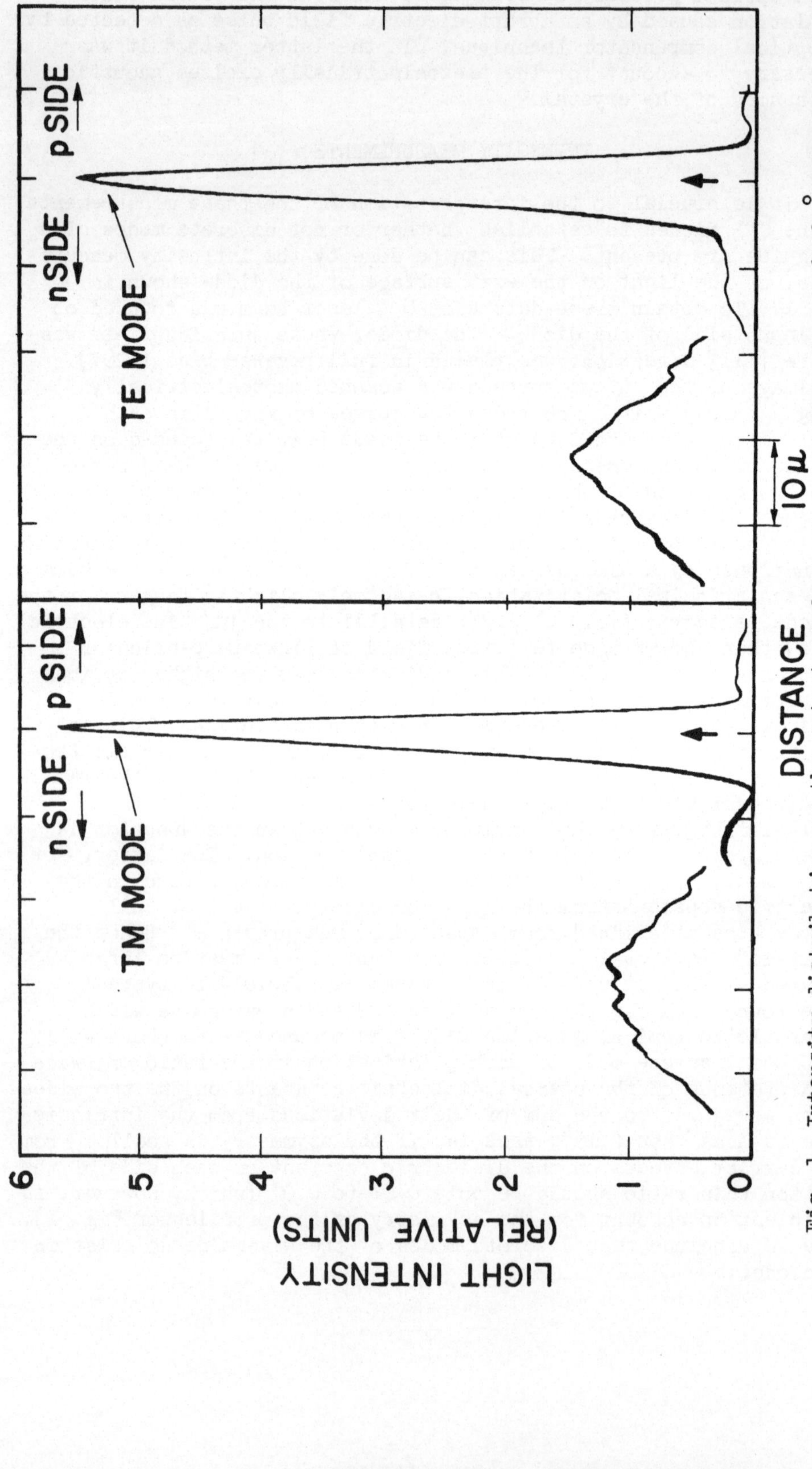

Fig. 1 Intensity distribution on the output surface of a GaP diode for a 6328 Å He-Ne laser beam focused on the input surface on the bulk n side of the crystal (low broad curves) and on the p-n junction (high narrow curves). The scales for the curves of each polarization are the same.

From the insensitivity of the intensity profile of the transmitted modes with respect to changes in the reverse bias it is concluded[7] that the "built-in" optical dielectric constant increase (or rather, that times the width over which it extends) is independent of applied bias. This "built-in" optical dielectric constant increase is mainly responsible for the mode confinement. The linear electro-optic effect, of course, gives a voltage dependent contribution to the optical dielectric constant. The width over which it extends is the electrical junction width as determined from capacitance measurements. This width, as is well known, varies with applied bias.

MODEL OF DIELECTRIC CONSTANT PROFILE

From the above observations, as well as the phase measurements to be presented, we are led to construct the model of the optical dielectric constant profile shown in Fig. 2. The coordinate x is measured perpendicular to the p-n junction plane. The planes $x = \pm w_1$ are the voltage independent boundaries between which the "built-in" dielectric constant K_D exists. The planes $x = \pm w_2$ form the electrical boundaries of the p-n junction. Since for the parameter values typical of p-n junction waveguides only the lowest order mode of each polarization can propagate, it is a good approximation to consider the spatial average of the junction electric field as extending uniformly between the junction boundaries.[4] For the same reason it is also a good approximation to take the dielectric constant changes as abrupt.[4] The dielectric constants K_1 and K_2 on the two sides of the waveguide are taken as different in order to produce the asymmetric profiles seen in Fig. 1. The two values of the optical dielectric constant shown between $x = \pm w_2$ represent the values for the two principal polarization states for a particular direction of propagation.

It should be emphasized that we are attributing the mode confinement observed to optical dielectric constant changes because changes in the absorption coefficient in the different regions are insufficient to account for the strength of the mode confinement observed.[4]

With the assumptions just outlined the discrete modes of the infinite planar dielectric waveguide can be found from Maxwell's equations by elementary methods.[7] The lowest order mode has only a single maximum in intensity inside the waveguide and falls off exponentially in the outermost regions. Solutions have been found[7] for the junction electric field in the [111], [110], and [100] crystalline directions since checking the symmetry of the effects is an important test of the model.

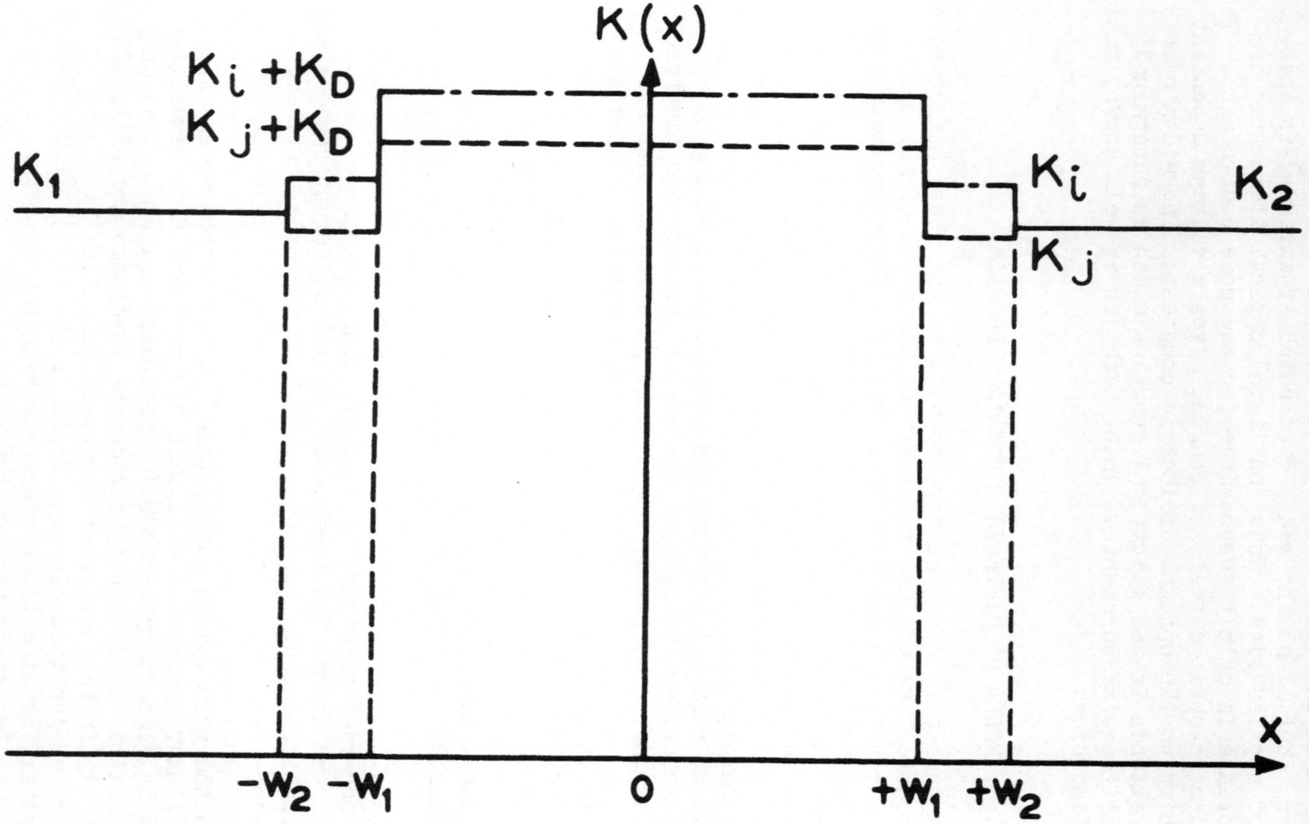

Fig. 2 Variation of two dielectric constant components for the waveguide model used in the analysis of the experiments versus the coordinate measured perpendicular to the junction plane.

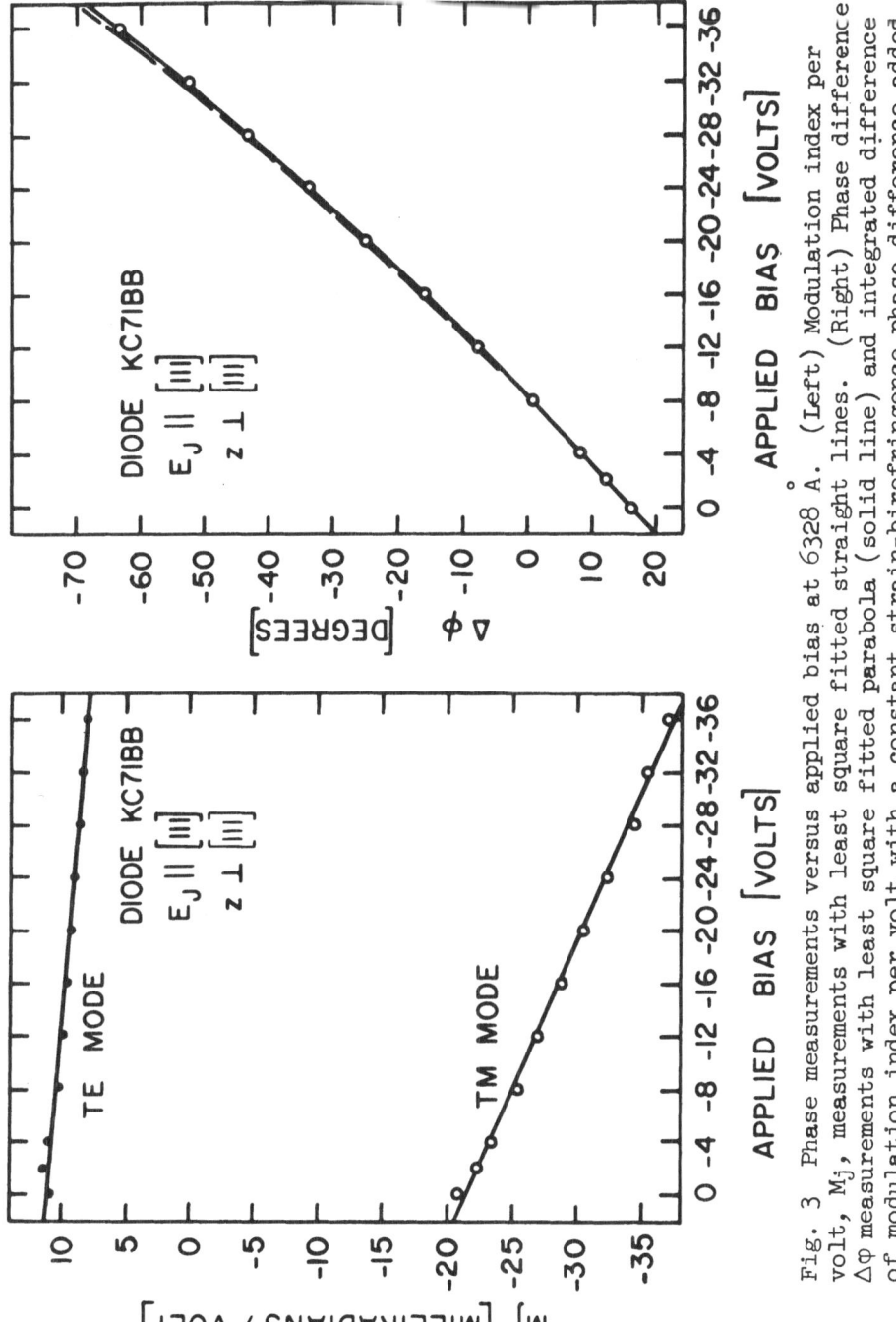

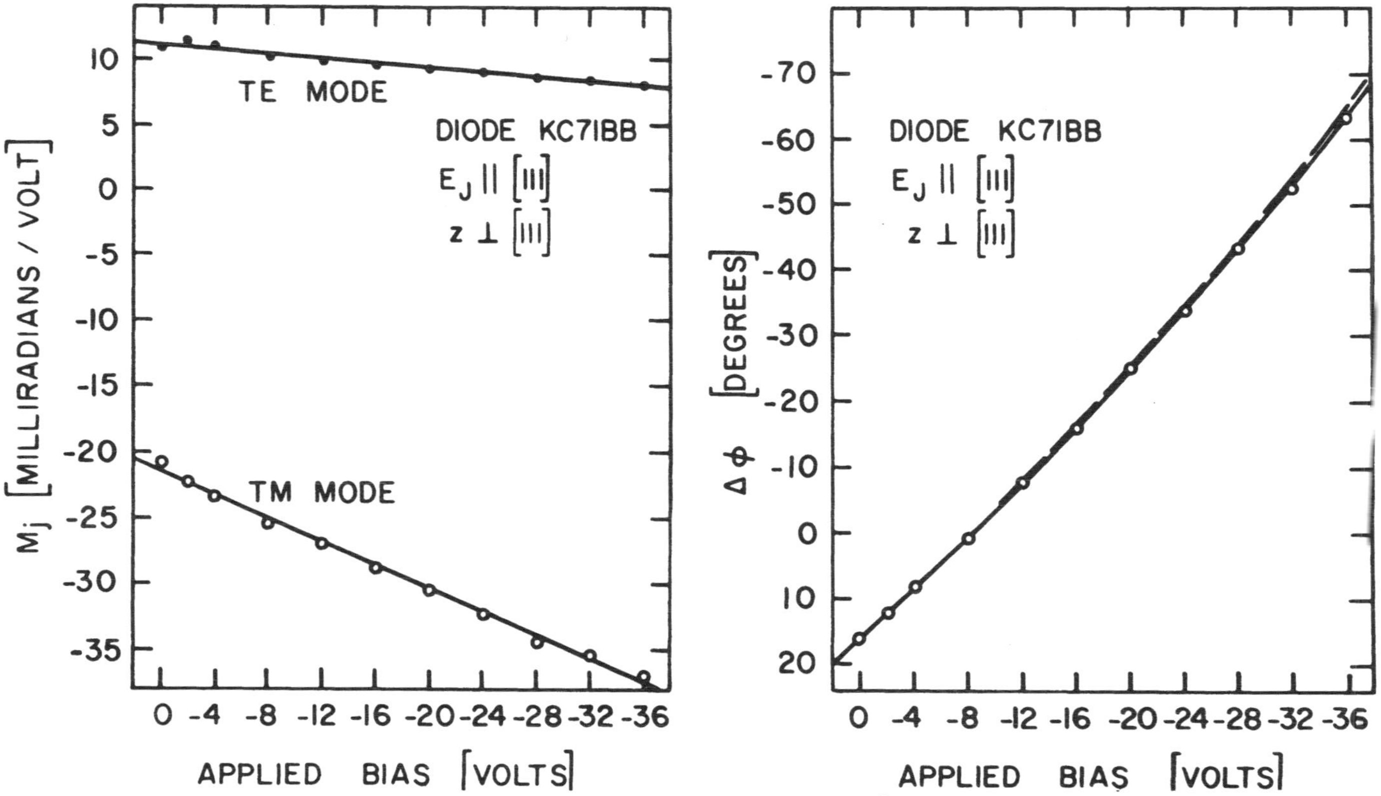

Fig. 3 Phase measurements versus applied bias at 6328 Å. (Left) Modulation index per volt, M_j, measurements with least square fitted straight lines. (Right) Phase difference $\Delta\varphi$ measurements with least square fitted parabola (solid line) and integrated difference of modulation index per volt with a constant strain-birefringence phase difference added to agree with the $\Delta\varphi$ data (dashed line).

PHASE MEASUREMENTS

Phase measurements of two types were carried out to confirm the above model.[7] One type was measurement of the change of phase with applied voltage for each of the two modes of polarization, while the other type was measurement of the phase difference between these two modes. The former were made by an optical heterodyne technique and the latter by an optical compensator method.

The model outlined above predicts[7] that the change of phase with voltage should contain a constant term and a term linear in the applied voltage and that the phase difference should contain a term linear in the applied voltage and a term quadratic in the applied voltage (except for the case of the junction field in the [110] direction). Experimental determination of the coefficient of the constant term in the former case or of the linear term in the latter case yields values for the product of the "built-in" fractional dielectric constant increase times the width over which it extends. This combination forms the one unknown parameter of the above model. All of the factors in the coefficient of the linear term of the former type measurement and in the quadratic term in the latter type measurement are known. However, experimental values of these coefficients can be used to determine values of the electro-optic coefficient for comparison with the value measured in bulk crystals.

Typical results for a diode with its junction field in the [111] direction are shown in Fig. 3. The dotted curve on the plot of phase difference, $\Delta\varphi$, was obtained by integrating with respect to voltage the curves for the derivative of the phase with respect to voltage, M_j, and then taking their difference (and adding a constant to account for strain birefringence). The agreement between the two types of measurements is good and their agreement with the form expected from the model is also good. The value of the "built-in" fractional dielectric constant increase times the width over which it extends is found to be 2.0×10^{-8} cm. If this width is taken to be the zero bias junction width, then the fractional dielectric constant increase is 1.1×10^{-3}. The value of the electro-optic coefficient deduced agrees with the bulk value to within 20%.

The results for other diodes, some with different crystalline orientations, have been generally satisfactory except for phase modulation measurements for the TM mode in diodes having the junction field in the [100] direction. For this situation the linear electro-optic effect should give no contribution to the modulation and the model predicts only a very small effect from the voltage variation of the junction width. Figure 4 shows such results. Mode mixing arising from strain can be ruled out since to be of sufficient size would require misorientation of the optic axes by about 10°, an effect not present. Length changes of the crystal

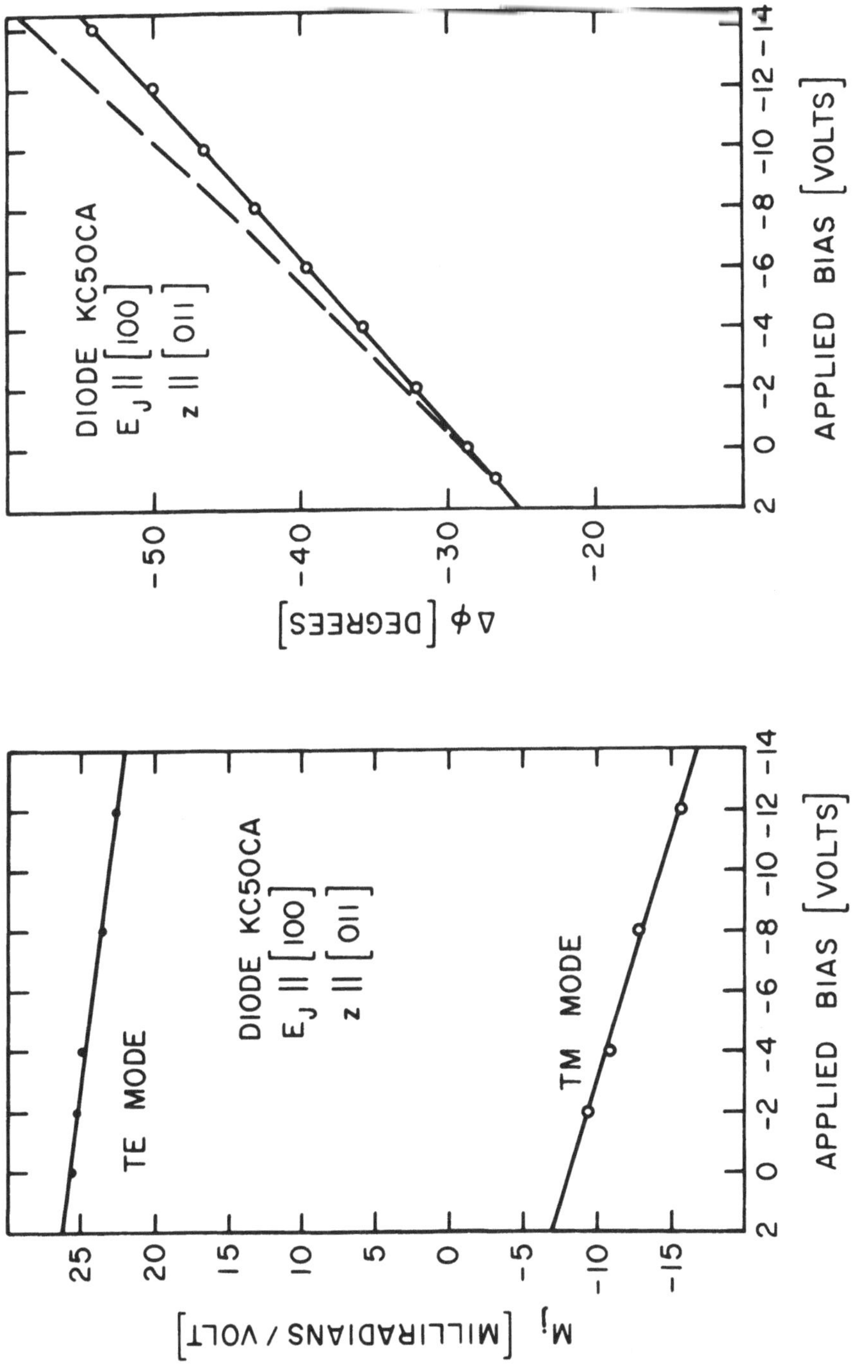

Fig. 4 Same caption as Fig. 3.

caused by the piezoelectric effect or perhaps also the electrostrictive effect can be ruled out since the bulk crystal surrounding the junction restricts such a change to no more than 1% of that required to explain the results of Fig. 4.

The only effect that seems capable of explaining this anomaly is the quadratic electro-optic (Kerr) effect. Values of the Kerr constants of $R_{11} \approx -4\times10^{-21}$ m^2/V^2 and $R_{12} \approx R_{44} \lesssim 10^{-21}$ m^2/V^2 are needed. These values generally improve the fit to experiment for the diodes of other orientations also. No other measurements of the Kerr coefficients are available to confirm this interpretation at present. They are, however reasonable values in terms of a generalized Miller's rule which says that third order anharmonic force terms are similar in a wide variety of materials. Based on values of this for Ge, Si, GaAs, and several perovskites a prediction of $|R_{ij}| \approx 10^{-21}$ m^2/V^2 results. For these reasons we attribute the origin of the anomaly to the Kerr effect but, in view of the measurement difficulties alluded to in the Introduction, we feel that the analysis in terms of a Kerr effect falls short of a proof of its presence here.

ORIGIN OF THE DIELECTRIC WAVEGUIDE

The origin of the larger dielectric constant in the junction is an interesting question. One possibility is the depression of the dielectric constant in the bulk crystal by the plasma resonance of the free carriers. In the depletion layer of the junction this depression would be absent. The magnitude of this effect is easily calculable and is found to be at least 50 times too small. It should be noted that, although only the product of this dielectric constant increase and the width over which it extends is measured, and this width is not known *a priori* to be the zero bias junction width as assumed, any drastic increase of w_1 from this value in order to decrease the dielectric constant increase would cause the expected light intensity distributions to be much broader than the measured ones.

A number of characteristics of the dielectric constant increase can be derived if the width over which it extends is taken as the zero bias junction width. Its magnitude does not vary with voltage, crystal orientation, or junction formation technique (junctions formed by liquid phase epitaxy have a comparable dielectric constant increase to diffused junctions). It depends on wavelength λ as roughly $\lambda^{3/2}$ which seems to rule out a simple bandgap shift as an explanation. Finally, there does not seem to be any pronounced dependence on the doping levels although the data are not completely conclusive. At present we do not know of any mechanism to explain these facts.

CONCLUSIONS

Our observations of light transmitted along the plane of reverse biased p-n junctions in GaP in conjunction with our studies of electromagnetic models of the junction waveguide have led us to the following conclusions: (1) an optical waveguide surrounding the junction occurs naturally. (2) The waveguide arises from a higher refractive index in the junction plane. (3) The asymmetry in this dielectric waveguide is sufficiently small to allow propagation of discrete modes. (4) For parameter values characteristic of the GaP p-n junctions studied, only the lowest order discrete mode of each polarization can propagate. (5) The product of the fractional optical dielectric constant increase and the width over which it extends is independent of bias. (6) The linear electro-optic (Pockels) effect creates a birefringent layer coincident with the junction depletion layer which expands with voltage.
(7) Certain anomalies in the phase measurements are best explained as arising from a small quadratic electro-optic (Kerr) effect, though the agreement falls short of a proof of its presence.

REFERENCES

1. D. F. Nelson and F. K. Reinhart, Appl. Phys. Lett. 5, 148 (1964).
2. F. K. Reinhart, J. Appl. Phys. 39, 3426 (1968).
3. D. F. Nelson, IEEE J. Quant. Electron. QE-3, 667 (1967).
4. D. F. Nelson and J. McKenna, J. Appl. Phys. 38, 4057 (1967).
5. J. McKenna, Bell Syst. Tech. J. 46, 1491 (1967), and a second paper to be published.
6. D. F. Nelson and E. H. Turner, J. Appl. Phys. 39, 3337 (1968).
7. F. K. Reinhart, D. F. Nelson and J. McKenna, to be published.

EPR SPECTRA OF CHROMIUM DOPED SPINEL

Alfred Kahan and Benjamin R. Capone

Air Force Cambridge Research Laboratories

Bedford, Massachusetts

Abstract. This article discusses the paramagnetic spectra of a chromium ion in natural and nonstoichiometric magnesium aluminate spinel. Methods of calculations are given and parameter values are derived from the angular variation of the magnetic resonance field.

The importance of lasers as a research tool in solid state physics and their widespread application in current optical technology needs no emphasis. Research on basic properties of laser and potential laser host crystals is justified in terms of the serious problems existing with available host crystals which limit and inhibit the further applications of laser systems. These difficulties involve the question of laser damage, that is, the deterioration and catastrophic failure of ruby and glass rods in high power level oscillator systems. These failures are well documented in the literature and need not be discussed in this paper.

The companion problem, or the alternate solution to laser damage, is the quest for novel laser host materials. Historically, the first, and probably still the most widely applied of all optical lasers, is chromium doped aluminum oxide, or ruby. Extensive efforts were expanded by various organizations on the improvement of this system and considerable resources invested in the development of an alternative to the aluminum oxide host crystal. With respect to potential laser host lattices, the program at AFCRL involves the growth and investigation of the static and dynamic properties of chromium and other transition metal and rare earth ion doped single crystals which could, under favorable conditions, lase. In particular, the materials under investigation include lithium and sodium germanate crystals and glasses, as well as magnesium aluminate spinels. The basic properties of these materials are reported in a series of presentations and publications.[1-6] Accordingly, they will be omitted from this report.

The oral presentation at the Institute consisted in a discussion of the structural, optical, and EPR properties of $Li_2Ge_7O_{15}:Cr^{3+}$ and $MgAl_2O_4:Cr^{3+}$. The present paper will limit itself to a basic review of some aspect of these investigations; namely, the electron paramagnetic spectrum of a chromium ion in an octahedral environment. Magnesium aluminate spinel is taken as an illustration, but calculations and procedures apply to chromium ions in other host lattices as well.

The general electron magnetic resonance spectrum of a chromium ion in magnesium aluminate spinel is similar to chromium in other oxide host lattices. The application of a magnetic field splits the $^4A_2$ ground state, and chromium with a spin of $S = 3/2$ yields $2S+1$, or 4 energy levels, with possibly six transitions between these levels. Three of these transitions are allowed while, owing to the selection rule $\Delta m = 1$, three are forbidden. In reality, in some materials one can clearly observe all six transitions, whereas in others even some of the allowed ones cannot be recorded.

For calculation purposes, the data are evaluated by applying the standard spin Hamiltonian applicable to chromium,[7-9]

$$H_s = \beta\,(g_x H_x S_x + g_y H_y S_y + g_z H_z S_z) + D\,[S_z^2 - 1/3\,S\,(S+1)] + E\,(S_x^2 - S_y^2) \qquad (1)$$

where the first term is the contribution of the Zeeman energy, and the D and E terms are the effects of the crystalline field on the fine structure. The second and third terms of the Hamiltonian are independent of the magnetic field, and the D and E parameters determine the energy value of the ground state splitting. Introducing the raising and lowering operators, $S_\pm = S_x \pm iS_y$, and defining $\gamma_\pm = \gamma_x \pm i\gamma_y$ and $\gamma_i = (\beta/2)g_i H_i$, one finds that the spin Hamiltonian can be expressed as

$$H_s = \gamma_- S_+ + \gamma_+ S_- + 2\gamma_z S_z + D(S_z^2 - 5/4) + (E/2)(S_+^2 + S_-^2) \qquad (2)$$

The energy eigenvalues W are found by operating on the eigenfunction set $|jm\rangle$, and applying relationships

$$\langle m'|m\rangle = \delta_{m'm}$$

$$S_z|m\rangle = m|m\rangle$$

$$S_{\pm}|jm\rangle = [j(j+1) - m(m\pm1)]^{1/2} \; |j,m \pm 1\rangle.$$

The secular determinant is then derived from the eigenvalue equation

$$\| \langle n|H_s - W|m\rangle \| = 0 \tag{3}$$

with n, m = 3/2, 1/2, -1/2, -3/2. Performing the operations, one obtains

$$\begin{vmatrix} 3\gamma_z+D-W & \sqrt{3}\gamma_- & \sqrt{3}E & 0 \\ \sqrt{3}\gamma_+ & \gamma_z-D-W & 2\gamma_- & \sqrt{3}E \\ \sqrt{3}E & 2\gamma_+ & -\gamma_z-D-W & \sqrt{3}\gamma_- \\ 0 & \sqrt{3}E & \sqrt{3}\gamma_+ & -3\gamma_z+D-W \end{vmatrix} = 0 \tag{4}$$

This determinant can be solved for W_i either by expanding and solving for the roots of the fourth order algebraic equation

$$\begin{aligned} W^4 - [10(\gamma_x^2 + \gamma_y^2 + \gamma_z^2) + 2(D^2+3E^2)]\,W^2 + [8D(\gamma_x^2+\gamma_y^2-2\gamma_z^2) \\ - 24E(\gamma_x^2 - \gamma_y^2)]W + [9(\gamma_x^2+\gamma_y^2 + \gamma_z^2)^2 + (D^2+3E^2)^2 \\ - 18E^2\,(\gamma_x^2 + \gamma_y^2 - \gamma_z^2) \\ + 2D^2(\gamma_x^2 + \gamma_y^2 - 5\gamma_z^2) + 24DE(\gamma_x^2 - \gamma_y^2)] = 0 \end{aligned} \tag{5}$$

or applying inversion and eigenvalue routines directly to Equation (4). The proper parameters are those for which the eigenvalues of this equation satisfy the magnetic field resonance frequency condition $\nu = W_i - W_j$ for all experimental points at all angles of rotation. One can also solve for the eigenvectors and calculate the intensity ratios as well. The computations are most easily accomplished by digital computer techniques, especially if many data points are involved. The object is to obtain the g-values, D and E parameters, and then compare and interpret the results in comparison to the behavior of Cr^{3+} in other host lattices. An exposition of these calculations as applied to ruby can be found in reference 10.

The most detailed work on the EPR spectrum of natural spinel,

$MgAl_2O_4:Cr^{3+}$, was completed by Stahl-Brada and W. Low.[11] Basically, they found that there are four octahedrally coordinated Cr^{3+} ions in the unit cell, distorted along the <111> directions. For this cubic crystal system, $g_x = g_y$ and $E = 0$, the coefficients of the secular equation simplify, and only three parameters enter in fitting the experimental data. The ions are located along the <111> and, for computational purposes, it is most convenient to obtain the experimental data with the magnetic field rotated parallel to the (110) plane of the crystal. In this configuration, the symmetry axis of two ions contained in this plane are 70°32′ apart and, consequently, a repeating spectrum occurs shifted by this angle and symmetric, both around the [110] and [001] directions. The angular variation of the two other ions, lying along the body diagonal of the unit cell outside the plane of rotation is described by $\theta = \cos^{-1}\left(\frac{\cos\theta'}{\sqrt{3}}\right)$, where θ' is the angle in the (110) plane between the magnetic field vector and the [111] direction. These two ions are situated symmetrically about the (110) plane and, for a perfectly oriented sample, the contribution of both ions as a function of angle of rotation should be superimposed at all angles of rotation.

In the experimental investigations of reference 11, only one transition was observed, but the symmetry relationship between the various ions was confirmed and the angular variations did follow the predicted behavior. The parameters, based on X-band (9.4 Gc/sec) frequency data, were computed as: $g_\perp = 1.989$, $g_\parallel = 1.986$, and $D = 0.495\ cm^{-1}$. Subsequently, these measurements were repeated by Atsarkin[12,13] at X- and K-band frequencies and the parameters, calculated from the field values at 0° and at 90° utilizing the data from both frequencies, were obtained as: $g_\perp = 1.98$, $g_\parallel = 1.985$, and $D = 0.923\ cm^{-1}$. While the discrepancies in the g-values can be attributed to variations in samples and experimental differences, the deviation in the D-value is more fundamental and the question, "Which value is right?" is a valid one. At first sight, it is obvious that Atsarkin's calculations should be right, as these are based on three independent experimental points, the minimum necessary to calculate three parameters. However, the solution is not necessarily unique, there is no indication of the accuracy of the parameter values, and the correctness of the angular behavior is not verified.

In order to clarify this situation, we computed the angular variation of the magnetic resonance fields based on both sets of parameters at X-band frequency. The results are shown in Figure 1. The calculations show that, within experimental accuracy, the main observed transition is insensitive to D, even at non-zero and non-ninety degree values; the resulting curves are superimposed; and

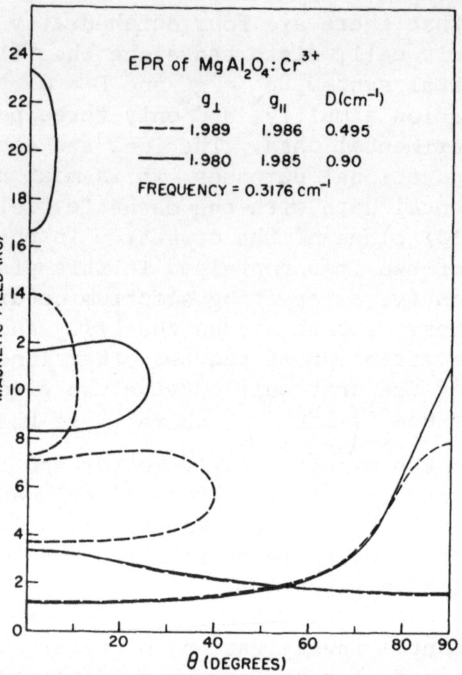

Figure 1. Calculated EPR spectra of Chromium doped magnesium aluminate spinel at X-band frequency, based on two different sets of parameters.

in Figure 1, for emphasis and clarity, the two curves are slightly displaced. Thus, at X-band, the parameter values cannot be uniquely determined from this transition only. For both sets of parameters, the value of 2D exceeds the resonance frequency; as a consequence, some of the transitions are multivalued, and one obtains folded spectra characteristic of such a situation.

In addition to the central transition, one other transition is unfolded and, at 90°, the curves based on the two sets of parameters are separated by approximately 3 kG. The experimental observation of this point would enable the determination of D without invoking any K-band data. This point has been observed by Berger,[14] and the calculated D-value based on the two transitions at 90° is 0.89 cm^{-1}, in close agreement with Atsarkin. However, Berger also measured at 90° the value of the lower transition at 0.769 cm^{-1} (23.08 Gc/sec) and, if all three ninety degree points are taken into account and a least-square fit applied to the computations, the calculated D-value becomes 0.916 cm^{-1}. This computer exercise shows that one should not quote an error on the calculated parameter value, usually given to a third decimal place, if only one or two points are applied in computing the particular value.

The comparative calculations were also performed at a frequency of 1.182 cm^{-1} (35.46 Gc/sec) with results shown in Figure 2. At this frequency, the main transition based on the two sets of parameters separates at all angles and the measurement of the angular variation of the resonance field can uniquely determine and differentiate between the two sets of values. Note that at this frequency, one of the 2D values, the dashed lines, exceed the resonance frequency and, consequently, all transitions are unfolded while some of the solid lines are still multivalued. Thus, for natural spinel, a D-value of the order of 0.9 cm^{-1} is the proper one and, as shown in Figure 2, for this condition, a tremendous increase in magnetic field occurs for the central transition at angles below 20°.

The curves of Figures 1 and 2 were calculated from a computer program based on Eq. (5). For chromium and other S = 3/2 materials, one still can easily expand the resulting 4 x 4 determinant. Initially, our calculations were carried out on computer programs obtained from the Quantum Chemistry Program Exchange, Chemistry Department, Indiana University, written by H. M. Gladney of IBM.

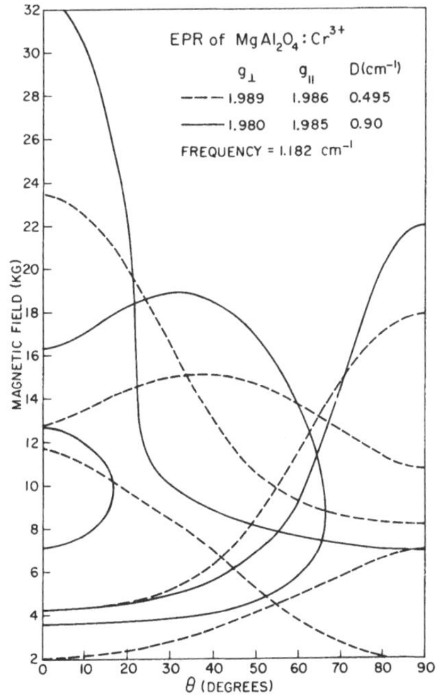

Figure 2. Calculated EPR spectra of chromium doped magnesium aluminate spinel at a frequency of 35.46 Gc/sec, based on two different sets of parameters.

These are general programs applicable to any spin value that will either calculate the least-square fitted parameters based on the experimental points, or determine from a set of parameter values the angular variation of the magnetic field, including intensities and eigenvectors of the transitions. However, owing to some adaptation difficulties together with some question about the capability of the program in calculating multivalued transitions, we did not have great confidence in the initially obtained values. Therefore, we decided to write the small program applicable to Cr^{3+} only. Consequently, at the time of experiment performance, we did not possess a fully trustworthy computer program for the analysis of EPR data.

Due to some broad resonance obscuring all significant data, our attempt to repeat the EPR experiment on a natural spinel sample was unsuccessful. However, our real purpose was to obtain data on synthetic spinels as a function of stoichiometry. A typical near-stoichiometric crystal spectrum at a frequency of 1.169 cm^{-1} (35.07 Gc/sec) is shown in Figure 3. With the magnetic field rotated in the (110) plane, one observes the central transition, the crystal symmetry connected spectra displaced at approximately 70° from the original, and the third set contributed by the two ions lying outside and symmetrical with respect to the plane of rotation. The shape of the basic curve is very similar to the graph of the central transition shown for X-band frequency in Figure 1 and to the angular variation of the spectra of reference 11. Our initial attempt to calculate the parameters from this experimental curve was unsuccessful.

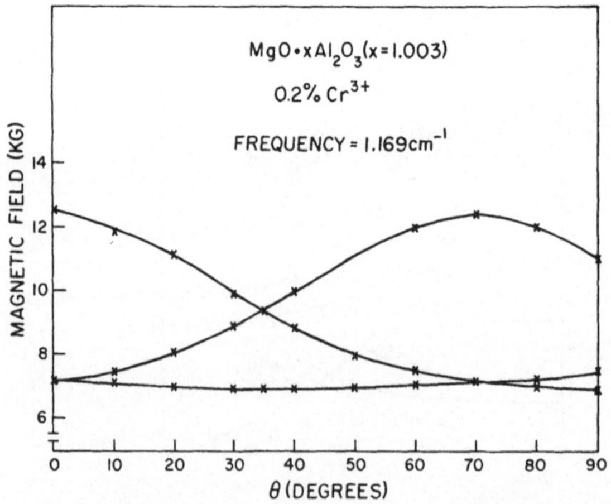

Figure 3. Experimental EPR data of nonstoichiometric magnesium aluminate spinel at 35.07 Gc/sec frequency.

The key to this problem is the behavior of spectra between 20 and 30°. If one collects data points every 10°, which we were doing, without having as a priori the calculated curves for natural spinel at K-band, the tendency would be to draw a smooth curve joining the points as in Figure 3, without realizing that two separate transitions are being connected. So, on the one hand, we were faced with spectra very much resembling the X-band data while, on the other, Gladney's computer program gave indication that the only way the data could be reasonably fitted would be by assuming that the 0° and 10° points belong to a different transition than the rest of the curve. This gave the clue for re-examination of the question concerning the proper D-value for natural spinel and for writing the simplified computer program. Finally, when the curves of Figures 1 and 2 were calculated, we realized the behavior of the main transition at high frequencies, with the corresponding increase in resonance field values below 20°. Figure 4 shows the calculated theoretical curves, the experimental points replotted from Figure 3, and additional points taken in the critical region of the transition. The particular parameters applicable to these curves, based on a least-squares fit of the eleven points, are: $g_{\perp} = 1.954$, $g_{\parallel} = 1.982$, and $D = 1.034$ cm^{-1}. In Figure 4, all

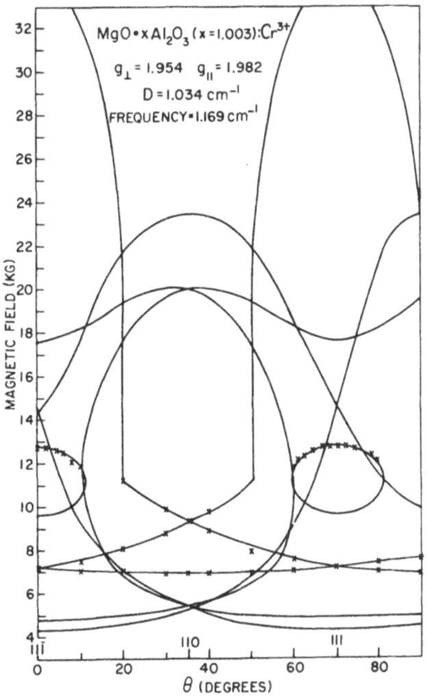

Figure 4. Experimental and calculated EPR spectra of nonstoichiometric magnesium aluminate spinel. Parameter values are indicated on the figure.

possible transitions for all three sets of ions are plotted even though experimentally only the central transition is observed for each set. Similar experiments have been performed on samples with different stoichiometric compositions but, as yet, no consistent variation of parameter values with stoichiometry has been obtained.

REFERENCES

1. Dugger, C.O., J. Appl. Phys. 38, 2345 (1967).
2. Lipson, H.G. and R.C. Powell, J. Appl. Phys. 38, 5409 (1967).
3. Powell, R. C., Phys. Rev. 173, 358 (1968).
4. Powell, R. C., J. Appl. Phys. 39, 4517 (1968).
5. Dugger, C. O. (to be published).
6. Lipson, H. G., A. Kahan, J.A.Adamski, E. Farrell, M.J.Redman and J. Kawamura, J. Crystal Growth, Vols. 3 & 4 (1968).
7. Bowers, K.D. and J. Owen, Rpt. Progr. Phys. 18, 304 (1955).
8. Low, W., Paramagnetic Resonance in Solids, Solid State Phys., Suppl. 2, Seitz and Turnbull, Eds., Academic Press, N.Y. (1960).
9. Al'tshuler, S.A. and B. M. Kozyrev, Electron Paramagnetic Resonance, Academic Press, N.Y. (1964).
10. Schulz-DuBois, E.O., Bell System Tech. J. 38, 271 (1959).
11. Stahl-Brada, R. and W. Low, Phys. Rev. 116, 561 (1959).
12. Atsarkin, V.A., J. Exptl. Theret. Phys. (USSR) 43, 839 (1962). English Trans. Sov. Phys. JETP, 16, 593 (1963).
13. Atsarkin, V.A., M.E.Zhabotinsky and A.V.Frantsesson, Quantum Electronics III, p. 759, edited by P. Grivet and N. Bloembergen, Columbia University Press, N.Y. (1964).
14. Berger, S. B., J. Appl. Phys. 36, 1048 (1965).

METHOD AND APPLICATION OF NEUTRON SCATTERING FOR THE STUDY OF FLUCTUATIONS IN CONDENSED MATTER

T. Springer
Institut für Festkörper- und Neutronenphysik der Kernforschungsanlage Jülich

and

W. Gläser
Institut für Angewandte Kernphysik des Kernforschungszentrums Karlsruhe

1. GENERAL PRINCIPLES OF THE METHOD; THE SCATTERING LAW

During the last 10 years the investigation of slow neutron scattering has developed into a successful method being widely used to study the dynamics of atoms in condensed matter. This lecture presents an introduction into some important features of the method. Then it discusses a few typical and present-day experiments which have been selected from the main fields of application, namely (1) phonon scattering in crystals, (2) scattering on fluctuations in liquids, and (3) scattering on magnetic fluctuations. This means that we restrict ourselves on the investigation of collective phenomena by coherent scattering.

The experimental quantity of interest is the scattering probability which can be described by a double differential cross section, namely the cross section per unit energy interval and per solid angle:

$$d^3\sigma/d\Omega dE = (k/k_0)A^2 S(\vec{\kappa},\omega) \qquad (1)$$

$\hbar\omega = \hbar^2(k_0^2 - k^2)/2m$ and $\hbar\vec{\kappa} = \hbar(\vec{k}_0 - \vec{k})$ are the energy and the momentum transfer during the scattering process in the sample. E_0, E are the energies, $\hbar\vec{k}_0$, $\hbar\vec{k}$ are the momentum vectors of the neutrons before and after scattering, respectively.

The scattering law $S(\vec{K},\omega)$ depends only on the dynamics and the structure of the sample, whereas the factor A depends on the interaction between the neutron and the scattering particles. This simple factorial representation follows from Fermi-Born's approximation which holds quite well within the experimental accuracy. Two types of the scattering law have to be considered. S_{coh}, if there is interference between the partial waves originating from the scattering particels, and S_{inc} if there is no interference at all. The latter case is important only for a few nuclei with spin, essentially for hydrogene. In the following, we consider only coherent scattering. In this case the quantity A is equal to the scattering length of the nucleus which is approximately equal to the nuclear radius if resonance contributions can be neglected.

The scattering law can be calculated by a Fourier transform over the time-dependent pair correlation function /1/, namely

$$S_{coh}(\vec{K},\omega) = \frac{1}{2\pi} \iint e^{i(\vec{K}\vec{r}-\omega t)} G(\vec{r},t)\, d^3\vec{r}\, dt \qquad (2)$$

The correlation function is given by

$$G(\vec{r},t) = N^{-1} \left\langle \sum_{i,j=1}^{N} d^3\vec{r}\,' \delta[\vec{r}+\vec{r}_i-\vec{r}\,'] \delta[\vec{r}\,'-\vec{r}_j] \right\rangle_T \qquad (3)$$

In the classical limit the correlation function measures the probability that, if an atom is at the origin at a time t = 0, an atom will be within a volume element at $\vec{r}$ at time t. For sufficiently large $\vec{r}$ or t this can be expressed by the autocorrelation function of the local density,

$$G(\vec{r},t) = \langle \rho(0,0)\rho(\vec{r},t) \rangle_T / \rho \qquad (4)$$

For magnetic scattering /2/ one has a spin autocorrelation function

$$G^{\alpha}(\vec{r},t) = \langle S^{\alpha}(0,0) S^{\alpha}(\vec{r},t) \rangle_T \qquad (5)$$

where $S^{\alpha}(\vec{r},t)$ is the spin component α at r and t. In this case, A has essentially to be replaced by $(1{,}91\ e^2/m_{el}c^2)F(K)$, where F is the form factor of the magnetic electron shell. The scattering cross section is then given by a summation over the components of G^{α} perpendicular to K.

It is noteworthy that the representation of eq. (1) holds, with the same function S_{coh}, for the scattering of x-rays (essentially with the classical electron radius as scattering length). For light scattering this formalism is not applicable because of the very large scattering amplitude, except for weak density fluctuations (e.g. thermally excited sound waves /3/). In the case of x-ray scattering no energy analysis is possible so far. Integrating (2) over all ω, $e^{-i\omega t}$ is transformed into a delta function of t, and the well-known instantaneous correlation function $G(\vec{r},0) = g(\vec{r})$ is obtained. The Fourier transformation (2) connects a space-time-plane with a K,ω-plane.

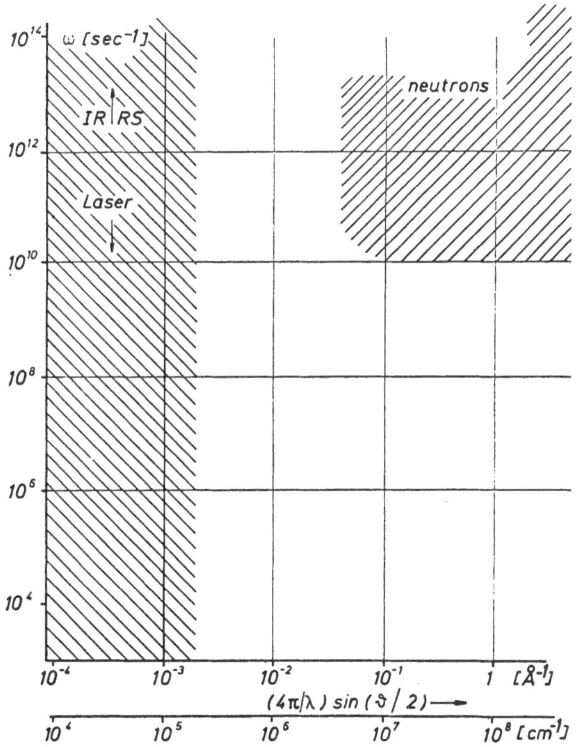

Fig. 1: Plane of frequencies ω and wave numbers K which can be analyzed by neutron and light scattering.

Fig. 1 shows the range of K and ω which can be covered by neutron scattering experiments as compared to the scattering of light (ϑ is the corresponding scattering angle, where $k = k_o$ was assumed). The obvious advantage of the neutrons is their short wave length λ and therefore the possibility to investigate fluctuations with wave lengths 1/K as short as the interatomic distances. The accessible ω range is essentially restricted by the available intensities.

The instrumental methods will be discussed only in principle. Neutrons with an energy E_o are separated from the reactor spectrum; after scattering they are analyzed with regard to E. Both steps can be performed either by means of Bragg reflections on single crystals, or by time-of-flight analysis using beam choppers. All combinations of both methods, applied to E_o as well as to E are possible and usual. The experimental resolution is determined by the angular spread of the neutrons before and after scattering, and by the energy width of E_o and E. Normally E_o is chosen in a range between 4.10^{-3} and 0,2 ev. In typical cases $\delta\hbar\omega$ is of the order of 10^{-4} ev up to a few 10^{-3} ev. The momentum resolution $\hbar\delta K$ can be of the order of 10^{-2} Å^{-1} under favourable conditions. It can be shown that the obtainable pulse rate at the detector is, for a given resolution, proportional to $(\delta\omega)^2$ and at least to $(\delta K)^2$ (see /4/). Therefore, an increase of the overall resolution e.g. by a factor of 3 would need (for fixed statistical accuracy of the results) an increase of the reactor power or of the experimental time by at least two orders of magnitude. Therefore, intensity and experimental time are the essential restrictions with regard to the obtainable resolution in nearly all cases, and not the quality of the spectrometers as such. In this connection one should have in mind that the flux of a reactor is 10^3 to 10^4 times smaller than that of a x-ray tube in a typical wave length interval.

A comprehensive description of the main subjects of slow neutron scattering can be found in a book /5/. A world-wide representation of the theoretical and experimental work in this field can be found in the proceedings of the corresponding IAEA conferences which are hold every few years /6/. A short review has been published about two years ago /7/.

2. SCATTERING ON PHONONS; INVESTIGATIONS ON METALS

The scattering law is especially simple if the

scatterer is a single crystal in which phonons can propagate /8/. If the momentum transfer is sufficiently small and the temperature is not too high, single phonon creation or annihilation is dominating and the scattering law is essentially a series of delta functions determined by the following equations

$$E_0 - E = \hbar^2(k_0^2 - k^2)/2m = \pm \hbar\nu \tag{7}$$

$$\vec{K} = 2\pi\vec{\tau} - \vec{q}(\nu) \tag{8}$$

$\vec{\tau}$ is a reciprocal lattice vector; $\vec{q}$ is the wave vector and ν the frequency of the phonon interacting with the neutron. The condition of "wave vector conservation" (8) is well known from diffuse X-ray scattering and from electron-phonon interactions in metals. It is sort of an interference condition for the double periodical structure existing in the crystal ($2\pi\vec{\tau}$ and $\vec{q}$).

Eqs. (7) and (8) have discrete solutions in all Brillouin zones corresponding to the different polarisations of the phonon. Therefore, for a given E_0 and $\vec{k}_0$, and for a certain orientation of the crystal one finds intensity peaks at certain values and directions of $\vec{k}$. From $\vec{k}$ one obtains $\hbar\nu$ (eq.7) and from the relation (8) one finds the vector $\vec{q}$. By changing the orientation of the crystal relatively to $\vec{k}_0$ and the direction of observation ($\vec{k}$) the whole reciprocal lattice can be covered with experimental points, each corresponding to a certain phonon $\nu(\vec{q})$. The intensity in the scattered neutron groups is proportional to the square of the dynamical structure factor $\sum_j(\vec{u}_j\vec{K})e^{i\vec{K}\vec{r}_j}$ where $\vec{u}_j$ is the polarization vector of the phonon, and where the sum goes over all atoms per unit cell. By proper choice of the orientation of $\vec{K}$ this can be practically made non-zero in all cases. Therefore, all vibration types are observable with neutron scattering throughout the Brillouin zone in contrast to optical spectroscopy. The structure factor allows in principle the determination of the polarization vector $\vec{u}_j$. However, reliable measurements of the peak intensities are difficult.

This method has been applied to a great variety of substances (cf. /6/), namely to a number of metals, semiconductors, ionic crystals, solid noble gases, and to one or two organic crystals, except substances with high neutron capture, e.g. Ag, Au, or certain rare earths, or with strong incoherent spin scattering (vanadium).

As an example of application of some topicality we discuss now investigations on metals in more detail. Dispersion curves $\nu(\vec{q})$ of metals measured in symmetry directions have been used to extract the interatomic force constants which in the Born- von Kármán theory describe the dynamics of the crystal lattice. In the fits of Born- von Kármán models to the experimental $\nu(\vec{q})$ curves, the force constants between a great number of neighbours in the lattice had to be taken into account to give a satisfactory description of the experimental results. The reason for the apparent long range forces between ions in metals is due to the influence of conduction electrons. It is therefore desirable to explain these forces more basicly by considering the interaction between the positive ions, between ions and conduction electrons, and between the electrons themselves.

It has been found advantageous to formulate the corresponding potentials in reciprocal rather than in normal space and a useful representation of the Fourier transform of the effective potential between the ions in simple metals is /9/

$$V_{eff}(\vec{q}) = V_{II}(\vec{q}) + V_{IE}^2(\vec{q})\,[\varepsilon^{-1}(\vec{q}) - 1]\,/\,C(\vec{q}) \tag{10}$$

Here V_{II} is the contribution of the direct Coulomb interaction of the ions which is calculated under the assumption of Gaussian charge distributions with the Ewald method /10/. V_{IE} describes the interaction of an ion with a conduction electron. This interaction is screened by the "response" of the sea of conduction electrons to the ionic charge. The last effect is taken into account by a q-dependent dielectric constant which can be calculated for a gas of free electrons /11/. The consideration of exchange and correlation of the conduction electrons is difficult. Several approximations have been proposed for this effect /12/, /13/ which change $\varepsilon(\vec{q})$ and determine $C(\vec{q})$ in eq. (10). An approximation for V_{IE} is the "pseudopotential" consisting of the attractive Coulomb potential of the bare ion and an additional repulsive term which can be approximated by a local potential in the case of simple metals. V_{IE} can be calculated then from first principles. However, the attempts made so far did not lead to quantitative agreement with measured integral quantities. Therefore, model potentials containing free parameters are currently in use. These parameters can be determined e.g. from spectroscopic data of the free atom /14/. Another approach is to use the phonon dispersion curves to determine open parameters /15/. A successful ansatz is

$$V_{IE}(\vec{q}) = \frac{-4\pi Z e^2}{q^2} + \gamma_1 [1+(qr_1)^2]^{-2} + \gamma_2 r_2 q [1+(qr_2)]^2 \qquad (11)$$

where the repulsive part of the pseudopotential is approximated by the radial part of hydrogen-like wave functions; γ_n and r_n are the model parameters and Ze is the effective charge of the ion. By inserting the expression (11) into eq. (10) and (10) into the equations of motion, these parameters can be determined by a least square fit to the phonon dispersion curves. Starting with eq. (11) the model potentials were derived for Na, Mg, and Al containing 4 experimentally determined parameters. By using the electronic band structure the Fermi surface, the cohesive energy, the crystal energy, and the electrical resistivity have been calculated in good agreement with experimental values /16/. It seems reasonable to use these potentials also for the calculation of other quantities which are influenced by the ion-electron interaction, e.g. for the calculation of the properties of lattice defects.

For the noble and transition metals the assumption of a small core of electrons bound to the ion, and of the free gas behaviour of the conduction electrons seem to fail. Nevertheless such an analysis might be a first approximation leading to a better understanding of the microscopic behaviour of these systems. This has been done in the case of copper by using three repulsive terms and taking the exchange interaction between the cores into account by a Born-Mayer potential /17/.

A similar approach replacing the more detailed model (11) by a Coulomb potential changed by a q-dependent form factor $G(q)$ has also been used to obtain an effective potential from phonon dispersion curves measured with neutrons /18/:

$$V_{eff}(q) = -\left(4\pi Z e^2 / q^2 \varepsilon(q)\right) G(q) \qquad (12)$$

This model has been used, e.g. for the analysis of dispersion curves of potassium and sodium /19/.

Using the model pseudopotential, the shape of the Fermi surface can be calculated by means of perturbation theory as mentioned above.

With further refinement of the experimental accuracy it is possible to detect fine irregularities in the dispersion curves which can be directly attributed to the influence of the Fermi surface. Such an irregularity is shown in fig. 2 for lead /20/. The qualitative explanation can be given as follows: The conduction electrons move very easily thereby screening the electrical field which is built up by the periodical displacement of the ions in a lattice vibration.

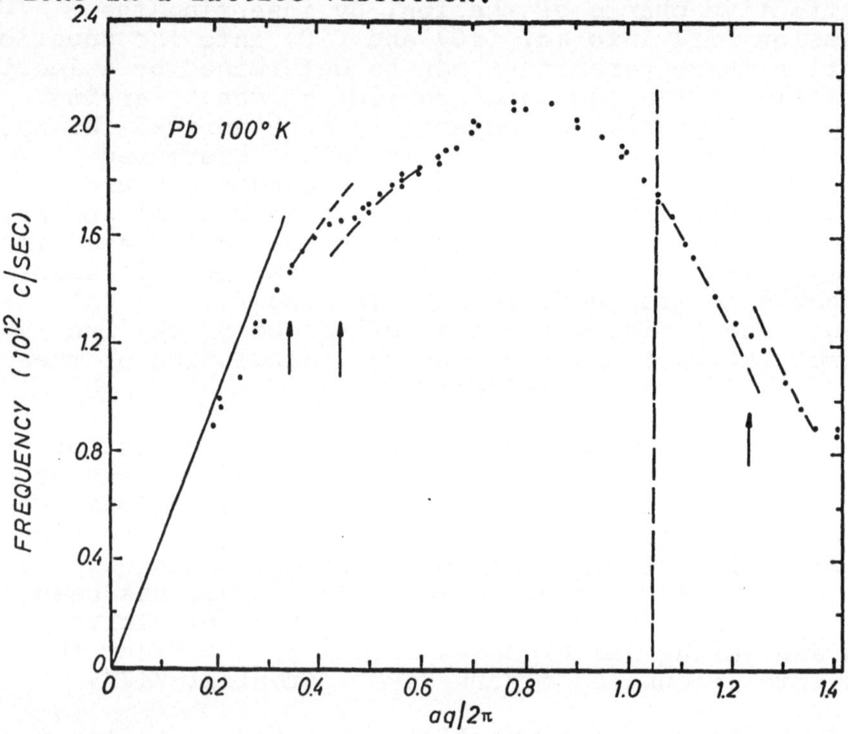

Fig. 2: Dispersion curve on lead showing a Kohn anomaly (Brockhouse et al./20/).

This screening is complete for a phonon with infinite wave length and it decreases as q increases. A special situation arises at $q = 2k_F$. Assuming a spherical Fermi surface for a free electron gas with diameter $2k_F$, $\varepsilon(q)$ has a logarithmic singularity in the slope at $2k_F$. An interaction between a phonon and an electron is only possible for pairs of wave vectors of the electron, k and k+q (k before and k+q after the interaction) if one of the states is occupied and one is empty. One can easily see that as soon as q exceeds $2k_F$ the number of electron states available for the interaction with the phonon decreases abruptly. Therefore one should expect an abrupt change in the restoring force connected with a change of the frequency ν /21/ (Kohn effect). For a more complicated surface this effect appears for such q-values which correspond to parallel pairs of tangential planes on the Fermi surface. A favourable case is lead. The Fermi surface of lead /22/ is given approximately by a sphere with deviations at the zone boundaris (fig.3). The anomalies have been actually found corresponding to the chords shown in the figure, in good agreement with conventional methods. Recently a very careful analysis has shown a number of further anomalies /23/.

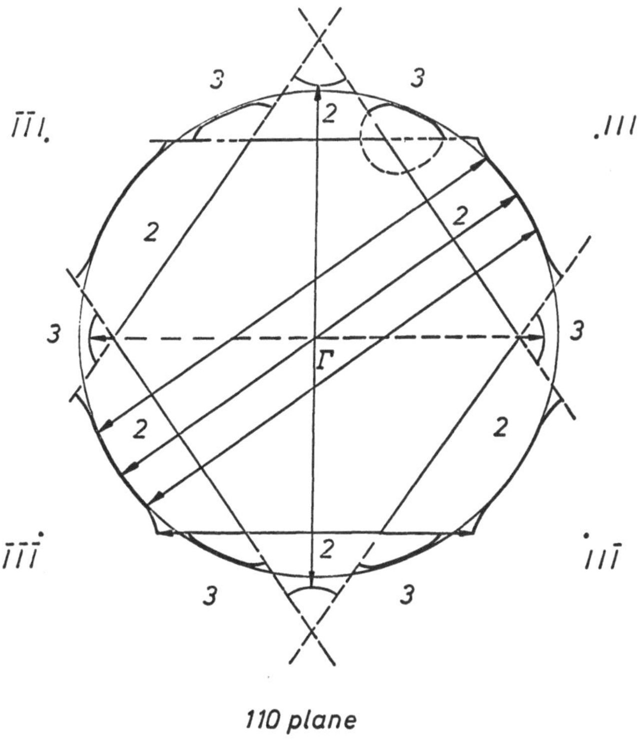

Fig. 3: Fermi surface of lead, showing chords corresponding to Kohn anomalies (from /20/).

Measurements on a series of Pb-Tl alloys have been performed /24/ changing the electron concentration from 3,4 to 4 electrons per atom. The expected increase of the Fermi radius was clearly indicated by the shift of the Kohn anomalies. Kohn anomalies have been seen in other metals, namely in Cu, Zn /25/, /26/, and in Al /27/. This method might be, in spite of being extremely tedious, very useful for experiments on disordered metals and on metals at elevated temperatures where other methods fail. The "strength" of the anomaly is determined besides the geometrical properties of the Fermi surface by the interaction strength /28/. So far, the effect is not quantitatively understood, however, and the experimental accuracy does not allow to draw quantitative conclusions. Also the width of a phonon line is determined by the interaction with the electrons besides the phonon-phonon interaction. In most cases the line widths are at best comparable with the resolution obtainable at modern triple axis spectrometers. Therefore, no relevant experiments on this subject are existing so far. Also x-ray scattering is able to detect the Kohn anomalies. In this case energy analysis is not possible; however, the diffuse thermal scattering intensity, depending on $\nu(\vec{q})$, should reveal these anomalies which have been actually detected on lead /29/.

3. FLUCTUATIONS IN LIQUIDS

In a liquid two kinds of atomic motions should be considered. The self-diffusion of individual atoms, and collective motions due to strong correlations existing between the atoms over large distances. The diffusivity of the atoms in liquids has been investigated, among many other methods, by studying the self part of the correlation function eq. (3), using incoherent neutron scattering which dominates in hydrogeneous liquids. (As an example we refer to the investigations discussed in ref. /30/). Here, we will restrict ourselves to the discussion of the collective modes which can be investigated by coherent inelastic scattering only.

As a very rough model for the theoretical discussion of such experiments one tries to describe the liquid like a polycrystal with phonons /31/, defining the reciprocal lattice vector $\vec{\tau}$ by the position of the peaks in the Fourier transform of the pair correlation function, and neglecting the actual width of τ and the width of the phonon energy $\hbar\nu$ which is expected to be large. Then one can apply the conservation laws eq. (7) and (8). In a liquid, $\vec{\tau}$ is distributed isotropically in reciprocal space. Therefore, scattering is only possible between the limits $2\pi\tau + q$ and $2\pi\tau - q$ to fulfill eq. (8). This would show up at a certain K as an abrupt change of the scattered intensity measured at fixed energy transfer, i.e. at fixed q.

In a similiar way, one would expect a step in the intensity measured at a fixed angle as a function of the energy transfer. Such breaks in the intensity distribution have been actually observed in several liquids, like aluminium, argon, and lead /32/, /33/, /34/. The existence of longitudinal "phonons" has been postulated from these results with a dispersion curve similiar to that of the solid phase. Transverse motions are not easily observable by this method because, according to theory, the described irregularity in the spectrum is only a smooth kink, not a step. More elaborate theories have been proposed which are improved with regard to the description of the liquid structure by a form factor /35/ instead of an artificial τ-vector.

The interpretation of the experiments has more lucidity if these are performed at sufficiently small K-values so that the quantity $\vec{\tau}$ in eq. (8) does not come into play, and one has

$$K = -q(\nu) \tag{13}$$

Therefore, one should find a dublett of broadened lines at

$$\hbar\omega = \pm h\nu \tag{14}$$

This is equivalent to the well-known Brillouin dublett for light scattering. Unfortunately, also in this case only "phonons" with longitudinal components are observable because $\vec{K}$ is parallel $\vec{q}$. Eqs. (13) and (14) have been used for the first time in the famous neutron scattering experiments on liquid He^4 where rather sharp lines have been observed. From these results a dispersion curve $\nu(q)$ of the excitations in He^4 has been determined in agreement with theory (see e.g. /36/). In normal liquids, Brillouin scattering of neutrons has been observed so far only on liquid CO /37/ and weakly pronounced on liquid lead /38/. The peaks are broad but their separation from the elastic middle peak is feasible. Fig. 4 compares the dispersion curves constructed from Brillouin scattering on liquid and on polycrystalline CO. Fig. 5 collects data taken from this method (crosses, from /38/) and from that described before ($\tau \neq 0$) for liquid lead. These experiments prove the existence of "phonons" (or better to say: of density fluctuations propagating as strongly damped waves) with wave lengths of the order of 5 to 15 Å and frequencies of the order of 10^{12} sec^{-1}. This is not unreasonable: It is known that the average jumping

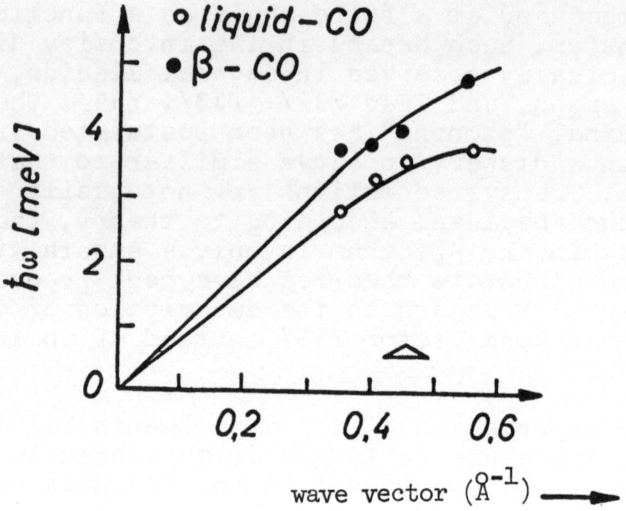

Fig. 4: Dispersion curve of "phonons" in liquid CO /37/, as compared to the solid.

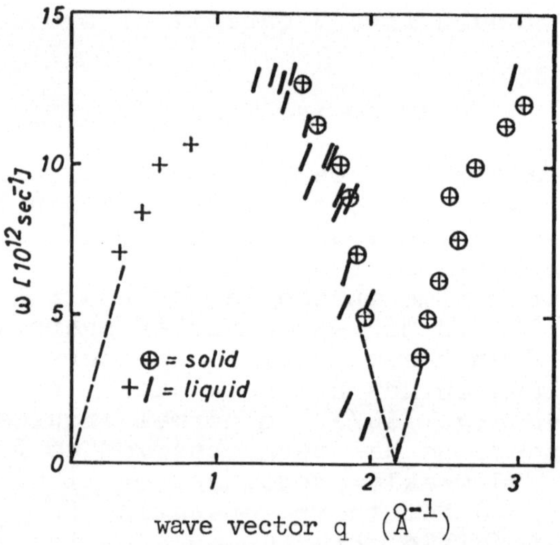

Fig. 5: Collected data for a longitudinal "phonon"-dispersion curve in liquid lead (from /39/).

times of atoms in such liquids are of the order of 10^{-11}sec (see e.g. /39/, /30/). This means that at these very high frequencies the liquid behaves very similar to a disordered solid. More experiments of this kind, especially on liquid metals would be of great interest.

The range of small K and ω values (see fig. 1) can be investigated by Brillouin scattering of light. This region is characterized by the hydrodynamical theory which assumes local thermodynamical equilibrium at every moment, and a relatively simple interpretation of the experiments is possible /40/. An interesting region is the gap between both methods in fig. 1 which might be partly covered by specialized and improved instruments at reactors with higher flux. A comprehensive review on the field of neutron scattering on liquids has been presented at the Copenhagen Conference /39/.

4. SCATTERING ON MAGNETIC FLUCTUATIONS CLOSE TO THE CURIE TEMPERATURE

At temperatures well above the Curie temperature T_c the magnetic moments of atoms in paramagnetic solids are oriented nearly independently from each other. Approaching T_c from above strong correlations appear between the magnetic moments. These correlations are equivalent to long range fluctuations of the magnetization. The correlation range spreads out widely as T_c is approached, until at and below T_c nearly complete alignement of the magnetic moments is established. At temperatures sufficiently below T_c well defined excitations, namely the spin waves, can exist.

Spin waves are able to interact with neutrons, and essentially the same conservation laws, eqs. (7) and (8), hold. Using these relations spin wave dispersion curves $\nu(\vec{q})$ have been measured in the same way as described in sect. 2 for many ferro and anti-ferro magnetic solids. We will not describe the spin wave experiments in detail, and refer therefore to the existing literature, e.g. /5/, /6/, /41/. Here we will rather discuss the magnetic fluctuations close to T_c, a field of great theoretical interest at present. These fluctuations are not necessarily of a simple periodical type and we have to refer to the general formulations eqs. (2) and (5). The long range correlations between magnetic moments are responsible for strong magnetic scattering. It takes place at small angles (1° and below) because the range of the correlations (order of 50...100 Å) is large as compared to the applied neutron wave lengths. This "Curie scattering" is analogous to the well-known critical opalescence of a gas close to its critical point. (For comprehensive reviews on the field of critical phenomena in general see for example /42/, /43/, /44/).

First, we discuss the static magnetic correlation function G(r,0) which is determined from the integrated experimental scattering intensity S(K) without ω-analysis. Molecular field theory gives a simple expression, namely

$$g(r) = \frac{v_0\, s(s+1)}{4\pi r_1^2\, r}\, e^{-r/\xi} \tag{15}$$

where v_0 is the atomic volume; r_1 is slowly varying with T. ξ is the correlation range which varies with T as $\varepsilon^{-\nu}$ with $\varepsilon = (T-T_c)/T_c$ and $\nu = 1/2$. The quantities r_1 and ξ can be easily calculated from the space moments of the exchange integral. The corresponding scattering law is a Lorentzian in K, namely

$$S(K) = \frac{2\, s(s+1)}{3 r_1^2 (K^2 + \xi^{-2})}$$

The half width $K_{1/2} = 1/\xi$ descreases as T approaches T_c. Simple thermodynamical arguments show that $S(0) = k_B T \chi(T)/\mu^2$ where $\chi(T)$ is the magnetic susceptibility, with $\chi(T)$ proportional to $\varepsilon^{-\gamma}$. On sees that a 1/S(K)-plot gives, over K^2, a straight line from which one can easily extract $\xi(T)$ and $\chi(T)$. This simple theory which is rooted in the work of Ornstein and Zernike in 1920, should not hold close to T_c where the fluctuations become very strong. A semi-empirical approach has been proposed /45/ writing $1/r^{1+\eta}$ in eq. (15) instead of 1/r with $\eta > 0$.

Many experiments have been performed to investigate the deviations from the simple Ornstein-Zernike formula (15). Results for Curie scattering on iron /46/ have been used to extract the temperature dependency of χ and ξ. A log-log representation of χ and ξ gives the critical exponents $\nu = 0{,}64 \pm 0{,}02$ in disagreement with the Ornstein-Zernike formula and $\gamma = 1{,}30 \pm 0{,}04$. Molecular field theory gives $\gamma = 1$; for a Heisenberg model with high spin one finds $\gamma = 1{,}33$. Similiar investigations have been performed on other paramagnets like cobalt and nickel /47/. So far it was not possible to find relevant deviations from (15) e.g. a finite value of η beyond the error limits, and further experiments are needed. Curie scattering has been investigated on nickel by changing the temperature and measuring different scattering angles simulataneously. From these and similiar experiments /47/, /48/ it has been found that the maximum of critical scattering shifts to higher temperatures as K increases (fig. 6). This can not be explained by theories of the type described above and no explanation is available so far.

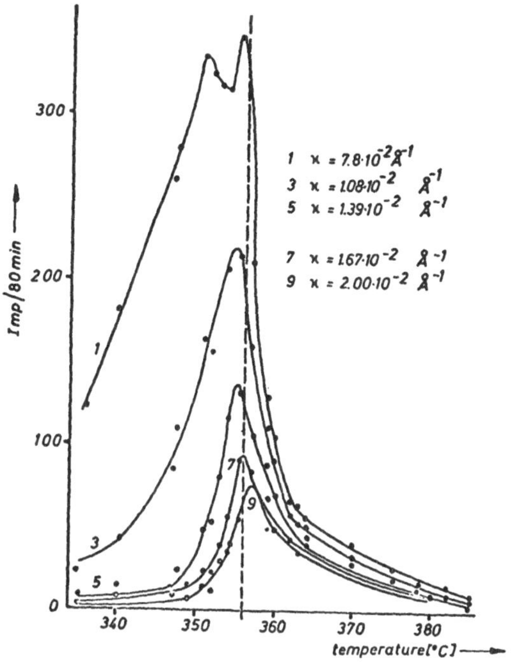

Fig. 6: Critical magnetic scattering on nickel at fixed scattering angle (or K) as a function of temperature /48/.

An important approach to interpret the critical experiments is the concept of the so-called scaling laws /42/, /49/. A number of physical quantities, like the correlation length ξ, the susceptibility χ, the magnetization M, or the specific heat C_H seem to have singularities at T_c which can be described very well by power laws of $\varepsilon=(T-T_c)/T_c$ like $\varepsilon^{-\nu}$, $\varepsilon^{-\gamma}$, ε^{β} and ε^{α}, respectively. By guessing a functional connection between certain thermodynamical variables, simple linear relations between these critical exponents can be established. The functional guess is made in such a way that it fits existing rigorous theories. These relations between the critical exponents $\nu, \gamma, \beta, \alpha$ are fulfilled quite well in many cases. They bring order into the enormous variety of critical experiments /42/. At a later stage they might give new physical insight into the collective behaviour of strongly coupled systems.

The situation is very complicated with regard to the time behaviour of the correlation function. No established theory of the collective motion of the magnetic moments close to T_c is existing. One approach used as a guide-line is the description of the movement of the spins by means of a diffusion-type equation /2/, /50/ for the local magnetization $\langle M(r,t)\rangle$:

$$\partial \langle M\rangle / \partial t = \Lambda_s \nabla^2 \langle M\rangle \qquad (16)$$

The corresponding scattering law for fluctuations following this equation of motion should be a Lorentzian in ω with an energy half width

$$\Gamma = \hbar K^2 \Lambda_s(T) \qquad (17)$$

Λ_s is the so called spin-diffusion constant which should behave like

$$\Lambda_s = \ell(T)/\chi(T)$$

$\ell(T)$ should vary slowly with T, but this cannot be proved a priori /51/. This description is equivalent to that which is known for the fluctuative behaviour of a gas close to T_c according to the hydrodynamical theory /40/. It follows a heat diffusion equation, i.e. a density disturbation $\delta\rho$ propagates via heat conduction. Close to T_c the same equation (16) holds for $\delta\rho$ and the corresponding quantity Λ_g is given by

$$\Lambda_g = \alpha(T)/c_p(T) = \ell'(T)/\chi(T)$$

Again the heat conductivity α and $\ell'(T)$ should vary slowly with T. Here χ is the compressibility. At T approaching T_c both theories predict that the fluctuations slow down and that the inelasticity vanishes at T_c. This has been actually confirmed by experiments at small K and ω on gases with optical Laser spectroscopy /52/ on SF_6 in the region where the hydrodynamical theory should hold.

The half width Γ of inelastic neutron scattering on paramagnets close to T_c has been investigated several times /53/, /46/, /54/. Fig. 7 shows a typical energy distribution centered around $\omega = 0$, measured on iron at a fixed scattering angle (or, approximately, at fixed K). These experiments are difficult because the half-width of inelasticity, Γ, is of the order of or smaller than the resolution of the spectrometers ($\sim 10^{-4}$ ev). In spite of these difficulties it has been demonstrated that Γ does not change appreciably at T_c. This means that the speed of the magnetical critical fluctuations is not reduced when T_c is approached /51/.

In this connection there is another question of importance, namely the existence of fluctuations very close to T_c which are periodical in space and time. These should be observable in the wings of the energy distribution discussed before. So far, no direct observation has been reported. However, indirect indications do exist. The end points of the wave vectors after scattering, $\vec{k}$, allowed

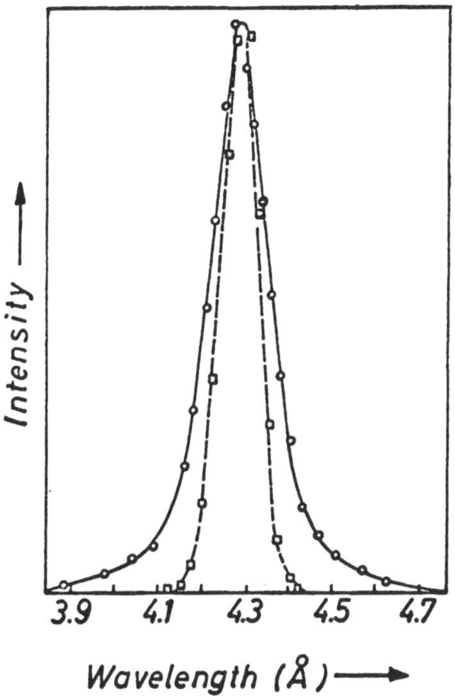

Fig. 7: Energy distribution of critical scattering on Fe as a function of the neutron wave length after scattering /46/. Inner curve represents the experimental resolution (Passell et al.).

by the conservation laws (7) and (8) are situated at a surface in reciprocal space whose shape is determined by $\vec{k}_0$ and by the dispersion law $\nu(\vec{q})$. Assuming $\nu = D(T)\,q^2$ (like for spin waves at temperatures well below T_c), this surface consists of two contacting spheres situated at the end point of the vector $\vec{k}_0$ (e.g. /55/). Their radii are $k_\pm = k_0/(1 + 2mD(T)/\hbar)$. $D(T)$ is supposed to decrease as T approaches T_c. Therefore, at a fixed scattering angle (i.e. direction of $\vec{k}$) a certain value of D and T exists where these spheres just touch the vector $\vec{k}$. This induces an abrupt change of the scattered intensity at a certain value of T which could explain the second peak shown in fig. 6. The corresponding step has also been observed in the scattered intensity at fixed T as a function of ϑ /56/. This evidence might be considered as an indirect prove of the existence of magnon-like excitations within a few degrees below or even above T_c. Direct experiments measuring the energy distribution with a resolution better than 10^{-4} ev would be of great interest in this region.

References

/1/ Van Hove, L., Phys. Rev. 95, 249 (1954)
/2/ Van Hove, L., Phys. Rev. 95, 1374 (1954)
/3/ Pecora, R., J. Chem. Phys. 40, 1604 (1964)
/4/ Maier-Leibnitz, J., Nukleonik 8, 61 (1966)
/5/ Egelstaff, P.A., (Ed.), "Thermal Neutron Scattering", Academic Press, London (1965)
/6/ "Inelastic Scattering of Neutrons in Solids and Liquids", International Atomic Energy Agency, Vienna, 1961, 1963, 1965; 1968 (in print)
/7/ Maier-Leibnitz, H., and Springer, T., Ann. Rev. Nucl. Sci., 16, 207 (1966)
/8/ Weinstock, R., Phys. Rev. 65, 1 (1944)
/9/ Toya, T., J. Res. Inst. Catalysis 6, 183 (1958)
/10/ Fuchs, K., Proc. Roy. Soc. A151, 585 (1935)
/11/ Ziman, J. M., Principles of the Theory of Solids, Cambridge University Press (1965)
/12/ Hubbard, J., Proc. Roy. Soc. A243, 336 (1958)
/13/ Sham, L.J., Proc. Roy. Soc. A283, 33 (1965)
/14/ Heine, V., Abarenkov, I., Phil. Mag. 9, 457 (1964)
/15/ Harrison, W.A., Pseudopotential in the Theory of Metals, Benjamin, New York (1966)
/16/ Schneider, T., Stoll, E., SNIS - SM 104/4+) see also Greene, M.P., Kohn, W., Phys. Rev. 137, A513 (1964)
/17/ Gläser, W., Schneider, T., Physikertagung 1966 München, Vorabdruck, p. 390
/18/ Krebs, K., Hölzl, K., Report EUR 362e (1967) and Solid State Comm. 5, 159 (1967)
/19/ Cowley, R.A., Woods, A.D.B., and Dolling, G., Phys. Rev. 150, 487 (1966)
/20/ Brockhouse, B.N., Arase, T., Caglioti, G., Rao, K.R., Woods, A.D.B., Phys. Rev. 128, 1099 (1962)
/21/ Kohn, W., Phys. Rev. Letters 2, 393 (1959)
/22/ Anderson, J.R., Gold, A.V., Phys. Rev. 139, 1459 (1965)
/23/ Stedman, R., Almquist, L., Nilsson, G., Raunio, G., Phys. Rev. 163, 567 (1967)
/24/ Ng, S.C., Brockhouse, B.N., SNIS SM - 104/53+)
/25/ Nilsson, G., SNIS SM - 104/40+)
/26/ Iyengar, P.K., Venkataraman, G., Gameel, Y.H., Rao, K.R. SNIS SM - 104/124+)
/27/ Stedman, R., Nilsson, G., Phys. Rev. Letters 15, 634 (1965)
/28/ Taylor, P.L., Phys. Rev., 131, 1995 (1963)

/29/ Paskin, A., Weis, R.J., Phys. Rev. Letters 9, 199 (1962)
/30/ Larsson, K.E., Dahlborg, U., Physica 30, 1561 (1964), Larsson, K.E., Queroz do Amaral, L., Ivanchev, N., Ripeanu, L., Bergstedt, L., Dahlborg, U., Phys. Rev. 151, 126 (1966)
/31/ Egelstaff, P.A., AERE-R 4101 (1962)
/32/ Larsson, K.E., Dahlborg, U., Jovic, D., Inelastic Scattering of Neutrons, 2, 117 (IAEA Vienna 1965)
/33/ Sköld, K., Larsson, K.E., Phys. Rev. 161, 102 (1967)
/34/ Cocking, S., Egelstaff, P.A., Phys. Letters 16, 130 (1965)
/35/ Singwi, K.S., Physica 31, 1257 (1965)
/36/ Henshaw, D.G., and Woods, A.D.B., Phys. Rev. 121, 1266 (1961)
/37/ Dorner, B., Mika, K., Stiller, H.H., SNIS SM-104/15+)
/38/ Dorner, B., Plesser, Th., Stiller, H.H., Disc. of the Faraday Soc., 43, 160 (1967)
/39/ Larsson, K.E., SNIS SM-104/200+)
/40/ Mountain, R.D., Rev. Mod. Phys. 38, 205 (1966)
/41/ De Gennes, P.G., in "Magnetism" (ed. Rado, G.T. and H. Suhl) Acad. Press, New York, 3, 115 (1963)
/42/ Kadanoff, L., Götze, W., Hamblen, D., Hecht, R., Lewis, E.A.S., Palcianuskas, V.V., Martin, R., Swift,J., Aspens, D., Kane, J., Rev. Mod. Phys. 39, 395 (1967)
/43/ Heller, P., Rept. Progr. Phys. 30, II, 794 (1967)
/44/ Fixmann, M., Adv. Chem. Phys. 6, 175 (1964)
/45/ Fisher, M.E., Burford, R.J., Phys. Rev. 156, 583 (1967)
/46/ Passell, L., Blinkowski, K., Brun, T., Nielsen, P., J. Appl. Phys. 35, 933 (1964)
/47/ Bally, D., Popovici, M., Totia, M., Grabcev, B., Lungu, A.M., SNIS SM-104/55+)
/48/ Stump, N., Maier, G., Phys. Letters 12, 625 (1967)
/49/ Widom, B., J. Chem. Phys. 43, 3892 and 3898 (1965)
/50/ Kadanoff, L.P., Martin, P.C., Anals of Physics 24, 419 (1963)
/51/ Kawasaki, K., Phys. Rev. 145, 224 (1966)
/52/ Saxman, A., Benedek, G.B., (1967), to be published, cf. [43], p. 805
/53/ Ericson, M., Jacrot, B., J. Phys. Chem. Sol. 13, 235 (1960)
/54/ Gordon, J., Kisdi-Kozó, E., Pál, L., Vizi, L., SNIS SM-104/9+)
/55/ Frikkee, E., Riste, T., Proc. Intern. Congr. Magnetism, 299 (1964)
/56/ Stringfellow, M.W., to be publ. in Proc.Phys. Soc. (London) 1968

+) SNIS SM... is used for the reports from the "Symposium on Inelastic Scattering of Neutrons", Copenhagen 1968, to be published by the IAEA (Vienna).

ON THE CALCULATION OF PHONON DISPERSION CURVES IN INSULATORS

B. Gliss

Institut für Theoretische Physik

Frankfurt, Germany

Abstract

The basic features of some wellknown phenomenological force constant models are reviewed. Their interrelation with attempts for ab initio calculations of the phonon-spectrum is discussed. A formal derivation of the dispersion curves in the harmonic Born-Oppenheimer approximation and results of an application of the method to LiF are given.

I INTRODUCTION

Considerable progress in the understanding of phonon dispersion curves has been achieved along several different lines during the last years.

While for ionic crystals efficient force constant models have been developed that enable quantitative comparison with experimental results from neutron and X-ray measurements [1-5), the simple metals have been treated by the self-consistent field method [6-8) correlating the electronic structure and the vibrational motion of the atoms.[+]

[+] The term atom here denotes the entity composed of the atomic nucleus and tightly bound electrons that can be considered to move rigidly with the nucleus during a vibrational motion. It does not mean atom in the chemical sense.

Most numerical applications have, however, been limited to the case where the electronic structure of the crystal can be described in the single-particle- and pseudopotential[9] approximation. Here the smooth part of the electronic wave function is taken to be plane-wave like, while the pseudopotential operator is approximated by an empirical local potential in momentum space. The electronic contribution to the harmonic force constants is then calculated by first order perturbation theory, the perturbing force being the change in the potential because of the displacement of the atoms from equilibrium.

Clearly, this method cannot be taken over to ionic crystals without change. The difficulties associated with the determination of the self-consistent potential an electron is subjected to in ionic crystals have not been overcome with reasonable accuracy to enable a true ab initio calculation of phonon dispersion curves from electronic states. Only in the limiting case of long waves (zero pseudomomentum) have the elastic constants that determine the slopes of the acoustic branches of the dispersion curves been computed from free-ion states orthonormalized up to the square of the overlap matrix [10-13]. Instead, phenomenological microscopic force constant models were constructed that introduce effective coordinates for the electronic degrees of freedom and assume harmonic forces between the ion cores and the electronic shells and among the electronic shells themselves. The functional form of the equations of motion for these models has in some instances been made plausible from more rigorous consideration.[14-16]

In this article we shall comment on the interrelation between the phenomenological force constant models and attempts for ab initio calculations of the dispersion curves. We shall also give a reformulation of the harmonic Born-Oppenheimer[17] method in terms of electronic Green functions thereby making the approximations involved in any practical calculation more explicit. Finally we shall report on numerical results obtained when an approximate calculation was attempted for LiF.[18]

Throughout this paper the following effects will be neglected:

1) Anharmonicity

2) The temperature dependence of the interatomic force constants

3) Nonadiabatic effects beyond the Born-Oppenheimer approximation+.

We shall limit the discussions mainly to ionic crystals. Here assumption (3) seems plausible from energy arguments. (2) must be taken into account when comparing theoretical predictions with experiment or with phenomenological models fitted to experiment. The anharmonic effects have been estimated by Cowley and Cowley[19]. In alkali halides they induce a small shift ($\approx$ 8% at room temperature for KBr) of the phonon frequencies against those at zero temperature.

Furthermore we shall assume that only the higher electronic energy levels take part in the reaction of the electronic system to ionic displacements. This seems to be justified for closed shell systems.

In the second paragraph we shall review the basic notations of lattice dynamics in the harmonic approximation. We shall then comment on attempts to link microscopic force constant models to quantum mechanical calculations. In the fourth paragraph we shall reformulate the problems in terms of Green functions and shall finally report numerical results of a semi-empirical quantum mechanical calculation for LiF.

II BASIC NOTATIONS OF LATTICE DYNAMICS

We use the symbols introduced by Montroll, Maradudin, and Weiss[20] ++. Let $\underline{x}\binom{\ell}{\kappa}$ stand for the instantaneous position of the κ - th atom in the ℓ - th cell, and let the index zero denote the respective equilibrium position. The deviation from it is defined as

$$\underline{u}\binom{\ell}{\kappa} = \underline{x}\binom{\ell}{\kappa} - \underline{x}^{\circ}\binom{\ell}{\kappa}$$

Let $\underline{p}\binom{\ell}{\kappa}$ be the momentum conjugate to $\underline{u}\binom{\ell}{\kappa}$. Then the hamiltonian of any model theory can be brought to the form:

$$\mathcal{H} = \sum_{\ell\kappa} \frac{\underline{p}\binom{\ell}{\kappa}^2}{2m_\kappa} + \tilde{\Phi}\left(\dots \underline{u}\binom{\ell}{\kappa}\dots\right)$$

+ The diagonal correction term to the Born-Oppenheimer approximation is of order $\frac{m_{El}}{M_{Ion}}$ and shall be neglected.

++ Their article contains a detailed analysis of lattice dynamics and should be referred to for a review of the field.

We assume that $\tilde{\Phi}$ can be expanded into a power series in terms of the atomic displacements and that the second order (harmonic) term is the leading one.

$$\tilde{\Phi} = \tilde{\Phi}_0 + \sum_{\ell\kappa} \left(\frac{\partial\tilde{\Phi}}{\partial \underline{x}\binom{\ell}{\kappa}}\right)_0^{\tau} \cdot \underline{u}\binom{\ell}{\kappa} + \sum_{\substack{\ell\ell'\\ \kappa\kappa'}} \underline{u}\binom{\ell}{\kappa}^{\tau} \left(\frac{\partial}{\partial \underline{x}\binom{\ell}{\kappa}} \otimes \frac{\partial}{\partial x\binom{\ell'}{\kappa'}} \tilde{\Phi}\right)_0 \underline{u}\binom{\ell'}{\kappa'}$$

Here a matrix notation has been used with τ denoting the transpose of a matrix and $\otimes$ symbolizing the direct produrct between two matrices. The expansion coefficients have to fulfill certain symmetry and stability requirements (for a review of the subject see Ludwig[21]). From stability the term linear in the displacements must vanish identically and

$$\tilde{\Phi}\binom{\ell\ell'}{\kappa\kappa'} \equiv \left(\frac{\partial}{\partial x\binom{\ell}{\kappa}} \otimes \frac{\partial}{\partial x\binom{\ell'}{\kappa'}} \tilde{\Phi}\right)_0$$

should be positive definite. The invariance of the potential energy under translations of the crystal as a whole gives rise to a sum rule:

$$(2.1) \qquad \sum_{\ell'\kappa'} \tilde{\Phi}\binom{\ell\ell'}{\kappa\kappa'} = 0$$

Other requirements (for instance invariance of $\tilde{\Phi}$ against the space group operations of the crystal) further restrict the number of independent matrix elements of $\tilde{\Phi}\binom{\ell\ell'}{\kappa\kappa'}$. Invariance under infinitesimal rotations and homogeneous deformations relates the harmonic coefficients to the anharmonic ones, so that the motion of a purely harmonic crystal is in itself inconsistent. For computational purposes the anharmonic effects are, however, usually neglected.

Since a constant in the hamiltonian does not influence the dynamics of the problem, the hamiltonian leading to linear equations of motion is:

$$\mathcal{H} = \sum_{\ell\kappa} \left\{ \frac{\underline{p}\binom{\ell}{\kappa}^2}{2m_\kappa} + \sum_{\ell'\kappa'} \underline{u}\binom{\ell}{\kappa}^{\tau} \tilde{\Phi}\binom{\ell\ell'}{\kappa\kappa'} \underline{u}\binom{\ell'}{\kappa'} \right.$$

The periodicity of the crystal is exploited by the Ansatz:

$$(2.2) \qquad \underline{u}\binom{\ell}{\kappa} = m_\kappa^{-\frac{1}{2}} \, \underline{u}\binom{q}{\kappa} \, e^{i(q^{\tau}\cdot\ell - \omega t)}$$

This yields the equation of motion:

$$(2.3) \qquad \omega^2 \underline{u}\binom{q}{\kappa} = \sum_{\ell'\kappa'} \tilde{D}\binom{q}{\kappa\kappa'} \underline{u}\binom{q}{\kappa'}$$

where the dynamical matrix is defined as

$$\tilde{D}\left(\begin{smallmatrix} q \\ \kappa\kappa' \end{smallmatrix}\right) = \sum_{\ell} \tilde{\Phi}\left(\begin{smallmatrix} 0\ell \\ \kappa\kappa' \end{smallmatrix}\right) e^{iq^T\ell} \left(m_\kappa m_{\kappa'}\right)^{-\frac{1}{2}}$$

and we have used the invariance of $\tilde{\Phi}$ under crystal translations:

$$\tilde{\Phi}\left(\begin{smallmatrix} \ell\ell' \\ \kappa\kappa' \end{smallmatrix}\right) = \tilde{\Phi}\left(\begin{smallmatrix} 0\ \ell'-\ell \\ \kappa\ \kappa' \end{smallmatrix}\right)$$

In order for (2.2) to be an elementary solution to the equations of motion, ω^2 and $\underline{u}\left(\begin{smallmatrix} q \\ \kappa \end{smallmatrix}\right)$ must fulfill the homogeneous equation (2.3). ω depends on the crystal momentum q and on a branch index enumerating the eigenvalues $\omega_j^2(q)$ of $\tilde{D}(q)$ and is called the frequency of the phonon $\left(\begin{smallmatrix} q \\ j \end{smallmatrix}\right)$. For further reference we note that the sum rule (2.1) reads in terms of the dynamical matrix:

$$(2.4) \qquad \sum_{\kappa'} \left(m_\kappa m_{\kappa'}\right)^{\frac{1}{2}} \tilde{D}\left(\begin{smallmatrix} 0 \\ \kappa\kappa' \end{smallmatrix}\right) = 0$$

So far we have not specified the interatomic potential $\tilde{\Phi}$. It is convenient to separate $\tilde{\Phi}$ explicitly into two parts:

$$(2.5) \qquad \tilde{\Phi} = \Phi^{\text{Coul}} + \Phi$$

where Φ^{Coul} denotes the direct interaction between the atoms due to their electrostatic charges while Φ is defined as

$$\Phi = \langle H^{EL} \rangle$$

H^{EL} is the hamiltonian characterizing the motion of the electrons in the presence of the charged nuclei:

$$(2.6) \quad H^{EL} = \sum_i \left\{ -\frac{\hbar^2}{2m}\frac{\partial^2}{\partial \underline{r}_i^2} - \sum_{\ell\kappa} V_\kappa\left(\underline{r}_i ; \cdots \underline{x}\left(\begin{smallmatrix} \ell \\ \kappa \end{smallmatrix}\right), \ldots\right) + \frac{e^2}{2} \sum_{j \neq i} |\underline{r}_i - \underline{r}_j|^{-1} \right\}$$

The quantum mechanical average of H^{EL} is taken with respect to the ground state of the electronic system for fixed nuclei, so that eigenvector and eigenvalue of H^{EL} depend on the atomic configuration.

The Coulomb contribution to the harmonic force constants can be handled by Ewalds ϑ-function transformation [22]. The basic quantity of interest in any model - or first-principles calculation is therefore:

$$\Phi\binom{\ell\ell'}{\kappa\kappa'} = \frac{\partial}{\partial x\binom{\ell}{\kappa}} \otimes \frac{\partial}{\partial x\binom{\ell'}{\kappa'}} \langle H^{El} \rangle \qquad (2.7)$$

This quantity is approximated by the classical microscopic force constant models. We shall try to evaluate it from a more fundamental aspect in IV.

III FORCE CONSTANT MODELS AND METHODS FOR QUANTUM MECHANICAL CALCULATIONS

Any straight-forward application of the Born-Oppenheimer formula (2.7) involves the solution of the electronic eigenvalue for all atomic configurations. Clearly, the number of atoms in a crystal is so large that this task can only be undertaken in special cases.

A particularly simple one is the motion of a substitutional impurity of negative charge and light mass (H^- -ion) in an alkali halide. Experiment shows that this special type of defect (the U-center) gives rise

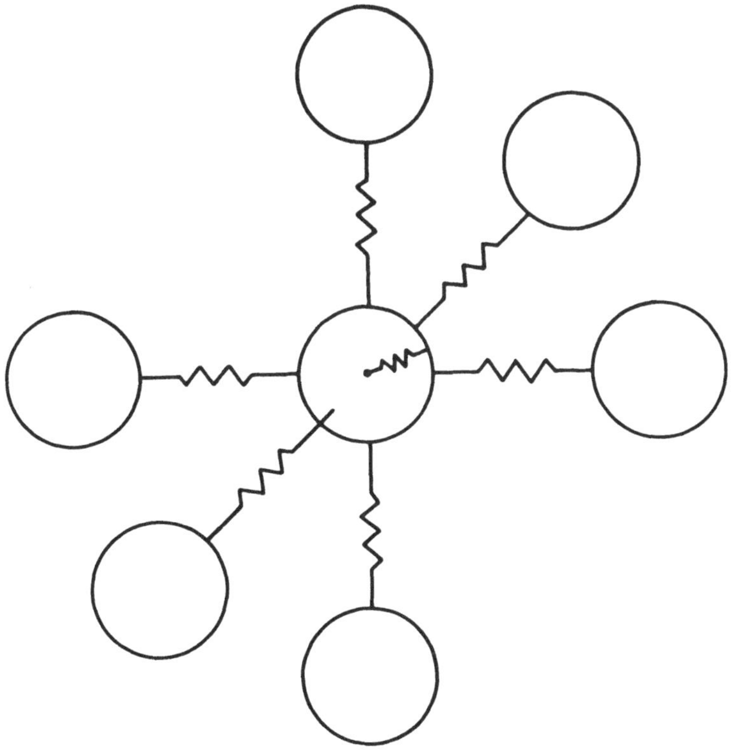

Fig. 1: The U-center

to a localized mode of vibration that involves to a first approximation the motion of the H^- in a rigid host lattice. Therefore the ground state energy of the electronic system depends only on the position of the H^- with respect to the neighbouring ion. It follows from symmetry considerations that the localized mode is directed along any of the three axes of a cube (see fig. 1). Its frequency can then be computed by calculating the ground state expectation value of H^{EL} as a function of the displacement of the H^- from equilibrium.

Even though the computational effort has been greatly reduced by the special properties of the system, it has only been recently that a variational calculation of the energy based on free-ion wave functions for the host-lattice-ions and a trial function for the electrons associated with the U-center has been performed[24,25]. The trial function is chosen in essentially two different ways. One corresponds to a hydrogen like s-type-function moving rigidly with the center the other allows for an admixture of a p-type function as the H^-- ion is displaced.

The calculation of $\langle H^{EL} \rangle$ when the ground state is approximated by a symmetrized combination of free atom wave functions moving rigidly with the displaced nuclei has first been performed by Heitler and London[25] for the interaction of two atoms. Later the cohesive energy of ionic crystals was treated by the same method+. If the wave function centered around different nuclei are properly orthogonolized, the Heitler London method yields results for the elastic constants, i.e. quantities determining the slopes of the acoustic branches of the phonon dispersion curves, that agree reasonably well with experiment.

The same physical idea, namely that the electronic orbits move rigidly with the displaced nuclei, underlies the Born- von Karman - or rigid-ion force constant model (from now on called RIM). Here the matrix elements of $\Phi\binom{\ell\ell'}{\kappa\kappa'}$ are sometimes derived from an assumed analytical form of the interatomic potential (e.g. Born-Mayer-potential), sometimes they are directly linked to experimental quantities by fitting given dispersion curves at a certain number of $\underset{\sim}{q}$ -points or relating them to other macroscopic data (e.g. elastic constants). Since the number of independent matrix elements increases rapidly

---

+ For a review of the subject see P.O.Löwdin[26]

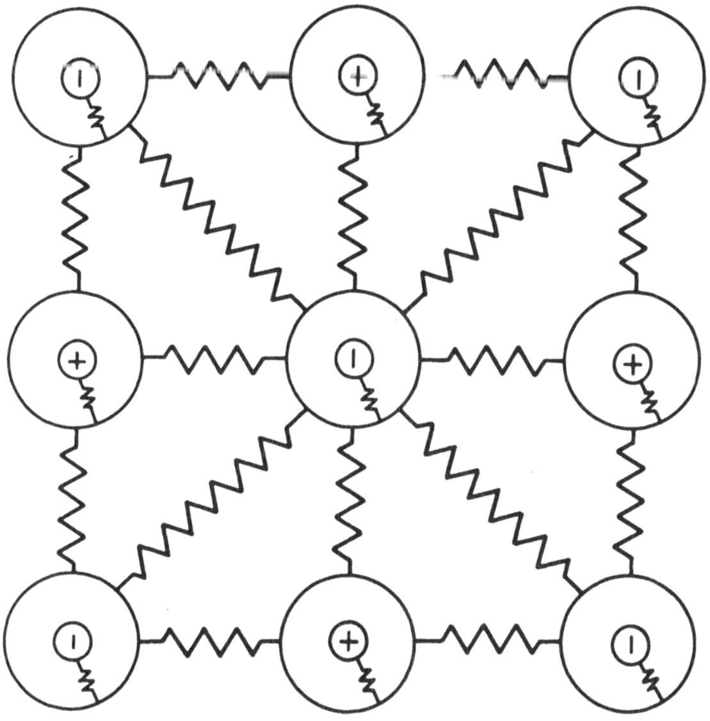

Fig. 2: Shell model for an alcali halide

with the distance between $\binom{l}{\kappa}$ and $\binom{l'}{\kappa'}$, only a limited number of neighbours of $\binom{l}{\kappa}$ is taken into account. This seems plausible, because in the Heitler London scheme the non-electrostatic forces between the ions decrease rapidly with distance.

An early quantitative calculation taking into account the Coulomb-forces derived from a central potential+ has been carried out by Kellermann [22].

When the dispersion cuves became better known through neutron diffraction experiments it became obvious, however, that the measured results could only be reproduced by using many disposable parameters that cannot be determined by other independent quantities [27] (e.g.elastic

---

\+ It should be noted that a rigorous Heitler London treatment allows for noncentral forces which lead to the nonfulfillment of the Cauchy relations for the elastic constants (Löwdin [11]) and a correction to the effective charge of the ions (Lundquist [28]).

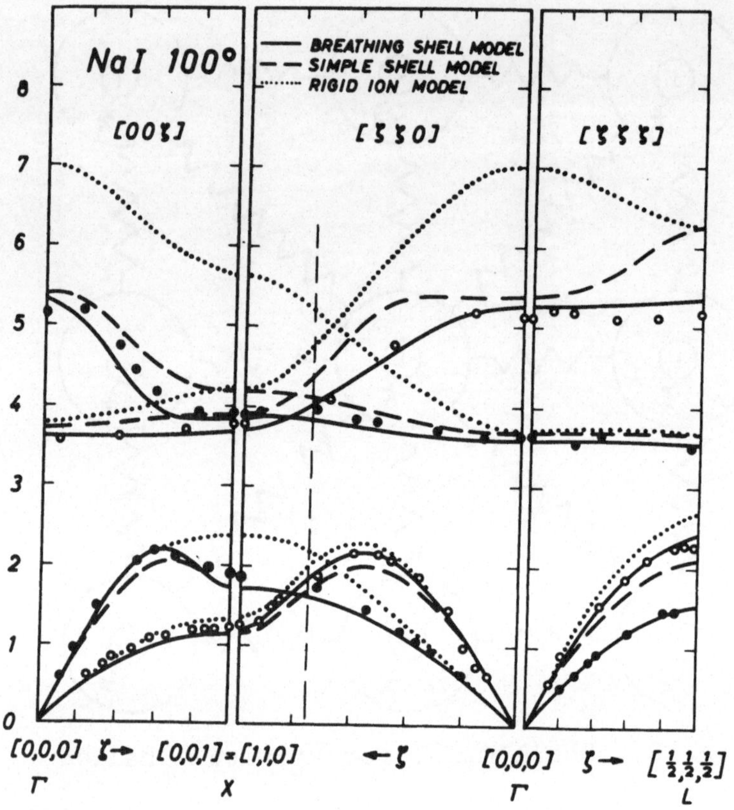

Fig. 3: Dispersion curves for NaI derived from various models (after ref. 3)

constants, dielectric constants). The characteristic difficulty a few parameter RIM calculation encounters for the alkali halides is the prediction of eigenfrequencies considerably higher than the experimental ones (see fig. 3).

It has been noted by many workers that this deficiency can be partly removed by taking the atomic polarizability into account. This introduces additional attractive force lowering the dispersion curves. Since the polarizability is associated with the deformation of the electronic charge cloud of the ions, it must be linked to the short range forces between these charge clouds.

A phenomenological model based on this idea is the shell model proposed by Dick and Overhauser [1] (see fig. (2)). The outer shell of the halide ion is assumed to move undeformed with respect to the ion core. Its coordinate is introduced as a new dynamical variable into the theory and the shell-shell and core-shell-forces are assumed to be harmonic springs. According to Cowley et al.[29], the equation of motion now reads:

$$(3.1) \qquad m\omega^2 \underline{U} = (R + ZCZ)\underline{U} + (T + ZCY)\underline{W}$$

$\underline{U}$ is a vector formed⁺ from the Fourier transforms of the core displacements. $\underline{W}$ is an analogous quantity for the relative core-shell-motion. The matrices R and T are Fourier transforms for the non-electrostatic forces between the cores and shells. The core- and shell-charges have been written down explicitly for the respective Coulomb interactions C. Equation (3.1) must be supplemented by an analogous equation for the motion of the shells. Here the adiabatic approximation is introduced by the statement that the motion of the electrons be force-free:

$$(3.2) \qquad 0 = (T^{T} + YCZ)\underline{U} + (S + YCY)\underline{W}$$

Solving for W in equation (3.2) and inserting in equation (3.1) the formal force constants equivalent to the shell model assumptions can be derived.

It is illuminating to perform this operation for a one-dimensional example of a linear chain with short range forces between nearest neighbours [30].

Let g be the core-shell- and f be the shell-shell-force constant, the equations of motion in real space are

$$m\omega^2 u(j) = g\,(u(j) - v(j))$$

$$0 = g\,[v(j) - u(j)] + f\,[2v(j) - v(j+1) - v(j-1)]$$

Using the Ansatz

$$u(j) = u(q)\, e^{iqaj}$$

$$v(j) = v(q)\, e^{iqaj}$$

where a is the lattice constant and q is in the first Brillouin zone $0 \le |q| \le \frac{\pi}{a}$ one gets:

⁺ It differs from the u used in eq. (2.2) by a mass-factor $m^{1/2}$.

$$m\omega^2 u(q) = g[u(q)-v(q)]$$

$$0 = g[v(q)-u(q)] + 2f\,v(q)[1-\cos aq]$$

We eliminate $v(q)$ and arrive at the dispersion relation

$$\omega^2(q) = 4\frac{f}{m}\frac{\sin^2\left(\frac{qa}{2}\right)}{1+4\frac{f}{g}\sin^2\left(\frac{qa}{2}\right)}$$

Passing to the rigid ion limit by letting the core-shell spring g become infinite one sees that the introduction of a displacable shell results here in a lowering of the dispersion branches at the end of the Brillouin zone.

While the shell model assumption improves the general agreement between calculated and measured dispersion curves (see fig. 3), it does not remove the discrepancy in the (1, 1, 1)-direction of the longitudinal optical mode without the introduction of further parameters. Furthermore, an increase in the number of disposable springs fitted to the dispersion curves can yield unphysical results for other physical quantities, e.g. negative polarizabilities of the alkali ions [27].

An analysis of the vibrational modes shows that in the L-O-(1, 1, 1)-mode the alkali ions move symmetrically against the halide ion in its equilibrium position. The relative core-shell displacement of the polarizable halide ion therefore is zero and the simple shell model gives the same result as the rigid ion model (compare fig. 3).

An improved model that overcomes this difficulty has been proposed by Schröder [3]. It takes the simplest (s-type) deformation of the electronic shell into account and has been termed the breathing shell model (BSM).

The equations of motion now contain a new scalar variable corresponding to the variation of the electron-shell radius and an additional constraint, i.e. the adiabatic condition for the scalar variable v. According to Schröder we get:

$$\begin{aligned} m\omega^2\underline{U} &= (R+ZCZ)\underline{U} + \underline{Q}v + (R+ZCY)\underline{W} \\ 0 &= (R+YCZ)\underline{U} \quad \underline{Q}v + (R+G+YCY)\underline{W} \\ 0 &= \underline{Q}^{+}(\underline{U}+\underline{W}) + Hv \end{aligned} \qquad (3.3)$$

Eq. (3.3) contains Cowley's formula for the special case of a simple shell model, i.e. no direct core-core-forces (R = T). There also appear two new matrices Q and H coupling the breathing motion to core-, shell-, and radial-displacements, respectively. When the matrix elements are expressed in terms of the various force constants, it can be shown that the motion of a scalar degree of freedom introduces one additional spring constant g' per breathing shell. Schröders results show that for most of the alkali halides good numerical agreement with experience can be reached, if g' is chosen equal to g^{+}. In a model with second nearest neighbour shell-shell-, no direct core-core-, and nearest neighbour core-shell interaction where only the halide ions are assumed polarizable, all parameters can then be determined from macroscopic quantities independent of neutron diffraction results. By this method the dispersion curves for LiF have been correctly predicted before they were measured by Dolling and coworkers [33].

So far we have only described the shell model and its extension as a plausible interpolation device between different sets of experimental data. There have, however, been attempts to link the shell model equations to the quantum mechanical calculations of the crystal energy.

Tolpygo and coworkers [14] have expressed the energy of a (homopolar) crystal up to second order in the overlap integrals. They introduced the dipole-moment associated with an atom as a macroscopic variable, i.e. a variable on the same footing as the nuclear displacements. Expanding the perturbed wave function in second order perturbation theory in terms of free-atom functions they varied the resultant wave function for given nuclear displacements and atomic dipole moments and were able to express the lattice energy in terms of these quantities. Since it does not contain the momenta conjugate to the dipole moments, the equation of motion immediately gives a relation between the dipole moments and the nuclear displacements and another one involving the force on the atoms. This is formally equivalent to the adiabatic condition (3.2) and the equation of motion (3.1) in the shell model treatment.

Another approach has been given by Cowley [15]. He in-

+ A quantum mechanical justification for this fact will be given in a forthcoming publication [31].

troduces the notion of subcells, i.e. regions of space associated with one particular atom in the crystal⁺, and uses the center of mass coordinate and the dipole moments of the subcell as dynamical variables for the lattice vibrations of a crystal.

No quantitative calculations starting from electronic wave functions have been based on these formal treatments. At this stage we shall therefore refer to the original literature.

The next paragraph sketches a formal treatment of the Born-Oppenheimer formula that could lead, if the one-electron-spectrum were completely known, to an ab initio calculation of the phonon frequencies of desired accuracy. An actual application involves, however, many assumptions and approximations based on experience, so that it should only be taken as a semiempirical device linking electronic properties to those of the lattice vibrations.

IV QUANTUM MECHANICAL TREATMENT OF THE HARMONIC BORN-OPPENHEIMER FORMULA

In the harmonic Born-Oppenheimer approximation the electronic contribution to the interatomic force constants is given by (2.7).

The derivative is to be taken with the lattice in its equilibrium configuration, and for zero temperature the quantum mechanical average is to be computed for the electronic ground state. Using the Hellman-Feynman theorem, equation (2.7) can be rewritten as

$$(4.1) \qquad \varphi\begin{pmatrix} \ell\ell' \\ \kappa\kappa' \end{pmatrix} = \frac{\partial}{\partial x\begin{pmatrix} \ell \\ \kappa \end{pmatrix}} \otimes \left\langle \frac{\partial H^{el}}{\partial x\begin{pmatrix} \ell' \\ \kappa' \end{pmatrix}} \right\rangle$$

where the only dependence of the electronic hamiltonian on the nuclear configuration is through the one-electron potential $V(\underline{r}, \{\cdots x\begin{pmatrix} \ell \\ \kappa \end{pmatrix} \cdots\})$ that an electron experiences in the presence of the ions.

⁺ It has, however, been pointed out [34,26] that the assignment of charge densities to particular atoms is in no way unique.

It is convenient to express (4.1) using the one-electron-Green-function+ (1 - G -function) defined by

$$(4.2)\qquad G(12) = (-i) \langle T\{\psi(1)\,\psi^+(2)\}\rangle$$

$\psi(1)$ and $\psi^+(2)$ are hermitian conjugate fermion field operators depending on the space time variable $1 \equiv (\underline{r}, t)$. T denotes chronological ordering. Using (4.2) we get for the force constant Φ :

$$(4.3)\qquad \Phi\left(\begin{smallmatrix} \ell & \ell' \\ \kappa & \kappa' \end{smallmatrix}\right) = (-i) \int d\tau \left\{ \frac{\partial G(11^+)}{\partial \underline{x}\left(\begin{smallmatrix} \ell \\ \kappa \end{smallmatrix}\right)} \otimes \frac{\partial V}{\partial x\left(\begin{smallmatrix} \ell' \\ \kappa' \end{smallmatrix}\right)} + G(11^+) \frac{\partial}{\partial \underline{x}\left(\begin{smallmatrix} \ell \\ \kappa \end{smallmatrix}\right)} \otimes \frac{\partial}{\partial \underline{x}\left(\begin{smallmatrix} \ell' \\ \kappa' \end{smallmatrix}\right)} V \right\}$$

where the notation $G(11^+) = \lim_{\epsilon \to +0} G(\underline{r}t, \underline{r}\,t+\epsilon)$ has been used. Let us assume that the electron potential can be written as a sum of single-ion contributions++ . Then the second term in (4.3) contributes only to $\Phi\left(\begin{smallmatrix} \ell & \ell \\ \kappa & \kappa \end{smallmatrix}\right)$ Since this quantity can be calculated from the $\Phi\left(\begin{smallmatrix} \ell & \ell' \\ \kappa & \kappa' \end{smallmatrix}\right)$ for $\left(\begin{smallmatrix} \ell \\ \kappa \end{smallmatrix}\right) \neq \left(\begin{smallmatrix} \ell' \\ \kappa' \end{smallmatrix}\right)$ (see eq. (2.1)), we shall concentrate on the first term in (4.3). It can be interpreted as the reaction of the electronic system due to a change in the electron-ion-potential which results from a change in the lattice configuration. In order to make its dependence on a change in the potential explicit, we consider the equation of motion for the 1-G-function and its adjoint:

$$(4.4a)\qquad L\,G(11') = \delta(1-1') - i \int d\tau_2\, v(\underline{r}-\underline{r}_2)\, G(121'2^+)\Big|_{t_1=t_2}$$

$$(4.4b)\qquad L^+ G(1''1) = \delta(1''-1) - i \int d\tau_2\, v(\underline{r}-\underline{r}_2)\, G(1''2^-1'2)\Big|_{t_1=t_2}$$

The single-particle operator L is defined as:

$$L = i\hbar \frac{\partial}{\partial t} + \frac{\hbar^2}{2m}\frac{\partial^2}{\partial \underline{r}^2} - V$$

$v(\underline{r}-\underline{r}')$ is the Coulomb interaction between the electrons, and G (1 2 3 4) is the 2-electron Green function (2-G-function):

$$G(1234) = \langle T\{\psi(1)\,\psi(2)\,\psi^+(3)\,\psi^+(4)\}\rangle$$

+ For an extensive account of Green-function techniques see Abricosov et al.[35]

++ see footnote on page 187.

We differentiate (4.4a) with respect to $x\binom{\ell}{k}$ and use (4.4b) to solve for $\frac{\partial G(1''1')}{\partial \underline{x}\binom{\ell}{k}}$. This gives:

$$(4.5)\quad \frac{\partial G(1''1')}{\partial \underline{x}\binom{\ell}{k}} = \int d1\, G(1''1) \frac{\partial V}{\partial x\binom{\ell}{k}} G(11') - i \int d1\, d\underline{r}_2\, v(\underline{r}-\underline{r}_2)\cdot$$

$$\cdot \left\{ G(1''1) \frac{\partial G(121'2^+)}{\partial \underline{x}\binom{\ell}{k}} - G(1''2^-12) \frac{\partial G(11')}{\partial \underline{x}\binom{\ell}{k}} \right\}$$

The first term arises from a simple perturbation treatment while the second one includes effects from the correlated motion of the electrons. To make its dependence on the single particle spectrum more evident, we introduce the mass operator of the system:

$$-i \int d\underline{r}_2\, v(\underline{r}-\underline{r}_2)\, G(121'2^+) \equiv \int d2\, M(12)\, G(21')$$

Then (4.5) yields:

$$(4.6)\quad \frac{\partial G(11')}{\partial \underline{x}\binom{\ell}{k}} = \int d1''\, G(11'') \left\{ \frac{\partial V(\underline{r}'')}{\partial \underline{x}\binom{\ell}{k}} G(1''1') + \int d2 \frac{\partial M(1''2)}{\partial x\binom{\ell}{k}} G(21') \right\}$$

Since the mass operator can be expressed as a series expansion in 1-G-functions and the electron-electron-interaction, eq. (4.6) is really a linear integral equation for $\frac{\partial G(11')}{\partial x\binom{\ell}{k}}$. Its solution depends on the one-electron-spectrum for undisplaced nuclei and the derivative of the electron ion potential.

V CONCLUDING REMARKS AND RESULTS

A practical application of (4.3) and (4.6) to the investigation of the phonon dispersion relations involves the following problems:

1) The 1-G-function, i.e. in a single-particle approximation the energy band-structure and the respective eigenfunctions, should be known.
2) The electron-ion-potential must be determined. It must be consistent with the potential used for the direct ion-ion-interaction, since both potentials are linked by the Born-Oppenheimer method.
3) The mass operator $M(11')$ must be approximated by a finite number of terms.

In an attempt to calculate the dispersion curves of LiF [17] the following approximations were made:

Only the Hartree-term in the mass operator was taken into account. This simplified formula (4.6) tremendously. The 1-G-function was then approximated in the single particle approximation. For the energy bands we used the Slater-Koster interpolation scheme [36] fitting the matrix elements of the effective Hamiltonian so that the resulting band structure agreed with the experimental data of Roessler and Walker [37]. The Bloch functions were approximated in the tigth binding scheme based on Clementi's [38] free-ion wave functions. The electron ion potential was chosen as a Coulomb-potential for the Li^+ -ion and as an empirical potential for the F^- -ion simulating the effect of that part of the electronic charge moving rigidly with the displaced ion. The parameters of the potential were fitted to the Rest-strahlen-frequency and the Lyddane-Sach-Teller relation at zero wave vector.

The procedure described involves approximations that are dictated by the numerical complexity of the problem$^+$. It cannot be considered as a rigorous ab initio calculation of phonon dispersion curves. In particular the electron-ion potential and the electronic band structure have not been determined by a self consistent calculation.

The method has, however, yielded sensible results for the dispersion curves of LiF, at least in the (1, 1, 1)-symmetry direction (see fig. 4). Here the results of the breathing-shell model calculation are reproduced, essentially. Since this model fits the neutron diffraction results of Dolling et al. [33] at room temperature, we have used a zero temperature BSM-calculation for comparison. Agreement is worse in the (1, 0, o)-direction. A detailed analysis shows that here the calculations depend much more critically on the functional form of the electron-ion potential.$^{++}$

$^+$ An alternative approach utilizing the APW method has been suggested recently [39,40], and a phenomenological model based on this idea has been applied to the calculation of phonon dispersion curves for silicon [41]

$^{++}$ Only symmetric combinations of the matrix elements of $\frac{\partial V}{\partial x\binom{\ell}{\kappa}}$ enter the final formulae for the (1, 1, 1)-direction.

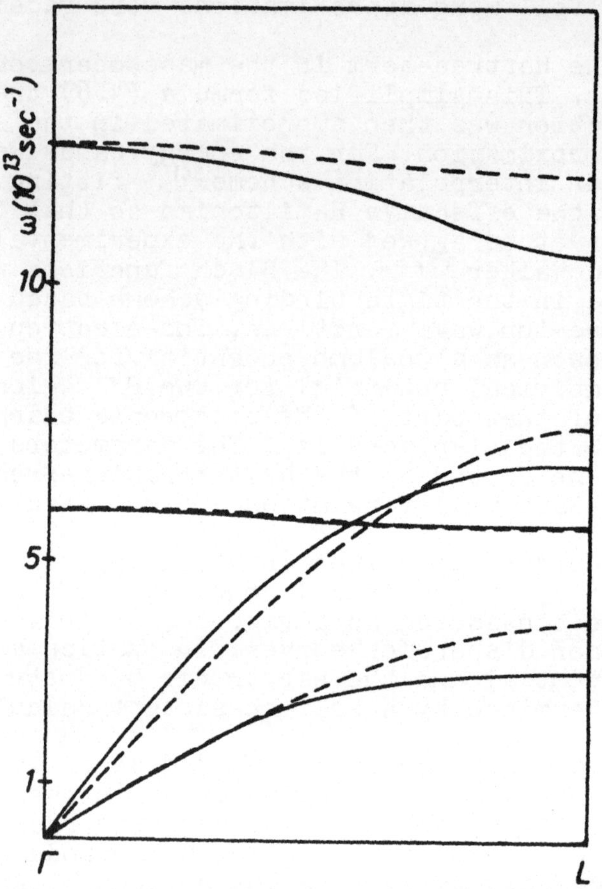

Fig. 4: Dispersion curves for LiF in the (1,1,1)-direction

Solid line: Quantum mechanical calculation

Broken line: BSM calculation

Several investigations of the band structure of the alkali halides are now available [42]. Therefore we hope that some of the discrepancies of our procedure can be removed in a future calculation.

It is a pleasure to acknowledge many discussions with Prof. H. Bilz on this work. Also discussions with Prof. F. Herman, S.O. Lundquist, and P. Rakah during the Chania Institute have been stimulating. The Deutsche Forschungsgemeinschaft and the Deutsche Rechenzentrum have enabled us to carry out the numerical calculations.

REFERENCES

1. B. G.Dick and A.W. Overhauser, Phys.Rev. 112, 90 (1958)
2. A.D.B.Woods, W. Cochran and B.N. Brockhouse, Phys. Rev. 119, 980 (1960)
3. U. Schröder, Sol. State Comm. 4, 347 (1966)
4. A.M. Karo and J.R. Hardy, Phys. Rev. 129, 2024 (1963)
5. A.M. Karo and J.R. Hardy, Phys. Rev. 141, 696 (1966)
6. L.J. Sham, Proc. Toy. Soc. (London) A 238, 37 (1965)
7. W.A. Harrison, Phys. Rev. 139 A, 179 (1965)
8. S.H. Vosko, R. Taylor, and G.H. Keech, Can.J.Phys. 43, 1187 (1965)
9. J.C. Phillips and L. Kleinman, Phys. Rev. 116, 287 (1959)
10. R. Landshoff, Zschr. Phys. 102, 201 (1936)
11. P.O. Löwdin, J. of Phys. 18, 365 (1950)
12. S.O. Lundquist, Ark. Fys. 6, 25 (1951 and
13. Ark, Fys. 9, 435 (1955)
14. K.B. Tolpygo and V.S. Mashkevitch, Soviet Physics JETP 5, 435 (1957)
15. R.A. Cowley, Proc. Roy. Soc, 268, 109 (1962)
16. K.B. Tolpygo, Sov. Phys. Sol. State 3, 685 (1961)
17. M. Born and R. Oppenheimer, Ann. d. Phys. 84, 457 (1927)
18. B. Gliss and H. Bilz, Phys. Rev. Letters 21, 884 (1968)
19. E.R. Cowley and R.A. Cowley, Proc. Roy. Soc. 292, 209 (1966)
20. A.A. Maradudin, E.W. Montroll, and G.H. Weiss, Sol. State Phys. Suppl. 3
21. W. Ludwig, Erg. der ex. Naturw. (Springer Tracts in Mod. Phys.) 43, 1 (1967)
22. E.W. Kellermann, Phil. Trans. Roy. Soc. (London) A 238, 513 (1940)
23. R.F. Wood and U. Öpic, Phys. Rev. 162, 736 (1967)
24. R.F. Wood and R.L. Gilbert, Phys. Rev. 162, 746 (1967)
25. W. Heitler and F. London, Zschr.Phys. 44, 455 (1927)
26. P.O. Löwdin, Adv. of Phys. 5, 1 (1956)
27. F. Herman, J. Phys. Chem. Solids 8, 405 (1959)
28. S.O. Lundquist, Ark. Fys. 12, 263 (1957)
29. R.A. Cowley, W. Cochran, B.N. Brockhouse, and A.D.B. Woods, Phys. Rev. 131, 1030 (1963)
30. H. Bilz in: S. Nudelman ed. "Optical Prop. of Solids" (in press)
31. H. Bilz and B. Gliss (to be published)
32. V. Nüßlein and U. Schröder, Phys. Stat. Sol. 21, 309 (1967)
33. G. Dolling, H.G. Smith, R.M. Nicklow, P.R. Vygayaraghavan, and M.K. Wilkinson, Phys. Rev. 168, 970 (1968)

34. W.I. Waller and S.O. Lundquist, Ark. Fys. 7, 121 (1954)
35. A.A. Abrikosov, L.P. Gorkov, and I.E. Dzyaloskinski, Methods of Quantum Field Theory in Statistical Physics, Prentice Hall, London (1963)
36. J.C. Slater and G.F. Koster, Phys. Rev. 94, 1498 (1954)
37. D.M. Roessler and W.C. Walker, J. Phys. Chem. Solids 28, 1507 (1967)
38. E. Clementi, IBM Journ. of Res. and Dev. 9, 2 (1965)
39. M.H. Cohen, R.M. Martin, and R.M. Pick (to be published)
40. S.K. Sinha, Phys. Rev. 163, 477 (1968)
41. R.M. Martin, Phys. Rev. Letters 21, 536 (1968)
42. B. Kunz, International Symposium on "Colour Centers in Alkali Halides", Rome, 179 (1968)

MANY-BODY EFFECTS IN SIMPLE METALS

Stig Lundqvist

Chalmers University of Technology

Göteborg, Sweden

I. INTRODUCTION. THE FERMI LIQUID

Many-body effects in metals have been extensively discussed and there is a vast literature on collective effects in metals, particularly plasmon effects, exciton problems, the metal-insulator transition, etc. Much less attention has been paid to what is considered as <u>one electron effects</u> in metals and the main objective of this review is to summarize some recent work in that area. We shall only discuss metals, i.e. those showing a predominantly free electron behaviour. Indeed, calculations of many-body effects have only been performed for either a uniform electron gas or for metals having an almost spherical Fermi surface such as Na. Practically nothing is known at the present time about many-body effects in metals having a complicated electron structure.

We shall mainly discuss results of calculations and not at all go into discussions of the detailed structure of the theory. The formulations used will contain the many-body effects to lowest order only and will be open to some criticism. It is quite obvious that more accurate work in the future will modify the numerical results, but there are strong reasons to believe that the qualitative features of the results reported in this review will remain.

The main questions to be answered are the following:

i) Why does one-electron theory generally work so well, i.e. why are the many-body corrections in most cases so small?

ii) What are the characteristic effects of the many-body interaction, i.e. when are the correction terms significant and what are the significant new effects due to the interaction?

These questions are indeed at least partly answered by the remarkable Landau theory of the Fermi liquid, which forms the basis for the modern theory of metals. In the Landau theory the physics is discussed in terms of a single particle distribution describing quasi-particles. Consider the change in energy corresponding to a small change $n^{(1)}(\underline{k}) = n(\underline{k}) - n^{0}(\underline{k})$ in the distribution $n^{0}(\underline{k})$ being the equilibrium distribution, one obtains to second order

$$E = \sum_{\underline{k}} E^{0}(\underline{k})n^{(1)}(\underline{k}) + \frac{1}{2}\sum_{\underline{k}\underline{k}'} f(\underline{k},\underline{k}')\, n^{(1)}(\underline{k})n^{(1)}(\underline{k}'), \qquad \text{I:1}$$

where $E^{0}(k)$ is the change in energy when we add one particle to the system in its ground state, while the change in energy when adding one particle in state $\underline{k}$ when we have a distribution $n(\underline{k})$ of quasi-particles is given by

$$E(\underline{k}) = E^{0}(\underline{k}) + \sum_{\underline{k}'} f(\underline{k},\underline{k}')\, n^{(1)}(\underline{k}') \qquad \text{I:2}$$

$f(\underline{k},\underline{k}')$ is the quasi-particle interaction, which makes the theory differ in a fundamental way from any single-particle theory. The last term in Eq. I:2 is numerically small compared to $E^{0}(\underline{k})$, but what counts in most problems is not the magnitude of $E(\underline{k})$ itself but the excitation energy $E(\underline{k}) - E_F$ where E_F is the Fermi energy and in such situations the interaction between quasi-particles will play an important role. The bare quasi-particle energy $E^{0}(\underline{k})$ and the quasi-particle interaction $f(\underline{k},\underline{k}')$ are formally defined according to Eq. I:1 as the first and second derivatives of the total energy with regard to the distribution function. $E^{0}(\underline{k})$ and $f(\underline{k},\underline{k}')$ cannot be calculated from the Landau theory itself but are the fundamental parameters of the theory.

In applications of the Landau theory one usually considers small disturbances of the distribution corresponding to a small fraction of the total number of particles, and the changes in the distribution are confined to the immediate neighbourhood of the Fermi surface. For an isotropic system the only information needed from $E^{0}(\underline{k})$ is given by the effective mass m^{*}, and, since $\underline{k}$ and $\underline{k}'$ are both very close to the Fermi surface, $f(\underline{k},\underline{k}')$ reduces to a function of the angle Θ between the vectors $\underline{k}$ and $\underline{k}'$, i.e. $f(\underline{k},\underline{k}') \rightarrow f(\Theta)$, and of the spin states associated with $\underline{k}$ and $\underline{k}'$. Furthermore, the interaction will only appear in certain integrals of the type given in Eq. I:2, and in typical applications they will only depend on one or a few of the expansion coefficients f_l in an expansion of $f(\Theta)$ in Legendre polynomials, $f(\Theta) = \sum_l f_l P_l(\cos\Theta)$.

Thus the interaction will appear in the final formulas through one or several of the coefficients $f_\ell = A_\ell + B_\ell \sigma \cdot \sigma'$, where the spin dependence for an isotropic, paramagnetic system is explicitly shown.

The Landau theory applies directly to the static properties of the system such as the electronic specific heat, paramagnetic susceptibility, compressibility etc. More important is the extension to time-dependent problems and transport properties. Application of the theory to wave-like excitations in external electric and magnetic fields, gives a spectrum of characteristic excitations, each with its specific dependence on the interaction parameters A_ℓ and B_ℓ . Many of those modes have not yet been studied experimentally, however, results have been obtained for the dispersion of plasma waves and recently also for spin waves and magnetoplasma waves.

So far theoretical calculations of the Landau interaction parameter from many-body theory have been very few. The variations of the single particle energy with small changes of the distribution is not very accurately known at present and further work has to be done. These aspects will not be further discussed in this review, which will deal with other applications of many-body theory, where some interesting results have recently been obtained. The reader is referred to ref. 1 for a comprehensive presentation of the Landau theory.

In section II the approximate form for the self-energy, which has been the basis of most calculations, will be discussed. In the following section we shall summarize the results of some calculations of quasi-particle properties for an electron gas and the conclusion one may draw for real metals. In particular we would like to draw attention to the old problem about the exchange and correlation potential to be used in energy band calculations. Section IV will present some results for the effect of electron-phonon interactions. Analog effects occur for the case of electron-electron interaction, although on a completely different energy scale; they will be discussed in section V. The implications of the results for some physical properties will be summarized in the concluding remarks in section VI.

II. THE SELF-ENERGY AND THE SPECTRAL WEIGHT FUNCTION

Suppose that we have solved the one-electron problem in the average potential of the lattice and obtained the energy as a function of wave number, $\varepsilon = \varepsilon(\underline{k})$. The next step would be to include the dynamical effects of the interaction and exchange effects, which cannot be described by an ordinary local potential. We consider for simplicity the case of uniform or almost uniform distributions of conduction electrons where specific effects of periodicity are of minor importance.

What is the proper generalization of an ordinary potential to include dynamical effects of the interaction? Direct physical arguments show that such a generalized potential must be non-local in space and retarded in time, i.e. of the form $\Sigma(x - x', t - t')$, (where the assumption of a uniform system has been explicitely used). In momentum-energy space we find, by making the Fourier transforms, a dependence on both momentum and energy $\Sigma(\underline{k}, \varepsilon)$, where ε is the energy variable. Σ is called the self-energy of the electron. Adding the self-energy to the average potential, we have to solve the equation

$$\varepsilon = \varepsilon(\underline{k}) + \Sigma(\underline{k}, \varepsilon) \qquad \text{II:1}$$

The self-energy includes all the interactions between the state considered and the rest of the system. This means that dissipative effects which lead to the decay of the state are included so that the self-energy must with necessity be complex,

$$\Sigma(\underline{k}, \varepsilon) = \Sigma_R(\underline{k}, \varepsilon) + i\,\Sigma_I(\underline{k}, \varepsilon),$$

The obvious question now arises how to represent the physics of the solution in a problem with a complex potential, which depends on both momentum and energy. The convenient quantity to use is the spectral weight function $A(\underline{k}, \varepsilon)$, defined by

$$A(\underline{k}, \varepsilon) = \frac{1}{\pi}\,\frac{|\Sigma_I(\underline{k}, \varepsilon)|}{[\varepsilon - \varepsilon(\underline{k}) - \Sigma_R(\underline{k}, \varepsilon)]^2 + |\Sigma_I(\underline{k}, \varepsilon)|^2}, \qquad \text{II:2}$$

In the special case that the imaginary part is independent of energy, i.e. $\Sigma_I(\underline{k}, \varepsilon) = \Gamma(\underline{k})$, the spectral function for a given $\underline{k}$ gives a Lorentzian spectrum and $\Gamma(\underline{k})$ determines the width of the line. With an imaginary part depending on the energy there are no restrictions on the shape of the spectral profile.

The spectral weight function is the general distribution function for electrons with regard to both momentum and energy, and from it one can calculate the density of states $N(\varepsilon)$ and momentum distribution $n(\underline{k})$ by integration, thus

$$N(\varepsilon) = \frac{1}{4\pi^3}\int d\underline{k}\; A(\underline{k}, \varepsilon); \quad n(\underline{k}) = \int_{-\infty}^{E_F} d\varepsilon\; A(\underline{k}, \varepsilon) \qquad \text{II:3}$$

We next wish to discuss the physical effects described by the self-energy and define the approximation, with which we shall be concerned.

In the simplest type of theory, the self-energy $\Sigma(\underline{k},\mathcal{E})$ is taken as energy-independent, which leads to just one quasi-particle peak in the spectral function $A(\underline{k},\mathcal{E})$. The Hartree approximation is obtained by taking $\Sigma(\underline{k},\mathcal{E}) = 0$, and the Hartree-Fock approximation by taking for $\Sigma(\underline{k},\mathcal{E})$ the Hartree-Fock exchange potential,

$$\Sigma(\underline{k},\mathcal{E}) = \frac{i}{(2\pi)^4}\int d^4q\, e^{i\delta q_0}\, v(q)\, G_0(\underline{k}+\underline{q}) =$$

$$-\frac{1}{(2\pi)^3}\int d^3q\, v(q)\, n(\underline{k}+q) \qquad \text{II:4}$$

where

$$G_0(k) = (k_0 - \mathcal{E}(\underline{k}) \pm i\delta)^{-1},$$

and

$$v(q) = \frac{4\pi e^2}{q^2}\,; \qquad n(\underline{k}) = \Theta\,(k_F - k). \qquad \text{II:5}$$

The Hartree-Fock approximation is, however, totally inadequate for describing the single-particle spectrum of metals, since it gives a far too large bandwidth and an erroneous structure of the density of states. Actually the so-called Hartree-Fock-Slater potential /2, which originally was given as an approximation of the Hartree-Fock potential, yields much better results.

Many-body theory tells us how to write exact expansions for the self-energy Σ. If the expansion is arranged after powers of the bare Coulomb potential $v(\underline{q})$, we find that the first-order term is identical to Eq. II:4. Physical intuition, however, tells us to expect an effective screened potential to appear in Σ, rather than the bare potential $v(\underline{q})$. A reasonable approximation for Σ has the form

$$\Sigma(k) = \frac{i}{(2\pi)^4}\int d^4q\, e^{i\delta q_0}\, v(q)\,\mathcal{E}_D^{-1}(q)\, G_0(\underline{k}+\underline{q}) \qquad \text{II:6}$$

where $\mathcal{E}_D^{-1}(q)$ is the dielectric function. This expression for Σ is described in many-body language as obtained by neglecting "vertex corrections". The dielectric function $\mathcal{E}_D(q)$ tells us the change in charge density $\delta\rho(\underline{r},t)$ induced by an infinitesimal external charge distribution $\rho^{ext}(\underline{r},t)$,

$$\delta\rho(\underline{r},t) = \int d^3r\,dt\left[\mathcal{E}_D^{-1}(\underline{r}-\underline{r},\, t-t) - \underline{1}\right]\,{}^{ext}(\underline{r},t), \qquad \text{II:7}$$

where $\underline{1}$ symbolizes a δ-function in space and time. Here $\mathcal{E}_D^{-1}(\underline{r},t)$ refers to the Fourier transform of $\mathcal{E}_D^{-1}(\underline{q},\mathcal{E})$ with respect to space and time.

It is possible to separate the expression for Σ in Eq.

II:6 to exhibit explicitly a screened exchange term and the effect of a "correlation hole". These effects have been anticipated long since by Wigner, Seitz, Peierls and others. To do this we introduce the spectral resolution of the inverse dielectric function

$$\mathcal{E}_D^{-1}(q, \ ) = 1 + \int_0^\infty \frac{d\mathcal{E}_1 \, 2\mathcal{E}_1 \, B(q,\mathcal{E})}{\mathcal{E}^2 - (\mathcal{E}_1^2 - i\delta)^2} \qquad \text{II:8}$$

$$B(q,\mathcal{E}) = \frac{1}{\pi} \operatorname{Im}(1 - \mathcal{E}_D^{-1}(q,\mathcal{E})) \geq 0.$$

We perform the energy integration in Eq. II:6 by closing the contour in the upper half-plane. We then have contributions from the poles in the Green's function G and in the inverse dielectric functions $\mathcal{E}_D^{-1}$,

$$\sum(\underline{k},\mathcal{E}) = -\frac{1}{(2\pi)^3} \int d^3q \, v(q) \, \mathcal{E}_D^{-1}(q, \mathcal{E}(\underline{k} + q) - \mathcal{E}) n(\underline{k} + \underline{q}) \qquad \text{II:9}$$

$$+ \frac{1}{(2\pi)^3} \int d^3q \, d\mathcal{E}_1 \, v(q) \frac{B(q,\mathcal{E}_1)_1}{\mathcal{E} - \mathcal{E}_1 - \mathcal{E}(\underline{k} + q)}.$$

By comparison with Eq. II:4 we see that the first term in Eq. II:9 is a screened exchange contribution. We now want to demonstrate that the second term has the significance of a correlation hole contribution, and do this by considering the classical self-energy problem.

The potential ϕ at a point $\underline{r} = \underline{v}t$ at time t from the "correlation hole" induced by a moving "classical" charge $\rho^{ext}(\underline{r},t) = \delta(\underline{r} - \underline{v}t)$ is in a linear approximation according to Eq. II:7.

$$\phi(\underline{v}t,t) = \int d^3r' dt \, v(\underline{v}t - \underline{r}') \left[\mathcal{E}_D^{-1} (\underline{r} - \underline{v}t, \ t - t) - 1 \right]$$
$$= \frac{1}{(2\pi)^3} \int d^3q \, v(\underline{q}) \left[\mathcal{E}_D^{-1}(q, \ \underline{q}\cdot\underline{v}) - 1 \right]. \qquad \text{II:10}$$

Using Eq. II:8 and the fact that $v(\underline{q})$ and $B(\underline{q},\mathcal{E})$ are even functions of $\underline{q}$, the expression for ϕ becomes

$$\phi(\underline{v}t,t) = -2 \frac{1}{(2\pi)^3} \int_0^\infty d^3q \, d\mathcal{E}_1 \, v(q) \frac{B(q, \mathcal{E}_1)}{\mathcal{E}_1 + \underline{q}\cdot\underline{v}}, \qquad \text{II:11}$$

We now see that the expression for ϕ in Eq. II:11 is the same as the last term in Eq. II:9 if we put $\mathcal{E}(\underline{k} + q) - \mathcal{E} = \underline{v} \cdot \underline{q}$, except for a factor 2. The factor 2 is cancelled by a factor $\frac{1}{2}$ arising from integration over the interaction strength. Since we deal with the self-energy of quasi-electrons, we can substitute

$$\mathcal{E}(\underline{k} + q) - \mathcal{E}(\underline{k}) = \underline{k} \cdot q/m + q^2/2m = \underline{v} \cdot q + q^2/2m. \qquad \text{II:12}$$

The last term in Eq. II:12 can certainly be neglected for swift particles.

III. SOME QUASI-PARTICLE PROPERTIES THE EXCHANGE AND CORRELATION POTENTIAL

The detailed structure of the spectral function will be discussed in the following two sections. In this section we shall only comment on properties that have to do with the dominating peak in the spectral function, which determines the quasi-particle properties. This peak is sharp at the Fermi surface and remains strongly peaked for a considerable range of momenta outside the Fermi surface as well.

Several calculations have been made using the electron gas model and we would like to refer in particular to the work by L. Hedin /3/ and T.M. Rice /4/. The only parameter in the electron gas model is the electron density, and one introduces a dimensionless parameter r_s, defined by the relation $4 (a_o r_s)^3/3 = V/N$, where a_o is the Bohr radius and V and N denote the volume and number of electrons respectively. The interesting range of metallic densities corresponds to r_s in the range 2 - 6. The calculations by Hedin and Rice include only the effect of electron-electron interaction; phonon effects will be discussed in section IV.

We shall briefly summarize some important results from the calculations by Hedin and Rice and refer to their papers for a comprehensive discussion.

i) The overall effect of the interaction is generally small because of the small range of the non-local interaction. Hedin has given results for $\sum(\underline{x} - \underline{x}', E_F)$ for electrons at the Fermi surface (ref. 3 Fig. 10). The range is typically smaller than the average distance between electrons and is considerably smaller than the range of the Hartree-Fock exchange potential. The fact that the interaction is strong only at distances considerably smaller than the average distances (the maximum magnitude of $\sum$ occurs at around 1/4 of the average distance) seems to be the key why the many-body modifications of some properties is so small.

ii) The effective mass correction because of electron-electron interaction is small. The correction is only a few percent. It is negative for small r_s. For large r_s the calculations are much less reliable. Various calculations agree that the correction is small but give different signs for the correction; it is more likely to be positive, i.e. a slight enhancement of the mass for large r_s, say $r_s > 4$.

iii) Hedin and Rice also give results for the Landau interaction. Since this has to do with the first order change in the self-energy, the results are much more sensitive to the approximation used. The results are probably reasonably accurate in the range $2 < r_s < 4$ to give the first few Legendre expansion coefficients, whereas the results for larger r_s are probably much less significant.

Most of the discussion of quasi-particles has been confined to Fermi surface properties. Indeed one often meets statements that quasi-particles are only well-defined at the Fermi surface and that the decay rate is so large that they are no longer well-defined excitations when we leave the immediate neighbourhood of the Fermi surface. Recent calculations by B.I. Lundqvist /5/, which will be reported in section V show on the contrary that the quasi-particle is a very well defined mode of excitation all the way to the bottom of the Fermi sea and in a considerable range outside the Fermi sea as well. This opens the range of applications to the general problem of energy band theory, optical properties in the ultraviolet and soft X-ray region etc. In this context we would like to comment on the old question of the effect in energy band calculations of exchange and correlations.

The limited usefulness of the Hartree-Fock theory in the theory of metals was observed long ago. The energy vs. k relation has infinite derivative at the Fermi level, and as a result the density of states vanishes at the Fermi level, which results in a number of strange properties completely at variance with experimental facts.

Slater /2/, suggested long ago an averaging procedure, resulting in a potential independent of k, depending only upon the local density. In the past ten years several proposals have been published how to incorporate in an average way the effects of exchange and correlation. A systematic method how to simulate the effect of correlation and exchange by the use of potential depending on the density alone has recently been developed by Kohn and Sham /6/. Their work resulted in a potential of the same form as the one introduced by Slater, but with a different numerical factor. It is of interest to compare these more or less ad hoc modifications with the results of systematic application of the many-body theory.

In order to find the quasi-particle peak in the spectrum according to Eq. II:2 we solve Eq. II:1 considering only the real part of Σ. The solution can be cast in the form

$$\Xi(\underline{k}) = \mathcal{E}(\underline{k}) + V_{ec}(\underline{k}), \qquad \text{III:1}$$

where $V_{ec}(\underline{k})$ can be interpreted as the effective exchange and correlation potential. In the Hartree-Fock theory $V_{ec}(\underline{k})$ corresponds to the H-F exchange potential; using the self-energy introduced in section II, $V_{ec}(\underline{k})$ gives the effect of the correlation hole plus screened exchange.

Results for $V_{ec}(\underline{k})$ were reported by Hedin (ref. 3 , Fig. 9) in the range $0 < |\underline{k}| < 1.5\ k_F$. The results show that there is only a moderate variation in $V_{ec}(\underline{k})$ over this range of momenta. Extrapolating this result to real metals we can conclude that there will be no appreciable distortion of energy band curves and no appreciable change in band width due to electron-electron interactions. The exchange and correlation potential therefore shows similar general characteristics as the potentials proposed by Slater and by Kohn and Sham, which are exactly constant for a uniform system.

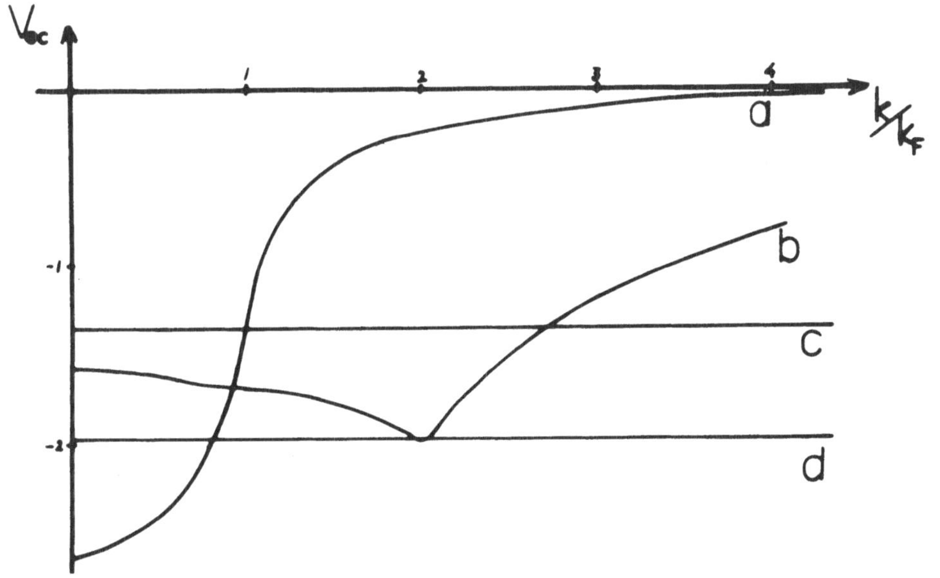

Fig. 1. The exchange and correlation potential V_{ec} in different approximations. a) in Hartree-Fock approximation b) from dynamical screened interaction c) the Kohn and Sham approximation d) from Slater exchange.

We observe that the conclusions just drawn are restricted to the momentum interval over which the calculations extend. For larger momenta we expect of course that the potential should approach zero, and because the screening starts to drop rapidly at around $k = 2k_F$, we expect that a gradual decrease in magnitude should set in around $k = 2k_F$. In Fig. 1 we give the results over an extended range of momenta for a density $r_s = 4$ (corresponding to Na). As a comparison we have also included the Hartree-Fock exchange potential as well as the potentials proposed by Slater and by Kohn and Sham.

The conclusion from this calculation is the following. There seems to be little distortion of the conduction band itself and the main effect is an almost constant shift. However, for highly excited states there is a gradual slow change (going asymptotically as k^{-1}). These effects should be observed to get the proper location of highly excited bands and in problems concerning low energy electron diffraction.

IV. THE ELECTRON-PHONON INTERACTION

In this section we will mention some of the effects of the electron-phonon interaction on the one-electron properties of metals. In order to have some intuitive understanding what to expect, we look into the basic physics of the electron-phonon interaction. When the ions move, most of the conduction electrons must adiabatically follow the motion of the ions because there are no available states of about the same energy into which they could scatter; only the electrons very close to the Fermi surface can scatter. The motion of the conduction electrons tend to screen out all deviations from charge neutrality. This results in a strong screening of the long-range bare electron-phonon interaction, and this screening produces an effective electron-phonon interaction. The dynamical effects of this interaction on the conduction electrons is confined to electrons within approximately a phonon frequency from the Fermi surface. This will lead to an enhancement of the density of states or effective mass on the Fermi surface. The spin susceptibility, however, will not be affected, because the electron-phonon interaction does not depend on spin and only samples a thin layer around the Fermi surface; therefore phonons will cause no change in the total energy if we have an unbalanced spin distribution.

The effects on the quasi-particle dispersion law can be obtained by using the appropriate modification of the formulas in section II, but they follow in a straightforward way using parturbation theory in the Brillouin-Wigner form, thus

$$E(\underline{k}) = \mathcal{E}(\underline{k}) + \sum_p |g_{k-p}|^2 \frac{1 - f_p}{E(k)-E(p)+\omega(k-p)} + \frac{f_p}{E(k)-E(p)-\omega(k-p)} \qquad \text{IV:1}$$

In Eq. 1 g_{k-p} is the electron-phonon matrix element and $(k-p)$ the phonon frequency.

The qualitative effect of the phonons is to flatten out the dispersion curve in the immediate neighbourhood of the Fermi surface, and this gives rise to an enhancement of the density of states, or, equivalently, the <u>thermal effective mass</u> and one obtains from Eq. 1

$$m_{ph} = m(1 + \lambda)$$
$$= N_o(E_F) \int \frac{d\Omega_{p-k}}{4\pi} \frac{2|g_{p-k}|^2}{\omega(p-k)} \qquad \text{IV:2}$$

where $N_o(E_F)$ is the density of states without electron-phonon interaction, $|\underline{p}| = |\underline{k}| = k_F$ and the integration extends over the full solid angle.

Ashcroft and Wilkins /7/ first calculated the corrections for Na, Al and Pb using Eq. 2 and several similar calculations have been published over the last few years. The most accurate values of λ are those deduced from tunneling data by McMillan and Rowell /8/. In table 1 we have given the theoretical values of λ by Ashcroft and Wilkins and some values obtained from superconductor tunneling.

<u>Table 1. Enhancement of the effective mass $m_{ph} = m(1 + \lambda)$</u>

	Na	Al	Pb	Sn	In	Ga
(theory)	0.18	0.49	1.05			
(tunneling)		0.38	1.12	0.60	0.71	0.60

It is obvious that the enhancement varies with temperature and that no enhancement is left at high temperatures, where the phonon system behaves like a fluctuating classical medium. A technical discussion of this effect requires the use of temperature Green functions, and we shall only state some results of the recent discussion of this effect.

There are essentially two effects of this temperature variation that can possibly be tested with experimental techniques now available. These possibilities have recently been discussed by G. Grimvall and J.W. Wilkins (private communications) and the brief report given here is from their work. One is the direct observation of the temperature variation of the effective mass in a cyclotron resonance experiment. The other effect is the corresponding additional variation in the specific heat. We write the specific heat in the low temperature regime in the form

$$C_e = \gamma_{tot} T = \left[\gamma_0 + \gamma(T) \right] T, \qquad \text{IV:3}$$

where $\gamma(T)$ incorporates all effects from the electron-phonon interaction. $\gamma(T)$ first increases with temperature, reaches a maximum and then gradually falls off towards the asymptotic value zero. The same qualitative behaviour holds for the enhancement factor $\lambda(T)$.

The temperature variation can be calculated if the electron-phonon matrix element and the phonon spectrum is known. Such a calculation has been performed by Grimvall ,9 , who also showed that the temperature variation of the effective mass was very accurately given by an Einstein model with an appropriately chosen Einstein temperature. In practice it turns out to be much better to use the combination of electron-phonon coupling strength and phonon spectrum, usually denoted $\alpha^2(\omega)\ F(\omega)$, which is obtained from tunneling data, because this is precisely the quantity needed in these calculations.

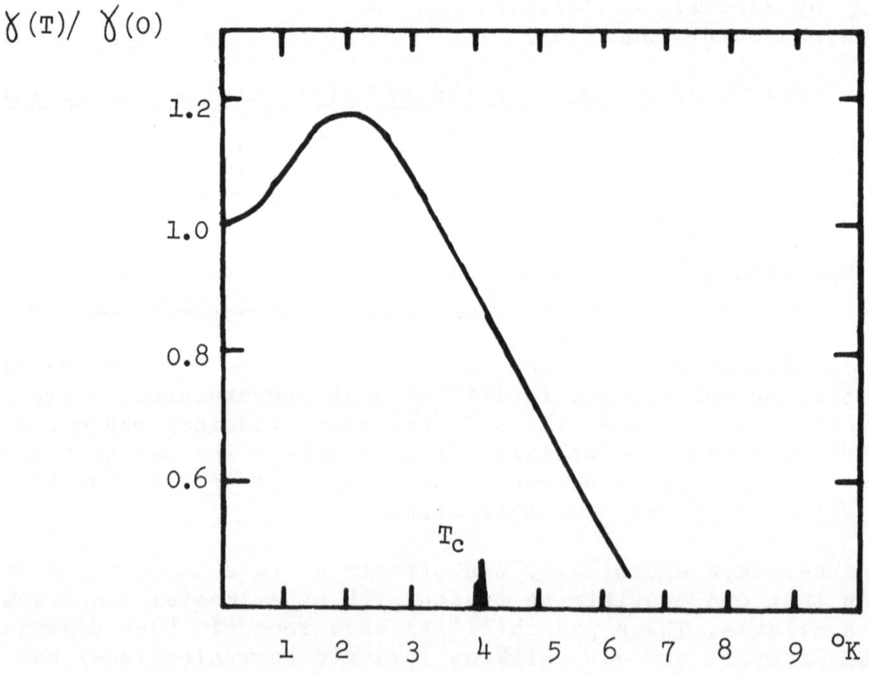

Fig. 2. Temperature dependence of the electron-phonon contribution to the electronic specific heat for mercure. $T_c = 4.16$ ^oK is the superconducting transition temperature.

There are considerable difficulties to observe the temperature variations in most metals with the techniques and accuracy now available. It seems, however, that the crucial quantity $\alpha^2(\omega)$ F(ω) for mercury has a strong peak at an energy corresponding to only 20 ^{o}K and Hg may therefore be a possible candidate (Pb may be a second possible choice). Calculations by Grimvall show that there is sufficient variation in the effective mass, e.g. the mass increases by 5 o/o from T = 1 ^{o}K to T = 3 ^{o}K and this change might be observed in a cyclotron resonance experiment. Similarly there is a considerable variation in the term γ(T) in the specific heat of mercury, which is illustrated in Fig. 2. This might be observed by measuring the difference in specific heat of the superconducting and normal (in an applied magnetic field) phases.

The final point we wish to discuss is the structure of the spectral function. Engelsberg and Schrieffer /10/ pointed out that in the neighbourhood of the Fermi surface the quasi-particle picture will no longer apply. They calculated the spectral function for an Einstein model and for a Debye spectrum and obtained a spectral function with a very complex structure in the region close to the Fermi surface. We refer to their paper for illustrative examples and a detailed discussion. The spectrum for Na has been calculated by Grimvall /9/, using a realistic model, including effects of umklapp, anisotropy in the phonon spectrum etc. For sodium no dramatic effects occur because of the rather weak electron-phonon interaction. For metals having a much stronger electron-phonon interaction there is indeed a pronounced structure in the spectral function similar to the examples given by Engelsberg and Schrieffer. The direct experimental verification of such a structure would be of considerable interest, but we are not aware of any available experimental technique that will resolve the structure of this spectrum.

V. THE ELECTRON-PLASMON INTERACTION

The problem in the preceding section was concerned with the interaction between an electron and the density fluctuations of the ions, i.e. the phonon system. The coupling constant is the screened electron-ion interaction and the characteristic energy involved is a typical phonon frequency. We now turn to a theoretically rather analog problem, namely, the interaction between an electron and the density fluctuations in the system of conduction electrons. The dominating mode of density fluctuations in this case is the plasmon, and accordingly the problem is qualitatively that of the coupled electron-plasmon system, where the remaining modes, the screened electron-hole pairs, mainly give rise to background and damping effects.

The electron-plasmon problem is in its mathematical structure equivalent to the polaron problem, i.e. the quasi-particle formed by strong coupling between an electron and the optical lattice modes. The effective electron-plasmon coupling constant is considerably larger than in the polaron problem and, more important, the frequency involved is not a phonon frequency but a plasmon frequency (5 - 20 eV). Accordingly, we have now to look for effects far away from the Fermi surface. Indeed, Fermi surface properties are only indirectly affected by the electron-plasmon effects.

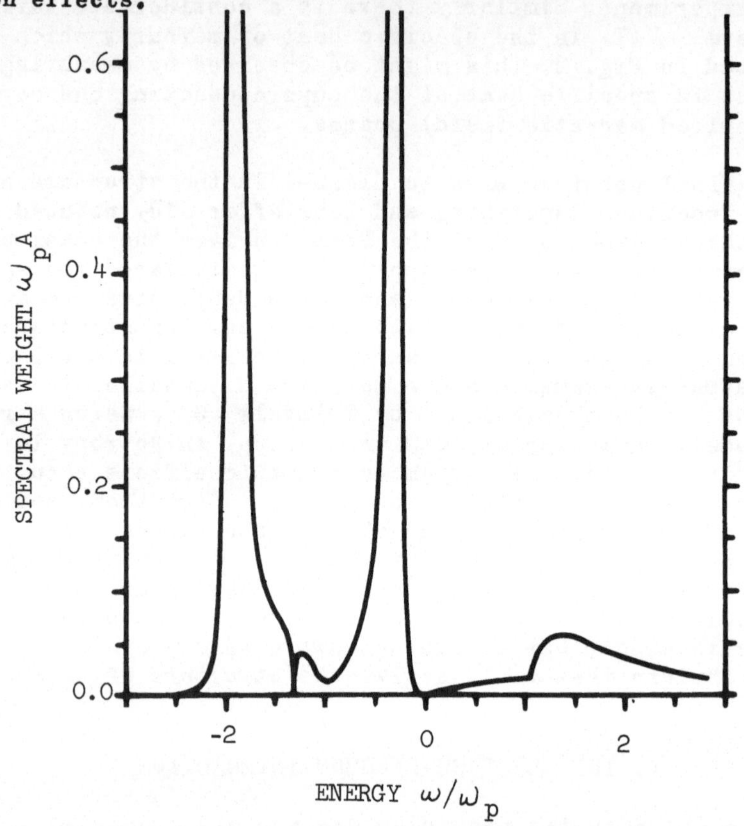

Fig. 3. Typical structure of the spectral weight function for hole states showing the plasmaron peak (left) the quasi-particle peak and some additional structure in the back-ground.

In order to study the effect of the electron-plasmon coupling we must look for structure in the spectral function quite faraway from the central quasi-particle peak. For a particle close to the Fermi level, the effect is two rather wide side bands on each side of the quasi-particle peak, corresponding to plasmon absorption and emission. For momenta inside the Fermi sea (hole states) the sideband on the negative energy side (choosing the zero of energy at the Fermi level) sharpens up when we go to smaller momenta and one gradually finds a strong peak that becomes even sharper than the quasi-particle peak. Thus one obtains two modes of excitation for holes with momenta less than, say 0.6 k_F as illustrated in Fig. 3. One corresponds to the ordinary quasi-particle peak and the new one to a hole strongly coupled to excited plasmons. This effects was first noted by L. Hedin, B.I. Lundqvist and the author 11 and has been extensively studied by B.I. Lundqvist 5 . A survey of the shape of the spectral function is given in Fig. 4.

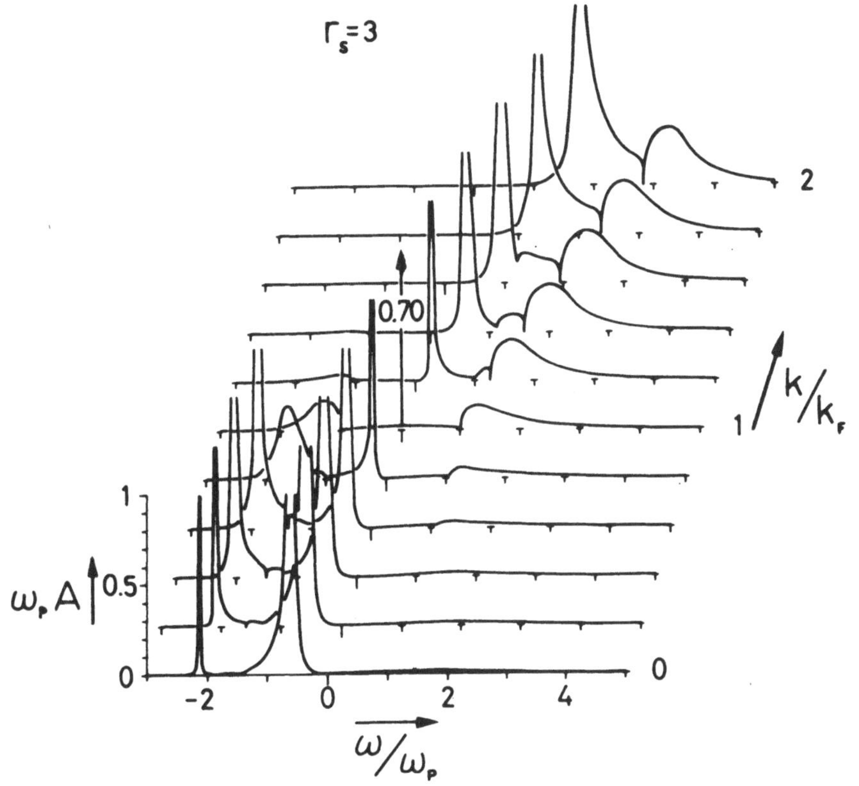

Fig. 4. Survey of the one-electron spectrum over a wide range of momenta from k = 0 to k = $2k_F$ and r_s = 3.

We refer to the original papers for a discussion of the theory, which is based on the approximation for the self-energy in section II. However, the effect can be understood in simple terms, just using perturbation theory as in the preceding section. The interaction couples the state $\underline{k}$ to other states with the same total momentum, a typical state containing one electron with energy $E(\underline{p})$ and one plasmon with energy $\omega(\underline{k} - \underline{p})$. Denoting the electron-plasmon coupling constant by g_{k-p} we find if we calculate the energy difference $E_{N+1}(\underline{k}) - E_N^0$ using perturbation theory that Eq. IV:1 applies in this case as well, with the appropriate reinterpretation of the symbols. Because the plasmon frequency varies only slowly with the wave number one will obtain a very strong resonance in the interaction term in Eq. IV:1 at ω_{pe}, the classical plasma frequency, and the rapid variation around the resonance give rise to two weakly damped solutions as illustrated in Fig. 3. The energy of the new mode differs from the sum by an appreciable amount as a result of the strong coupling.

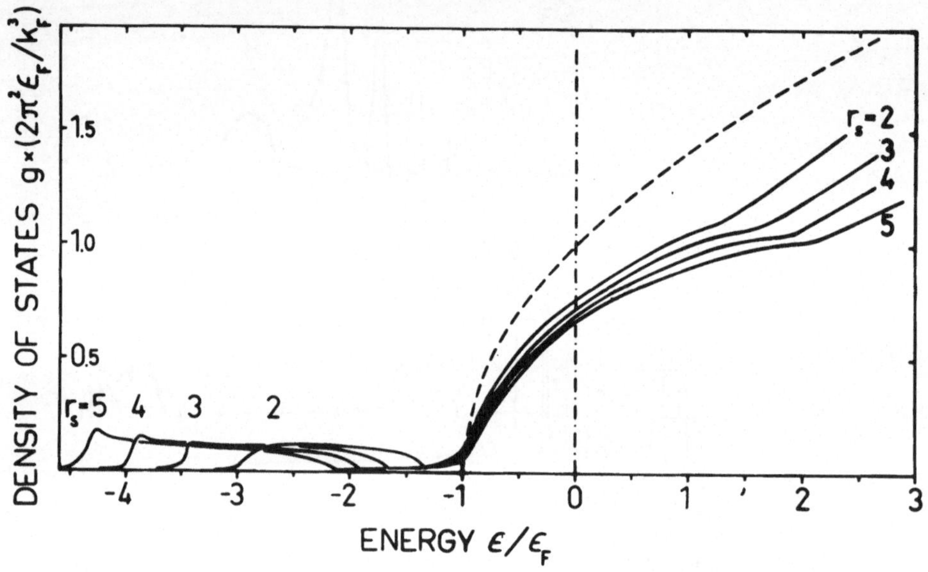

Fig. 5. The density of states in the density range $r_s = 2 - 5$.

The new state of hole and plasmons has been called a plasmaron state. It has an appreciable spectral strength and must therefore be considered as a strong mode of excitation of the system. It has not yet been experimentally verified. The plasmaron branch of the spectrum has an appreciable effect on the density of states of the system. The density of state curve has a characteristic tail as shown in Fig. 5, with a small peak at the low-energy end. This tail could be observed e.g. in soft X-ray emission and has been discussed by L. Hedin, 12. The experimental verification in a soft X-ray emission experiment seems to be marginal at the present time, however, recent work by J. R. Cuthill et al. ,13, gives some experimental evidence.

The existence of a new elementary hole excitation is of

course expected to show up in experiments in a characteristic way, however, the energy range, typically 5 - 40 eV below the Fermi level, makes the effect fall outside the range of most experiments. For example, one would expect an additional structure in the optical absorption spectrum and, of course, also in photoemission, in both cases in the far ultraviolet. The effects should be observable also in degenerate semiconductors, where it would fall in the infrared region of the spectrum. However, in this region the plasmons may mix strongly with the optical phonons and thus give rise to a different structure.

It should be noted that these strong effects of coupling between holes and plasmons occur not only for holes in the Fermi sea but for bound holes in inner shells as well, as has been found theoretically by B. I. Lundqvist (private communication). The calculations predict the existence of strong satellites of the plasmaron type shifted away from the quasiparticle peak. These electron-plasmon effects could be studied by X-ray photo-electron spectroscopy (XPS), which also seem to be the natural method to study the plasmaron effects in the conduction band.

VI. CONCLUDING REMARKS

We would like to give a brief summary and outlook referring to the material presented in this review. The simple quasiparticle properties are described by the quasi-particle dispersion law $E^{0}(\underline{k})$. This should be given accurately by ordinary energy band calculations if one introduces an appropriate exchange and correlation potential. Calculations show that this potential varies only very little over the Fermi sea, but that its variation with energy for highly excited states should be considered. The important many-body effects on Fermi surface properties is given by the Landau interaction $f(\underline{k},\underline{k}')$. Present calculations are probably not very accurate and especially not reliable for low-density metals and more experiments are needed as well to determine accurately the parameters in the Landau theory.

The effect of the electron-phonon interaction is confined to the small region around the Fermi surface, in which the electrons and phonons are strongly coupled. The gross features of the effect are well understood theoretically and experimentally, but more works needs to be done e.g. about the temperature effects. Future experiments may possible resolve the rather complicated structure of the spectral function in strong coupling systems.

The theoretical study of electron-plasmaron coupling has given new and interesting effects. The plasmons effects seem to occur in the conduction band as well as in the inner shells.

Thus, whereas effects of electron-electron interaction seems to be of secondary importance for Fermi surface properties, there seems to be strong dynamical effects in the far uv and X-ray region. This seems to open an interesting area for further theoretical and experimental work to investigate these effects in various properties such as optical absorption and photoemission in the far ultraviolet, soft X-ray emission, X-ray photoemission spectroscopy etc.

References

1 D. Pines and P. Nozières, Theory of Quantum Liquids I, Benjamin, New York, 1966.

2 J. Slater, Phys. Rev. 81, 385 (1951).

3 L. Hedin, Phys. Rev. 139, A796 (1965).

4 T. M. Rice, Ann. Phys. 31, 100 (1965).

5 B. I. Lundqvist, Phys. kondens. Materie 6, 193 (1967), ibid. 6, 206 (1967) 7, 177 (1968).

6 W. Kohn and L. Sham, Phys. Rev. 137, A1697 (1965), ibid 140, A1133 (1965).

7 N. Ashcroft and J. W. Wilkins, Phys. Letters 14, 285 (1965).

8 W. L. McMillan and J. M. Rowell, chapter in Superconductivity (editor R. D. Parks), Academic Press, New York (to be published).

9 G. Grimvall, Phys. kondens. Materie 6, 15 (1967); J. Phys. Chem. Solids 29, 1221 (1968), Phys. kondens. Materie (to be published).

10 S. Engelsberg and J. R. Schrieffer, Phys. Rev. 131, 993 (1963). See also J. R. Schrieffer, Theory of Superconductivity, Benjamin, New York (1964).

11 L. Hedin, B. I. Lundqvist and S. Lundqvist, Solid State Commun. 5, 237 (1967).

12 L. Hedin, Solid State Commun. 5, 451 (1967), Soft X-ray Band Spectra and the Electronic Structure of Metals and Materials, Academic Press, New York (to be published).

13 J. R. Cuthill, R. C. Dobbyn, A. J. McAlister and M. L. Williams (to be published).

LONGITUDINAL DIELECTRIC PHENOMENA

OF THE QUANTUM PLASMA

Norman J. Horing

Department of Physics, Stevens Institute of Technology

Hoboken, New Jersey 07030 USA

I. Introduction

This didactic lecture is intended to provide an elementary introduction to some of the physical phenomena which occur in the solid state quantum plasma. In this spirit we avoid explicit discussion of the preferred tools of theoretical research in this field, namely thermodynamic Green's functions [1], since an exposition of these elegant and powerful tools is not necessary for our present limited purpose. In fact we shall discuss many of the phenomena of the quantum plasma from a classical point of view. However, before embarking on this discussion, it is appropriate to remark briefly on an illuminating quantum mechanical formulation [2] of the problem. To start with one has the many electron Hamiltonian,

$$H = \sum_i \frac{P_i^2}{2m} + \frac{1}{2} \sum_{i \neq j} \frac{e^2}{|\vec{r}_i - \vec{r}_j|} = H^0_{el.} + H^{coul.}_{e-e} \quad . \tag{1}$$

The first term represents free electrons, and the second term represents the long-range Coulomb interaction among the electrons. The long-range of this interaction does in fact allow the electrons to behave in a collective manner, and to be specific, self consistent collective density oscillations can exist in the many electron plasma. The significance of these plasma oscillations (plasmons) as elementary excitations of the system has been brought into clear focus by the construction of a canonical transformation which puts H into an approximately equivalent Hamiltonian of the form [2],

$$H \rightarrow H^0_{el.} + H^0_{pl.} + H^{shield}_{e-e} + H_{el-pl} \, . \tag{2}$$

As above, H^o_{el} represents free electrons. Similarly H^o_{pl} represents free plasmons, the self consistent collective density oscillations which arise in consequence of the long range character of the Coulomb electron-electron interaction. The microscopic dynamical mechanism of the plasmon will be discussed further in the next section, and it will be seen that the plasma density oscillation (as well as its accompanying electric field oscillation which may be understood as a longitudinal photon) occurs with a natural frequency given by $\omega_p = (4\pi e^2\rho_0/m)^{1/2}$. Associated quantum plasmon energies involve multiplication by $\hbar$, so that H^o_{pl} involves a typical quantum mechanical zero point plasmon oscillator energy of $\hbar\omega_p/2$ summed over possible plasmon states contributing to the ground state of the many electron system. In addition, one has a residual short range shielded electron-electron interaction term $H^{shield}_{e\text{-}e}$. Finally, the fact that plasmons are not exact normal modes of the system (that is to say that they are damped and have finite lifetime as elementary excitations) is reflected in the presence of an electron-plasmon interaction term $H_{el.-pl.}$, which provides a mechanism for energy exchange between plasmons and electrons. Under equilibrium circumstances the plasmon yields energy to the electrons, and for sufficiently high wavenumber the plasmon is so heavily damped as to be meaningless as an elementary excitation; (this delimits possible plasmon states contributing to $H^o_{pl.}$ ground state in the discussion above). However, under appropriate drifting conditions the directionality of energy flow via this mechanism can be reversed, resulting in electrons yielding energy to the plasmons, and concomitant microinstability of the plasmons.

II. Microscopic Dynamics

Perhaps the first remark that should be made concerning the quantum plasma is that it is a *plasma,* and as such it is amenable to description in essentially the same terms of self-consistent microscopic dynamics as is the classical plasma. It is in the framework of this description that we shall discuss the dynamical mechanisms responsible for the phenomena briefly described in the Introduction.

The classical description of plasma dynamics proceeds [3] from a collisionless Boltzmann equation for the distribution function $f(\vec{r},\vec{v},t)$ which is linearized in the perturbing electric field, ($f \cong f_o + f_1$; $f_o \sim$ initial dist.; $f_1 \sim V \sim E$),

$$\frac{\partial f_1(\vec{r},\vec{v},t)}{\partial t} + \vec{v}\cdot\vec{\nabla}_{\vec{r}} f_1(\vec{r},\vec{v},t) - \frac{e}{m}\,\vec{\nabla}_{\vec{r}} V\cdot\vec{\nabla}_{\vec{v}} f_0(\vec{v}) = 0. \tag{3}$$

The perturbing electric field on an element of charge must be understood as that due to the Coulomb field of all other elements of charge, so that V must satisfy the Poisson equation, $(\rho_1 \sim \int d\vec{v} f_1)$,

$$\nabla_{\vec{r}}^2 V(\vec{r}, t) = -4\pi e \int d\vec{v} f_1(\vec{r}, \vec{v}, t). \tag{4}$$

The appearance of f_1 on the right hand side reflects the self-consistency of this description of microscopic plasma dynamics, which is known as the Vlasov-Boltzmann description. Solving this system of equations as an initial value problem, (note that whereas we are presently treating the microscopic plasma dynamics classically, we are making limited provision for nonclassical models by allowing that the initial distribution f_0 may include effects due to the Pauli principle, and other quantum effects when ambient magnetic fields modify f_0 through Landau quantization of orbits), one can succinctly re-express the solution for $f_1 \sim V$ in terms of a longitudinal dynamic conductivity σ by calculating the current $\vec{j} \sim \int d\vec{v}\, \vec{v}\, f_1 \sim \sigma \vec{\nabla} V$; or equivlently it can be expressed in terms of a longitudinal inverse dielectric function $\mathcal{K}$,

$$\mathcal{K}(\vec{p},\omega) = \epsilon^{-1}(\vec{p},\omega) = [1 + 4\pi i\, \sigma(\vec{p},\omega)/\omega]^{-1}. \tag{5}$$

Thus one finally obtains the description of the self consistent microscopic plasma dynamics in terms of a frequency (ω) and wave-number ($\vec{p}$) dependent inverse dielectric function given by

$$\mathcal{K}(\vec{p},\omega) = \left[1 + \frac{4\pi e^2}{mp^2} \int d\vec{v} \frac{\vec{p}\cdot\vec{\nabla}_{\vec{v}} f_0(\vec{v})}{\omega - \vec{p}\cdot\vec{v}}\right]^{-1}. \tag{6}$$

A small positive imaginary part must be associated with $\omega \to \omega + i\varepsilon$ to define the integral in the denominator in accordance with the proper prescription for solution of the initial value problem [4].

Now, the condition that the plasma sustain a longitudinal electric/density oscillation is that the inverse dielectric function shall have a pole (i.e. a zero of the dielectric function). Then the denominator of the r.h.s. of (6) must vanish, and one obtains the plasmon dispersion relation as,

$$1 = \frac{4\pi e^2}{mp^2} \int d\vec{v} \frac{\vec{p}\cdot\vec{\nabla}_{\vec{v}} f_0(\vec{v})}{\vec{p}\cdot\vec{v} - \omega}. \tag{7}$$

Considering a low wavenumber expansion of the r.h.s. of (7), one has

$$1 = \frac{4\pi e^2 \rho_0}{m} \frac{1}{\omega^2}\ ; \quad \omega = \omega_p = \sqrt{\frac{4\pi e^2 \rho_0}{m}}. \tag{8}$$

This result expressed in terms of the density $\rho_0 \sim \int d\vec{v} f_0$ is valid in both the nondegenerate and degenerate statistical regimes of f_0. Moreover, this result indicates a self sustaining normal mode type oscillation of longitudinal electric field/density which has a purely oscillatory time dependence $\sim e^{i\omega_p t}$. Actually this plasmon root of the dispersion relation is in fact not purely real. This is due to the prescription putting $\omega \to \omega + i\epsilon$ for solution of the initial value problem. Consequently the $\vec{v}$-integral above carries an imaginary part with it according to the formula

$$\frac{1}{\omega - \vec{p}\cdot\vec{v} + i\epsilon} = \frac{\text{Principal Part}}{\omega - \vec{p}\cdot\vec{v}} - i\pi\delta(\omega - \vec{p}\cdot\vec{v}). \tag{9}$$

This imaginary addition to the dispersion relation results in a complex root [4], whose real part is again ω_p for low wavenumber. However there is a nontrivial imaginary part γ which represents a temporal decay $\sim e^{i\omega_p t}\, e^{-\gamma t}$, so that the plasmon has finite lifetime as an elementary excitation. This reflects the fact that the plasmon is not an exact normal mode of the system, due to electron-plasmon interaction. In the case of equilibrium under consideration this "Landau damping" of the plasmon indicates loss of energy by the plasmon to electrons. The solution of the dispersion relation for the complex root with small imaginary damping constant γ yields the result [5]

$$\gamma = -\frac{\pi\omega_p^3}{2p^2}\, \overline{f_o'}\left(\frac{\omega_p}{p}\right) \tag{10}$$

where $\overline{f_o'}(v_{\parallel}) = \frac{1}{\rho_o}\frac{d}{dv_{\parallel}}\int d^2 v_{\perp} f_o(\vec{v})$, (and $v_{\parallel}$ and $v_{\perp}$ refer to velocity components parallel to $\vec{p}$ and perpendicular to $\vec{p}$ respectively). We shall not discuss further explicit evaluation of γ except to say that it is small so long as $p < p_D = \sqrt{m\omega_p^2\left(\frac{1}{KT} \text{ or } \frac{1}{E_F}\right)}$ and hence only plasmon states with corresponding low wavenumbers are meaningful as elementary excitations, others are heavily damped.

As indicated in the Introduction, the directionality of energy flow via the electron-plasmon interaction mechanism can be reversed by appropriate drifting of the carriers, as well as other appropriate distortions of f_o. This would manifest itself in a reversal of the sign of γ, corresponding to microinstability [5] of the plasmon, in which the plasmon gains energy at the expense of the electrons.

It should be pointed out that the general plasma wave-electron energy exchange mechanism incorporates a mechanism for plasma "memory". Recent classical plasma wave "echo" work [6] has shown that if one excites a first plasma wave and waits for it to damp out, and then excites a second plasma wave and waits for it to damp out also, then a plasma wave "echo" appears later after both have damped out. Such a "memory" phenomenon is characteristic of the time reversible microscopic dynamics of the plasma wave-electron energy exchange mechanism, and it clearly distinguishes the treatment of energy exchange in internal plasma dynamics from simple "collision time approximation" descriptions of energy exchange (such as impurity damping) which are devoid of such memory phenomena. (In this connection it should be noted that the overall effect of collisions is to destroy the "echo"). Thus far our discussion has treated the dynamical mechanisms responsible for H_{pl}^{0} and H_{el-pl} as they are embodied in the inverse dielectric function. The remaining term,

H_{e-e}^{shield}, can also be discussed in these terms since the "long time" static shielding which the plasma provides is given by the zero frequency limit of the inverse dielectric function,

$$\mathcal{K}(\vec{p},\omega=0)=\left[1-\frac{4\pi e^2}{mp^2}\int d\vec{v}\,\frac{\vec{p}\cdot\vec{\nabla}_{\vec{v}}f_o(\vec{v})}{\vec{p}\cdot\vec{v}}\right]^{-1}=\left[1+\frac{4\pi e^2(\partial\rho_o/\partial\zeta)}{p^2}\right]^{-1}=\left[1+\frac{p_D^2}{p^2}\right]^{-1} \quad (11)$$

(where we have recognized the Debye wavenumber as $p_D^2=4\pi e^2(\partial\rho_o/\partial\zeta)$, with ζ = chemical potential). Note that $\mathcal{K}(\vec{p},\omega=0)$ has a simple pole at $p=\pm ip_D$. The shielded electron-electron interaction potential arising from this result is well known as a spatially decaying exponential,

$$V\sim\frac{1}{r}e^{-p_D r} \quad (12)$$

This is the Debye-Thomas-Fermi static shielding law.

III. Magnetic Field Effects [8]

In accordance with our discussion of the self-consistent microscopic dynamics of the plasma on a classical basis, but making limited provision for nonclassical models by allowing that the initial distribution f_0 may include quantum effects (for example the Pauli principle would determine f_0 as a Fermi-Dirac distribution function for a semiclassical model, and inclusion of Landau quantization effects of a magnetic field in the initial distribution would give rise to a "quasiclassical" model), one can readily incorporate the effect of an ambient magnetic field into the basic equation (3) by adding a Lorentz force term $\sim\frac{e}{mc}\vec{v}\times\vec{H}\cdot\vec{\nabla}_{\vec{v}}f_o(\vec{v})$ on the left hand side. Following the same procedure as indicated above, the frequency and wavenumber dependent inverse dielectric function $\mathcal{K}(\vec{p},\omega)$ is obtained as

$$[\mathcal{K}(\vec{p},\omega)]^{-1}=1-\frac{4\pi e^2}{p^2}\int_0^\infty d\omega' f_o(\omega')\int_{c-i\infty}^{c+i\infty}\frac{ds}{2\pi i}e^{s\omega'}\left(\frac{2\pi}{ms}\right)^{3/2}s\,\cdot$$

$$\cdot\left[-1+i\omega\int_0^\infty dT e^{-i\omega T}\exp\left(-\frac{p_z^2T^2}{2ms}\right)\exp\left(\frac{\bar{p}^2}{m\omega_c^2 s}(\cos\omega_c T-1)\right)\right], \quad (13)$$

with the distribution function f_0 normalized as

$$\rho_o=\int d\vec{v}\,f_o\left(\frac{mv^2}{2}\right)=\frac{4\pi\sqrt{2}}{m^{3/2}}\int_0^\infty d\omega'\sqrt{\omega'}f_o(\omega').$$

It should be noted that the s-integral is an inverse Laplace transform with appropriate contour of integration, the magnetic field $\vec{H}$ is taken along the z-direction, giving rise to anisotropic behaviour of p_z and $\bar{p}=(p_x,p_y)$, and that the cyclotron frequency $\omega_c=\frac{eH}{mc}$ is the parameter describing magnetic field effects. If f_0

is taken to be a Maxwellian distribution, then the result (13) reduces to the purely classical result of I. B. Bernstein [7]. On the other hand, if one retains arbitrary f_0 and lets the magnetic field vanish ($\omega_c \to 0$), then one obtains equation (6) above.

The plasmon dispersion relation is again determined by the condition $[K(\vec{p},\omega)]^{-1} = 0$; that is

$$1 = \frac{4\pi e^2}{p^2} \int_0^\infty d\omega' f_0(\omega') \int_{c-i\infty}^{c+i\infty} \frac{ds}{2\pi i} e^{s\omega'} \left(\frac{2\pi}{ms}\right)^{3/2} s \cdot$$

$$\cdot \left[-1 + i\omega \int_0^\infty dT e^{-i\omega T} \exp\left(\frac{-p_z^2 T^2}{2ms}\right) \exp\left(\frac{\bar{p}^2}{m\omega_c^2 s}(\cos\omega_c T - 1)\right) \right] . \quad (14)$$

Considering a low wavenumber expansion of the r.h.s. of (14), one has (ϕ = angle between $\vec{p}$ and $\vec{H}$),

$$1 = \frac{\omega_p^2 \cos^2\phi}{\omega^2} + \frac{\omega_p^2 \sin^2\phi}{\omega^2 - \omega_c^2} . \quad (15)$$

This has two roots, indicating two principal plasmon modes (as opposed to the zero field limit in which there is just one),

$$\omega_\pm^2 = \frac{1}{2}(\omega_p^2 + \omega_c^2) \pm \frac{1}{2}\sqrt{(\omega_p^2 + \omega_c^2)^2 - 4\omega_p^2 \omega_c^2 \cos^2\phi} . \quad (16)$$

Of course if one takes account of higher order terms in the p-wavenumber expansion, small shifts in the positions of these roots will be found. In addition one finds that the higher order wavenumber terms added to the r.h.s. of (15) introduce additional poles at all integral multiples of ω_c for $\vec{p} \perp \vec{H}$; and these poles result in additional roots of dispersion relation which can be understood as plasmon resonances near each value $n\omega_c$ (so long as $\vec{p}$ is perpendicular to $\vec{H}$, or nearly perpendicular to $\vec{H}$).

The prescription putting $\omega \to \omega + i\epsilon$ for solution of the initial value problem again results in plasmon damping of the principal modes under equilibrium conditions (which is small in proportion to $e^{-\zeta/KT}$ when f_0 is a Fermi-Dirac distribution). However it is interesting to note that there is exactly *zero* damping of the plasmon resonance for $\vec{p} \perp \vec{H}$, even at finite temperatures. This is a direct consequence of the fact that the magnetic field forces the electrons into circular orbits, so that an electron interacting with the potential trough of a plasma oscillation and exchanging energy at a rate $-e\vec{\nabla}V\cdot\vec{v}$ finds that the energy gained from the plasmon during half of its circular orbit is lost during the other half of its circular orbit since $\vec{v}$ reverses sign during the orbit. This being the case, one may ask how nonzero damping re-emerges in the zero field limit. Physically, the answer is that the orbits straighten out with infinite (large) radius of curvature. Mathematically, one finds that as $\omega_c \to 0$, the many plasmon resonances near each value $n\omega_c$ are being packed very closely together, and phase averaging so that their meaning and identity as individual plasmon resonances is destroyed. Considering a consistent phase averaging of damping one finds that the single plasmon mode at ω_p (which survives the phase averaging of the resonances) does in fact have nonzero damping. Our earlier remarks concerning micro-instability of plasmons for appropriate nonequilibrium distribution functions f_0 are applicable in the case of nontrivial magnetic field also.

The static shielding law corresponding to equation (13) is *devoid* of magnetic field corrections. This is readily seen from the fact that the term involving ω_c in (13) vanishes for $\omega \to 0$. Physically, the reason that the magnetic field has no effect on static shielding is that it cannot do any work within the framework of a classical description of internal plasma dynamics. Therefore the magnetic field cannot provide any of the energy which would be required for a rearrangement of the static shielding charges, and so it cannot affect the shielding law and the zero field result (equation 12) is still valid. However, it should be noted that when the quantized nature of the electron energy spectrum is felt, and the magnetic field can modify the electron energy spectrum through Landau quantization, then magnetic field corrections will occur in the static shielding law in conjunction with quantum corrections.

IV. Quantum Effects [8]

In the Introduction, we have already indicated one important distinctly quantum mechanical effect arising in the quantum plasma, namely the zero point energy of the plasmon. Moreover, the fact that the plasmon has the character of a longitudinal photon indicates that plasmons obey Bose-Einstein statistics. In addition, there are quantum effects on the inverse dielectric function $K(\bar{p},\omega)$ which in turn affect the plasmon frequencies and their wavenumber dependencies, damping, and the static shielding law as well.

The random phase approximation (RPA) is the quantum generalization of the self-consistent description of microscopic plasma dynamics which we discussed above in terms of a Vlasov-Boltzmann equation. Since the RPA provides a quantized treatment of plasma electron dynamics as well as statistics, it clearly supercedes the limited provision we made for quantum statistics in f_0 above. Now the self-consistency of the RPA is specifically designed to take account of the long range aspects of the electron-electron Coulomb interaction. Moreover, for electrons interacting over very long distances, one may expect that the correlation effects of interaction are controlled by essentially classical dynamics with small quantum corrections. Since long distances correspond to low wavenumbers, this means that the quantized RPA description of the inverse dielectric function $K(\bar{p},\omega)$ embodying correlation effects of interaction must have a low wave wavenumber behaviour which is essentially classical with small quantum corrections. In fact this expectation is realized in the exact RPA inverse dielectric function (in the presence of an arbitrarily strong ambient magnetic field), which is given by $K(\vec{p},\omega)$ [8],

$$[K(\vec{p},\omega)]^{-1} = 1 - \frac{4\pi e^2}{p^2} \int_0^\infty d\omega' \frac{f_0(\omega')}{\hbar^3} \int_{c-i\infty}^{c+i\infty} \frac{ds}{2\pi i} \; e^{s\omega'} \frac{\pi^{3/2}}{(2\pi)^3} \sqrt{\frac{2m}{s}} \frac{mh\omega_c}{\tanh \hbar(\omega_c/2)s} \aleph, \quad (17a)$$

where

$$\aleph = \frac{2i}{\hbar} \int_0^\infty dT e^{-i\omega T} \left\{ \begin{aligned} & \exp\left(\frac{-p_z^2}{8ms}([2T - i\hbar s]^2 + \hbar^2 s^2)\right) \\ & \cdot \exp\left(\frac{\hbar \bar{p}^2}{2m\omega_c} \frac{\cos[(\omega_c/2)(2T - i\hbar s)] - \cosh[\hbar(\omega_c/2)s]}{\sinh[\hbar(\omega_c/2)s]}\right) \\ & - \exp\left(\frac{-p_z^2}{8ms}([2T + i\hbar s]^2 + \hbar^2 s^2)\right) \\ & \cdot \exp\left(\frac{\hbar \bar{p}^2}{2m\omega_c} \frac{\cos[(\omega_c/2)(2T + i\hbar s)] - \cosh[\hbar(\omega_c/2)s]}{\sinh[\hbar(\omega_c/2)s]}\right). \end{aligned} \right. \quad (17b)$$

As indicated above, the low wavenumber limit of this quantum RPA result yields just the plasmon dispersion relation of the Vlasov-Boltzmann equation which describes self consistent internal plasma dynamics classically, $1 = \frac{\omega_p^2 \cos^2\phi}{\omega^2} + \frac{\omega_p^2 \sin^2\phi}{\omega^2 - \omega_c^2}$, (low wavenumber, eq. (15)) and it also yields the same low wavenumber Debye-Thomas-Fermi static shielding law predicted by the Vlasov-Boltzmann equation (equation 12), which we have already seen to be devoid of magnetic field corrections as well as quantum corrections at low wavenumber. Of course quantum corrections occur as higher wavenumbers are taken into consideration. It should be noted that the occurrence of a factor $\hbar^3$ under $f_0(\omega')$ in equation 17a (as well as some other normalization factors) is actually spurious since f_0 is normalized here in accordance with

$$\rho_0 = 2 \int_0^\infty d\omega' \frac{f_0(\omega')}{\hbar^3} \int_{c-i\infty}^{c+i\infty} \frac{ds}{2\pi i} \; e^{s\omega'} \frac{\pi^{3/2}}{(2\pi)^3} \sqrt{\frac{2m}{s}} \frac{m\hbar\omega_c}{\tanh \hbar(\omega_c/2)s}, \quad (18)$$

which also incorporates Landau quantization effects of the magnetic field.

Notwithstanding the essentially classical character of the low wavenumber behaviour of $\mathcal{K}(\vec{p},\omega)$, one may expect quantum corrections to arise in conjunction with consideration of the higher order wavenumber terms embodied in the r.h.s. of equation (17a) for $\mathcal{K}(\vec{p},\omega)$. The higher order wavenumber corrections bearing quantum corrections will affect the plasmon dispersion relation, damping, and static shielding as well. The exact nature of these corrections depends on which statistical regime is under consideration, and what magnetic field strength is involved. Since quantum effects are most prominent and interesting in the degenerate statistical regime, we will limit our discussion to the case of degeneracy, but will briefly consider quantum effects for a variety of field strengths. (Throughout the following discussion it should be borne in mind that low wavenumber phenomena, such as the principal plasmon modes, are essentially devoid of quantum effects in accordance with the discussion above.)

In the case of zero magnetic field, an exact evaluation of the r.h.s. of equation (17a) in the case of degeneracy reveals the presence of log-singularity terms of the form $\ln|\omega \pm \hbar p p_F/m \pm \hbar p^2/2m|$, where p_F is the Fermi wavenumber determined by the Fermi energy $E_F = \zeta$ according to the relation $\zeta = E_F = \hbar^2 p_F^2/2m$. Apart from the small effect that such terms have in contributing quantum corrected wavenumber dependent shifts to the location of the principal plasmon mode at $\omega \sim \omega_p$, one finds that the accompanying imaginary part $(\omega \to \omega + i\epsilon)$ indicates very heavy plasmon damping for wavenumbers higher than those for which the argument of the log-singularity vanishes, $\omega = \hbar p p_F/m + \hbar p^2/2m$. This heavy damping in the case of degeneracy may be understood in terms of plasmon decay into an electron-hole pair for sufficiently high momentum $\hbar p$, and it stands in contrast to the very light plasmon damping for lower wavenumbers which is proportional to $e^{-\zeta\beta} \to 0$ in the case of degeneracy. It should also be noted that the presence of the log-singularity terms in the plasmon dispersion relation also introduces the possibility of a zero sound type resonance root. The log-singularity terms also introduce an important static shielding effect. Putting $\omega \to 0$ one obtains log-singularity terms of the form $\ln|p - 2p_F|$, whose effect in the long distance static shielding law is to produce a term of the form $V(r) \sim (d^2/r^3)\cos 2p_F r$. This is the well known Friedel-Kohn "wiggle", and it is of particular importance because of its long range character (in that it does not suffer a spatial exponential decay, as does the Debye-Thomas-Fermi static shielding law). Thus, interesting and important quantum effects emerge at higher wavenumber $p \sim 2p_F$, (despite the relative unimportance of quantum effects at low wavenumber).

In the case of a magnetic field of intermediate strength ($\zeta \sim E_F > \hbar\omega_c > KT$), one can expect de Haas - van Alphen (DHVA) oscillatory effects to be prominent among quantum corrections. Physically, DHVA effects arise in conjunction with the Landau quantization of energy levels. Analytically, DHVA oscillatory terms arise from the isolated singularities of the inverse Laplace transform s-integrand of equation (17a) corresponding to the zeros of $\sinh[\hbar(\omega_c/2)s]$. At zero wavenumber, the dispersion relation for the two principal plasmon modes in magnetic field (equation 15) stands unmodified. However, DHVA quantum corrections do indeed accompany the wavenumber dependent shifts of the locations of the two principal plasmon modes (which can be calculated from a wavenumber power expansion of the r.h.s. of equation 17a). In addition, the plasmon resonances which occur near each value $n\omega_c$ for $\vec{p} \perp \vec{H}$ at higher wavenumbers bear DHVA oscillatory quantum corrections. The static shielding law in the case of intermediate field strength is capable of bearing magnetic field corrections in conjunction with quantum corrections since the magnetic field modifies the electron energy spectrum through Landau quantization. Indeed, the wavenumber power expansion of the r.h.s. of equation (17a) for zero frequency results in magnetic field quantum corrections of the DHVA type which make the Debye-Thomas-Fermi static shielding law spatially anisotropic. It is of even greater interest to consider intermediate magnetic field strength corrections to the high wavenumber ($p \sim 2p_F$) Friedel-Kohn "wiggle", but the analytical complexity defies brief description. We shall be content to point out that the r-dependent shielded potential involves the magnetic field quantum correction parameter $\frac{\hbar\omega_c}{\zeta}$ in the form of a "mixed" parameter $\frac{\hbar\omega_c}{\zeta} p_F r$, where $p_F r \gg 1$ and $\frac{\hbar\omega_c}{\zeta} \ll 1$. When the smallness of $\frac{\hbar\omega_c}{\zeta}$ is dominant, and $\frac{\hbar\omega_c}{\zeta} p_F r \ll 1$, then the Friedel-Kohn "wiggle" retains its long range form. However, at larger distances when $p_F r \gg 1$ dominates, and $\frac{\hbar\omega_c}{\zeta} p_F r \gg 1$, then the long range of the Friedel-Kohn "wiggle" is destroyed. In essense this happens because large "mixed" parameter means that $\frac{\hbar\omega_c}{\zeta}$ is *effectively* large, in which case the static shielded potential $V(\vec{r})$ effectively feels the internal dynamics associated with the quantum strong field limit, in which all electrons are confined to the lowest Landau level. It is characteristic of the quantum strong field limit that the log-singularity responsible for the "wiggle" is replaced by its one-dimensional counterpart $\ln|p_z - 2p_F|$. This anisotropic log-singularity yields a static shielding law which is still oscillatory along the direction of the magnetic field, but which suffers an exponential decay in the transverse direction $\sim\cos(2p_F r_z)\, e^{-2p_F|\overline{r}|}$. Hence the long range of the Friedel-Kohn "wiggle" is destroyed when the "mixed" parameter is large, $\frac{\hbar\omega_c}{\zeta} p_F r \gg 1$. Moreover, there are no DHVA oscillatory corrections for large mixed parameter in view of the fact $V(\vec{r})$ effectively feels the internal dynamics associated with the quantum strong field limit, in which all electrons are confined to the lowest Landau level so that one does not have Landau levels flicking past the Fermi level.

The basic nature of the quantum strong field limit has already been indicated. All electrons are confined to the lowest Landau level, and $\hbar\omega_c > \zeta \sim E_F > KT$. Clearly one may expect maximal magnetic field quantum effects in this situation. However the zero wavenumber principal plasmon mode dispersion relation (equation 15) again stands unmodified, as one should expect from our earlier discussion. It is the consideration of higher wavenumber parts of the r.h.s. of equation (17a) which reveals important magnetic field-quantum effects. An exact evaluation of equation (17a) in the quantum strong field limit yields the result,

$$[K(\bar{p},\omega)]^{-1} = 1 - \frac{m\omega_p^2}{2\hbar p^2}\sqrt{\frac{m}{2p_z^2\zeta}}\exp\left(\frac{-\hbar\bar{p}^2}{2m\omega_c}\right)\sum_{n=0}^{\infty}\frac{1}{n!}\left(\frac{\hbar\bar{p}^2}{2m\omega_c}\right)^n \cdot \left\{\log\frac{\omega - n\omega_c - \hbar p_z^2/2m + \sqrt{2p_z^2\zeta/m}}{\omega - n\omega_c - \hbar p_z^2/2m - \sqrt{2p_z^2\zeta/m}} + (\omega \to -\omega)\right\} . \tag{19}$$

A wavenumber power expansion of the r.h.s. of equation (19) yields wavenumber shifts of the locations of the two principal plasmon modes which bear magnetic field-quantum corrections. It is of even greater interest to maintain the anisotropic log-singularities of equation (19) intact, and investigate their effects on the plasmon spectrum. One then finds that the p_z-dependent log-singularities imply the existence of two plasmon resonances near each value $n\omega_c$, one of which is undamped and the other is mildly damped. These remarks hold for propagation $\vec{p}$ not perpendicular to the magnetic field $\vec{H}$, ($p_z \neq 0$). In the limit of perpendicular propagation ($p_z = 0$) there is just one undamped resonance near each value $n\omega_c$. In regard to the static shielding law, the $\omega \to 0$ limit of the $n = 0$ log-singularity takes the form $\ln|p_z - 2p_F|$. This anisotropic result gives the principal contribution to the quantum strong field counterpart of the Friedel-Kohn "wiggle", and the strong anisotropy destroys the long range character of the "wiggle" as we indicated above.

V. Thermodynamics

The quantity of central concern in regard to statistical thermodynamics is the grand partition function $Z = e^W$ which is defined by

$$Z \equiv e^W \equiv \mathrm{tr}\, e^{-\beta(H-\zeta N)} \tag{20}$$

where $\beta \equiv (KT)^{-1}$ and ζ = chemical potential. The Hamiltonian is composed of a free electron part $H^o_{el.}$ and an electron-electron Coulomb interaction part $H^{coul.}_{e-e}$ whose strength we allow to be variable in proportion to a coupling strength parameter λ;

$$H = H^o_{el.} + \lambda H^{coul.}_{e-e} \quad \begin{pmatrix} \lambda = 1 \text{ for full coupling strength} \\ \lambda = 0 \text{ for zero coupling strength} \end{pmatrix} . \tag{21}$$

Differentiating log Z = W with respect to λ one has

$$\frac{\partial W}{\partial \lambda} = -\beta \frac{\mathrm{tr} H^{coul.}_{e-e} e^{-\beta(H-\zeta N)}}{\mathrm{tr} e^{-\beta(H-\zeta N)}} \equiv -\beta \langle H^{coul.}_{e-e} \rangle \tag{22}$$

where we have noted that the thermal average of a quantity X is just given by

$$\langle X \rangle \equiv \frac{\mathrm{tr} X e^{-\beta(H-\zeta N)}}{\mathrm{tr} e^{-\beta(H-\zeta N)}} . \tag{23}$$

If this relation is now integrated with respect to λ, one obtains

$$W_{\lambda=1} - W_{\lambda=0} = -\beta \int_0^1 \frac{d\lambda}{\lambda} \langle \lambda H^{coul.}_{e-e} \rangle . \tag{24}$$

On the left we have the difference in W between full coupling strength and zero coupling strength, and this is expressed on the right hand side as a coupling strength parameter integral over the thermal average of Coulomb interaction energy with coupling strength parameter λ, $\langle \lambda H^{coul.}_{e-e} \rangle$. One may expect that the inverse dielectric function $\mathcal{K}(\vec{p}, \omega)$ plays a central role in the determination of $\langle \lambda H^{coul.}_{e-e} \rangle$ since it describes the internal dynamics of the plasma both quantum mechanically and classically. This expectation is in fact realized, but it is beyond the scope of our present purpose to present the concomitant derivation. Therefore we shall just write down the part of the final formula which explicitly exhibits the role of $\mathcal{K}(\vec{p}, \omega)$ in the determination of $W_{\lambda=1} - W_{\lambda=0}$;

$$W_{\lambda=1} - W_{\lambda=0} \sim -\frac{\hbar\beta(\text{volume})}{2} \int_0^1 \frac{d\lambda}{\lambda} \int_{-\infty}^{+\infty} \frac{d\vec{p}}{(2\pi)^3} \int_0^\infty \frac{d\omega}{2\pi} \left(\coth \frac{\hbar\omega\beta}{2}\right) a_\lambda(\vec{p}, \omega) \tag{25a}$$

where $a_\lambda(\vec{p}, \omega)$ is defined in terms of $\mathcal{K}_\lambda(\vec{p}, \omega)$ as

$$a_\lambda(\vec{p}, \omega) = \frac{1}{i}\left(\mathcal{K}_\lambda(\vec{p}, \omega - i\epsilon) - \mathcal{K}_\lambda(\vec{p}, \omega + i\epsilon)\right) . \tag{25b}$$

The subscript λ indicates that the quantities must be evaluated at coupling strength parameter λ. The factor $\coth \frac{\hbar\omega\beta}{2}$ is associated with a Bose-Einstein distribution function which arises in connection with the fact that the inverse dielectric function acts as a propagator for longitudinal photons which obey Bose-Einstein statistics; moreover it also yields zero point plasmon energy. For example, if one considers $\mathcal{K}(\vec{p}, \omega)$ in the vicinity of the low wavenumber pole at $\omega \to \omega_p$ which describes the plasmon mode in the absence of a magnetic field, it is readily seen that $\mathcal{K}(\vec{p}, \omega)$ has the form

$$\mathcal{K}(\vec{p}, \omega) = \frac{z}{\omega - \omega_p} \quad \text{where} \quad z = \omega_p/2 . \tag{26}$$

Then

$$a(\vec{p}, \omega) \; \frac{1}{i}\left(\frac{z}{\omega-\omega_p-i\epsilon} - \frac{z}{\omega-\omega_p+i\epsilon}\right) \tag{27a}$$

$$= 2\pi z\,\delta(\omega-\omega_p) = \pi\,\omega_p\,\delta(\omega-\omega_p) \;. \tag{27b}$$

Recognizing that $\omega_p \sim \lambda^{1/2}$, we have

$$a_\lambda(\vec{p}, \omega) = \pi\omega_p\lambda^{1/2}\delta(\omega-\omega_p\lambda^{1/2}) \tag{27c}$$

and then the ω-integral above is immediate;

$$W_{\lambda=1} - W_{\lambda=0} \sim -\frac{\hbar\beta\,(\text{volume})}{2}\int_0^1 \frac{d\lambda}{\lambda}\int_{-\infty}^{\infty}\frac{d\vec{p}}{(2\pi)^3}\frac{\omega_p}{2}\lambda^{1/2}\coth(\hbar\omega_p\beta\lambda^{1/2}) \;. \tag{28a}$$

At zero temperature $\hbar\omega_p\beta = \hbar\omega_p/kT \to \infty$, so coth $(\hbar\omega_p\beta\lambda^{1/2}) \to 1$, and the right hand side becomes

$$\sim -\hbar\beta\frac{(\text{volume})}{4}\,\omega_p\int_0^1\frac{d\lambda}{\lambda}\int_{-\infty}^{+\infty}\frac{d\vec{p}}{(2\pi)^3}\lambda^{1/2} \;. \tag{28b}$$

It must be borne in mind that the simple form $K(\vec{p}, \omega) = \frac{z}{\omega-\omega_p}$ is valid only at low wavenumbers so long as the plasmon is lightly damped. At higher wavenumbers the plasmon damping is heavy so that this simple form is invalid and the plasmon is meaningless as an elementary excitation. Therefore, the contributing plasmon states are delimited by a wavenumber cutoff on the p-integral which ensures convergence. Carrying out the integrations, and calculating plasmon contribution to correlation energy by means of the formula $E = -\partial W/\partial\beta$, one finds a result which corresponds to associating a zero point oscillator energy of $\frac{1}{2}\hbar\omega_p$ with each contributing plasmon state. (It should be noted that in the classical limit coth $\frac{\hbar\omega_p\beta}{2} \to \frac{2}{\hbar\omega_p\beta}$, and then one would find a result which corresponds to associating an equipartition energy of ½KT with each contributing plasmon degree of freedom, instead of zero point energy.)

The above example is clearly limited in its restricted consideration of the frequency-wavenumber region corresponding to plasmon behaviour of $K(\vec{p}, \omega)$. Actually the integrals defining the correlated part of W are extended over all frequency and wavenumber so that other regions of behaviour of $K(\vec{p}, \omega)$, such as damping and static shielding, will contribute to the overall result. Noting that the free particle part of W (W_0) has already been subtracted off, one might say that consideration of these regions corresponds to a reconstruction (in terms of statistical thermodynamics) of the canonical transformation described in the introduction

$$H \to H^{\circ}_{el.} + H^{\circ}_{pl.} + H^{shield}_{e-e} + H_{el.-pl.}$$

in which the physical phenomena associated with these terms of H are seen to arise from the various regions of behaviour of a single inverse dielectric function $K(\vec{p}, \omega)$ describing the microscopic internal dynamics of the quantum plasma in a completely unified manner.

Thus $\mathcal{K}(\vec{p}, \omega)$ serves as an important analytical link relating the microscopic dynamical description of the quantum plasma with the canonical transformation description, succinctly embodying the important features of both descriptions.

Footnotes & References

(1) The interested reader may find discussion of thermodynamic Green's functions in the following references:
  (a) P. C. Martin and J. Schwinger, Phys. Rev. 115, 1342 (1959)
  (b) L. P. Kadanoff and G. Baym, "Quantum Statistical Mechanics", W. A. Benjamin Inc., (1962).

(2) See D. Pines, "Elementary Excitations in Solids", W. A. Benjamin Inc. (1964), and references cited in Pines book.

(3) A. I. Akhiezer, "Collective Oscillations in a Plasma", Pergamon Press (1967). (Also see ref. 3 pg. 180).

(4) L. D. Landau, "Collected Papers", Pergamon Press (1965), page 445.

(5) J. D. Jackson, J. Nucl. Energy Part C 1 171 (1960).

(6) (a) R. W. Gould, T. M. O'Neil, and J. H. Malmberg, Phys. Rev. Letters 19 (1967) 219
  (b) R. W. Gould, Physics Letters 25A (1967) 559
  (c) T. M. O'Neil and R. W. Gould, Phys. Fluids 11 (1968) 134.

(7) I. B. Bernstein, Phys. Rev. 109, 10 (1958).

(8) Detailed discussion of magnetic field effects arising in connection with both classical and quantum descriptions of internal plasma dynamics can be found in:
  (a) N. J. Horing, Proc. Plasma Effects Symposium, Int. Conf. on Physics of Semiconductors, Paris, July 1964.
  (b) N. J. Horing, Annals of Physics 31, 1 (1965).
  (c) N. J. Horing, Phys. Rev. 136-2A, A494 (1964).
  (d) N. J. Horing, Proc. Int. Conf. on Physics of Semiconductors, Kyoto 1966, J. Phys. Soc. Japan, Vol. 21 Supp. (1966).

QUASICLASSICAL MODEL OF IMPURITY SHIELDING BY QUANTUM PLASMA IN MAGNETIC FIELD

Norman J. Horing

Department of Physics, Stevens Institute of Technology
Hoboken, New Jersey, 07030, U.S.A.

ABSTRACT

A "quasiclassical model" of the quantum plasma in a magnetic field is discussed here. This model is predicated on the assumption that the internal dynamics of the quantum plasma are essentially classical: However, the distribution function which weights statistical averages is taken to be the appropriate quantum distribution function which takes account of both the Pauli principle and Landau quantization of energy levels in a magnetic field.

The static shielding law of the quasiclassical model is evaluated for low-intermediate magnetic field strengths. The recent work of A.K. Das and E. de Alba is thereby generalized to nonzero temperature and intermediate field strength to take account of de Haas-van Alphen (DHVA) oscillatory phenomena. This spatially isotropic static shielding law is seen to be of the Debye-Thomas-Fermi (DTF) type. The physical information embodied in the static shielding law of the quasiclassical model is evaluated, limitations are noted, and possibilities for improvement are discussed.

Finally, the plasmon dispersion relation of the quasiclassical model with ambient magnetic field is given here.

I. Introduction

The R.P.A. (random phase approximation) dielectric function description of the quantum plasma provides a very convenient formulation for studying the internal dynamics of the plasma, which are comprised of two fundamentally important physical phenomena: plasma oscillations, and static shielding. Moreover, the dielectric function description also provides a fruitful point of departure for studying the thermodynamics (with the use of a coupling constant integral). The importance of the fundamental physical phenomena of plasma oscillations and static shielding in the determination of the thermodynamics of the quantum plasma is apparent in the fact that the latter involves contributions from the frequency wavenumber dependent inverse dielectric function in all regimes of both frequency and wavenumber. Calculations of thermodynamic quantities such as correlation energy, have in fact been technically complex. The inclusion of an ambient magnetic field has resulted in additional complexity, and very considerable difficulty in handling the analysis, even within the scope of the R.P.A.

In view of these considerations, it is desirable to review the physical content of the R.P.A. description of the quantum plasma in a magnetic field with the intention of abstracting from it an even simpler description bearing its important physical features-in the hope that the simpler description will be useful in overcoming the technical difficulties of carrying out the analysis of magnetic field corrections to the thermodynamics of the quantum plasma.

We have in fact investigated simpler descriptions such as the semiclassical model, the nondegenerate limit, the classical limit, and the quantum strong field limit, in earlier works.[1,2] The recent work of A.K. Das and E. de Alba[3] has called to our attention another relatively simple description, which we shall call the "quasiclassical model", in contra-distinction to the semiclassical model (details will be supplied below). We will report here on the physical features of this model, at least in so far as static shielding is concerned, and in so doing, we shall generalize the work of Das and de Alba to nonzero temperature, and intermediate magnetic field strength to take account of de Haas-van Alphen (DHVA) oscillatory phenomena.

The quasiclassical model presently under consideration is predicated on the assumption that the internal dynamics of the quantum plasma may be described in purely classical terms, so that a classical collisionless linearized Vlasov-Boltzmann equation is used. However, the initial distribution function (which weights the statistical averages over the classically described internal dynamical processes) is taken to be the appropriate quantum distribution function which takes account of both the Pauli principle, and Landau quantization of energy levels in magnetic field. It is in this last respect that the quasiclassical model is superior to the semiclassical model (which does take account of the Pauli principle, but neglects Landau quantization in the initial distribution function). Thus the quasiclassical and semiclassical models differ in regard to their prescriptions for quantum statistical averaging. Apart from this, they are identical in that they employ the same classical description of the internal plasma dynamics, namely the Vlasov-Boltzmann equation. In this work we shall investigate the ramifications of the quasiclassical model in regard to static shielding, thereby generalizing the results of Das and de Alba as indicated above. We shall also evaluate the physical information embodied in the model, noting its limitations and possibilities for improvement. Furthermore, the plasmon dispersion relation appropriate to the quasiclassical model will be written down here, but the analysis of the plasmon modes and resonances will not be carried out.

II. Shielding Law of the Quasiclassical Model

The explicit use of the Vlasov-Boltzmann equation in connection with the magnetic field dependence of the static shielding law of the quasiclassical model may be found in the work of Das and de Alba. In this work, we will view it as an appropriately chosen limit of the full R.P.A. result, which has already been found to be given by[1],

$$V(\vec{r},\infty) = \int \frac{d\vec{p}}{(2\pi)^3}\, e^{i\vec{p}\cdot\vec{r}}\, \frac{4\pi e/p^2}{1-(4\pi e^2/p^2)\, \mathcal{I}m\, I(\vec{p},\omega=0+i\epsilon)} , \tag{1}$$

where

$$\begin{aligned} &1-(4\pi e^2/p^2)\, \mathcal{I}m\, I(\vec{p},\omega=0+i\epsilon) \\ &= 1-(4\pi e^2/p^2)\int_0^\infty d\omega\, \frac{f_0(\omega)}{\hbar^3}\int_{-i\infty+\delta}^{i\infty+\delta} \frac{ds}{2\pi i}\, e^{s\omega}\, \frac{\pi^{3/2}}{(2\pi)^3}\sqrt{\frac{2m}{s}}\, \frac{m\hbar\omega_c}{\tanh(\hbar\omega_c s/2)}\, K_0 , \end{aligned} \tag{2}$$

and

$$K_0 = -s\int_{-1}^{1} dT\, e^{(\hbar^2 p_z^2/8m)s(T^2-1)} \exp\left\{\frac{\hbar \bar{p}^2}{2m\omega_c}\, \frac{\cosh(\hbar\omega_c sT/2)-\cosh(\hbar\omega_c s/2)}{\sinh(\hbar\omega_c s/2)}\right\}. \tag{3}$$

Now, the full R.P.A. result above can easily be reduced to the quasiclassical model by replacing $\mathcal{K}_0$ by its limit as $\hbar \to 0$. This results in the removal of quantum dynamical effects. Apart from this, quantum statistical effects arising from the Pauli principle and Landau quantization are left intact in equation (2), and then the resulting static shielding law may be calculated from equation (1). Thus, the static shielding law of the quasiclassical model calls for the replacement,

$$\mathcal{K}_0 \to \lim_{\hbar \to 0} \mathcal{K}_0 = -s \int_{-1}^{1} d\tau\, e^0 = -2s\,. \tag{4}$$

Along with this, it should be remarked that the appropriate expression for density is given by[2], (ref. 2, equa. III 5a),

$$\rho = 2 \int_0^\infty d\omega\, \frac{f_0(\omega)}{\hbar^3} \int_{-i\infty+\delta}^{i\infty+\delta} \frac{ds}{2\pi i}\, e^{\omega s} \frac{\pi^{3/2}}{(2\pi)^3} \sqrt{\frac{2m}{s}}\, \frac{m\hbar\omega_c}{\tanh(\hbar\omega_c s/2)}\,. \tag{5}$$

Noting that

$$\frac{\partial \rho}{\partial \xi} = 2 \int_0^\infty d\omega\, \frac{f_0(\omega)}{\hbar^3} \int_{-i\infty+\delta}^{i\infty+\delta} \frac{ds}{2\pi i}\, e^{\omega s} \frac{\pi^{3/2}}{(2\pi)^3} \sqrt{\frac{2m}{s}}\, \frac{m\hbar\omega_c s}{\tanh(\hbar\omega_c s/2)}\,, \tag{6}$$

it is then readily verified that for the quasiclassical model, one obtains

$$1 - (4\pi e^2/p^2)\, \mathrm{Im}\, I(\vec{p}, \omega = 0 + i\epsilon) = 1 + (4\pi e^2/p^2)\, \partial\rho/\partial\xi\,. \tag{7}$$

The static shielding law of the quasiclassical model is therefore given by (magnetic field effects included),

$$V(\vec{r}, \infty) = \int \frac{d\vec{p}}{(2\pi)^3}\, e^{i\vec{p}\cdot\vec{r}}\, \frac{4\pi e}{p^2 + 4\pi e^2\, \partial\rho/\partial\xi}\,. \tag{8}$$

This shielding integral is the familiar Debye-Thomas-Fermi (DTF) type shielding integral, and it is readily evaluated as,

$$V(\vec{r}, \infty) = \frac{e}{r} \exp\left\{-r\sqrt{4\pi e^2 \cdot \partial\rho/\partial\xi}\right\}\,. \tag{9}$$

Clearly the result is spatially isotropic, and the associated DTF shielding length is given by r_D ,

$$r_D = \left[4\pi e^2 \cdot \partial\rho/\partial\xi\right]^{-1/2} . \tag{10}$$

The familiarity of these results is a reflection of the fact that the simplified classical description of internal dynamics in the quasiclassical model gives rise to essentially classical results for shielding. However, there is in fact an improvement in that $\partial\rho/\partial\xi$ (as given above) embodies the statistical effects of Landau quantization as well as the Pauli principle. Therefore it will yield monotonic as well as DHVA - oscillatory magnetic field corrections, and temperature dependence.

The evaluation of $\partial\rho/\partial\xi$ will be carried out here for low-intermediate magnetic field strength, $\hbar\omega_c << \xi \sim E_F, \hbar\omega_c \sim 1/\beta = KT$. In this case there are two classes of magnetic field corrections:

(a) Monotonic magnetic field corrections which arise from the branch cut of the s-integrand of equation (6) for $\partial\rho/\partial\xi$.

(b) DHVA oscillatory magnetic field corrections which arise from the isolated poles of the s-integrand of equation (6) for $\partial\rho/\partial\xi$.

These may be evaluated directly by contour integration. Alternatively, one may use similar results obtained by contour integration for ρ (equation 5), and then perform the ξ-differentiation. In fact the contour integration for ρ has been carried out (ref. 2, appendix I), and we have both monotonic and DHVA parts,

$$\rho = \rho_{Mon.} + \rho_{DHVA} . \tag{11}$$

Here,

$$\rho_{DHVA} = \frac{m^{3/2}(\hbar\omega_c)^{1/2}}{\pi\beta\hbar^3} \sum_{n=1}^{\infty} \frac{\cos\left[(2\pi n/\hbar\omega_c)\xi - 3\pi/4\right]}{n^{1/2}\sinh(2\pi^2 n/\hbar\omega_c\beta)} , \tag{12}$$

and therefore we obtain the temperature dependent result

$$\frac{\partial\rho_{DHVA}}{\partial\xi} = -\frac{2m^{3/2}}{\beta(\hbar\omega_c)^{1/2}\hbar^3} \sum_{n=1}^{\infty} \frac{n^{1/2}\sin\left[(2\pi n/\hbar\omega_c)\xi - 3\pi/4\right]}{\sinh(2\pi^2 n/\hbar\omega_c\beta)} . \tag{13}$$

$\rho_{Mon.}$ will be calculated directly, since we need a little more detail about it than is supplied in the indicated reference. Noting that $\rho_{Mon.}$ is given by the branch cut contribution (subscript Γ)

of the contour integral for ρ (equation 5), and retaining the first two terms of the Laurent expansion for $[\tanh(\hbar\omega_c s/2)]^{-1}$ we find,

$$\rho_{Mon.} = 2\int_0^\infty d\omega \frac{f_0(\omega)}{\hbar^3}\int_\Gamma \frac{ds}{2\pi i} e^{\omega s}\frac{\pi^{3/2}}{(2\pi)^3}\left(\frac{2m}{s}\right)^{3/2}\left(1+\frac{(\hbar\omega_c)^2 s^2}{12}\right)$$

$$= \left(1+\frac{(\hbar\omega_c)^2}{12}\frac{\partial^2}{\partial\zeta^2}\right)2\int_0^\infty d\omega\frac{f_0(\omega)}{\hbar^3}\int_\Gamma\frac{ds}{2\pi i}e^{\omega s}\frac{\pi^{3/2}}{(2\pi)^3}\left(\frac{2m}{s}\right)^{3/2}. \quad (14)$$

Recognizing that the integral expression on the rhs is just the density at zero field, ρ^0,

$$\rho^0 = 2\int_0^\infty d\omega\frac{f_0(\omega)}{\hbar^3}\int_\Gamma\frac{ds}{2\pi i}e^{\omega s}\frac{\pi^{3/2}}{(2\pi)^3}\left(\frac{2m}{s}\right)^{3/2} = \left(\frac{m}{2\pi}\right)^{3/2}\frac{2}{\Gamma(5/2)}\frac{\zeta^{3/2}}{\hbar^3}, \quad (15)$$

one finds that,

$$\rho_{Mon.} = \left(1+\frac{(\hbar\omega_c)^2}{12}\frac{\partial^2}{\partial\zeta^2}\right)\rho^0 = \left(1+\frac{1}{16}\left[\frac{\hbar\omega_c}{\zeta}\right]^2\right)\rho^0. \quad (16)$$

Taking account of the ζ-dependence of ρ^0, one then obtains the result,

$$\frac{\partial\rho_{Mon.}}{\partial\zeta} = \frac{3}{2}\frac{\rho^0}{\zeta}\left(1-\frac{1}{48}\left[\frac{\hbar\omega_c}{\zeta}\right]^2\right). \quad (17)$$

The evaluation of $\rho_{Mon.}$ has been carried out at zero temperature, since it is clear that any temperature dependence of the branch cut contribution $\rho_{Mon.}$ enters through the tail of the Fermi function and hence must be dominated by the very small factor $e^{-\zeta\beta}$, and it may therefore be neglected.

III. Results, and Evaluation of the Quasiclassical Model

In summary, $\partial\rho/\partial\zeta$ has a DHVA part given by equation (13), and a monotonic part given by equation (17). Translated into a result for $r_D = [4\pi e^2 \cdot \partial\rho/\partial\zeta]^{-1/2} \sim K_D^{-1}$, equation (17) essentially reproduces the result of Das and de Alba (except for a factor of 2 in the magnetic field term, which seems to be due to the inclusion of spin in this analysis). Moreover, the new terms represented by equation (13) yield the temperature dependent DHVA oscillatory terms which should be accounted for in the low-intermediate magnetic field strength regime.

The overall spatial isotropy of the shielding law of the quasiclassical model indicates that it is somewhat less accurate than the shielding results reported in reference (1). Nevertheless, it seems to provide a reasonably good and sensible description of the DTF part of the shielding law with magnetic field corrections, and enjoys the advantage of being easy to interpret physically insofar as the self-consistency aspects of plasma shielding are concerned.

However, the quasiclassical model has a serious defect in that it is incapable of describing the long-range Friedel-Kohn "wiggle" part of the shielding law for quantum plasmas, and certainly can not describe the influence of very high magnetic fields (quantum strong field limit) on this part of the shielding law. (This problem is discussed in reference 1). Nevertheless, it may be possible to synthesize the full shielding law, including a description of the long-range Friedel-Kohn "wiggle" by incorporating an ordinary Hartree-Fock type term along with the quasiclassical model, since such a term should give at least a reasonable qualitative description of the behaviour of the Friedel-Kohn "wiggle" in very high magnetic fields.

Although our attention in this work has been confined to an investigation of the ramifications of the quasiclassical model in regard to static shielding, we shall briefly indicate the appropriate dispersion relation for further investigation of the model in regard to the plasma oscillation spectrum. This dispersion relation is given by,

$$1 = \frac{4\pi e^2}{p^2}\int_0^\infty d\omega \frac{f_0(\omega)}{\hbar^3}\int_{-i\infty+\delta}^{i\infty+\delta}\frac{ds}{2\pi i}\, e^{s\omega}\frac{\pi^{3/2}}{(2\pi)^3}\sqrt{\frac{2m}{s}}\,\frac{m\hbar\omega_c}{\tanh(\hbar\omega_c s/2)}\cdot 2s\,\cdot$$

$$\cdot\left\{-1 + i\Omega\int_0^\infty d\tau\, e^{-i\Omega\tau}\exp\left(\frac{-p_z^2\,\tau^2}{2ms}\right)\exp\left(\frac{\bar{p}^2}{m\omega_c^2 s}(\cos\omega_c\tau - 1)\right)\right\}. \quad (18)$$

Although this result can be derived using a Vlasov-Boltzmann equation, we obtain it from the full R.P.A. plasmon dispersion relation (see reference 2, equation III. 34,35) by putting $\hbar \to 0$ in the dynamical part, while retaining quantum effects (due to Landau quantization and the Pauli principle) in the statistical part. The classical nature of the dynamical part is obvious from the absence of $\hbar$ in the brace {...} quantity in equation (18). The retention of quantum effects in the statistical averaging may be seen by inspection of the pre-brace factors of the s-integrand of this equation. The use of this dispersion relation to investigate the plasma oscillation spectrum in the quasiclassical model should provide further insight regarding the capacity of the model to describe magnetic field effects on the fundamental physical phenomena of the quantum plasma.

Acknowledgements

We wish to express our appreciation to Dr. A. K. Das for his kindness in making available a prepublication copy of his manuscript.

References and Footnotes

1.-N. J. Horing, Proc. Int. Conf. Physics of Semiconductors, Kyoto, Japan, 1966, J. Phys. Soc. Japan, Vol. 21 Supp., 1966, pp. 704 ff.

2.-N. J. Horing, Annals of Physics, Vol. 21, No. 1, Jan. 1965, pp. 1 ff.

Note: All notation and symbol usage here is in accordance with that of references 1 and 2.

3.-A. K. Das and E. de Alba, private communication, to be published.

BAND APPROACH TO THE TRANSITION METAL OXIDES

J. Feinleib

Lincoln Laboratory,* Massachusetts Institute of Technology, Lexington, Massachusetts 02173,U.S.A.

In these lectures we shall consider the conduction processes in the oxides of the transition metals. Rather than try to catalog the properties of this wide range of compounds which is the subject of several recent reviews,[1,2] we will try to look at the physical processes that make these materials, which appear very similar in structure, to exhibit an extremely rich range of physical properties: metals with 10^{-5} Ω-cm resistivity, and insulators with 10^{15} Ω-cm resistivity, ferromagnets, antiferromagnets and diamagnets. I have selected a few materials which exemplify the different properties and will discuss them in the light of the theories that are currently applied, and which have been discussed by Frank Herman, Hans Frederikse and others during this Summer Institute.

As is well known the transition metal oxides (TMOs) present a paradox to the band theorists because they appear to violate a very basic rule: namely, that the electronic energy bands for crystals in which the atoms have partly filled shells should give metallic conduction. This paradox is most apparent for the compounds of the transition metals because here the d shell, which can contain ten electrons, is known to be only partly filled in many compounds which are good insulators. In the TMOs of the first series from Ti to Cu, when there

---

*Operated with support from the U. S. Air Force.

are less than 10 d electrons, these electrons can distribute themselves among the d orbitals in such a way that some electrons remain unpaired and a net spin moment results. This experimental fact supports the point of view taken by many who deal with the magnetic properties of the TMOs, namely that the unpaired electrons on the transition metal ion are essentially localized. In this view the wave function of the electron is determined primarily by the interaction with its own ion core potential and its electrostatic repulsion from the other d electrons on the same site. The effect of the crystal is to perturb these zeroth order states through the, essentially, local crystal field set up by the neighboring ions. The periodic potential of the crystal plays a minor role, and the interaction with the electrons on neighboring sites is only a small perturbation which is observable primarily in the long range magnetic ordering in the crystal. The evidence for such a view comes mainly from two types of experiments: optical absorption and magnetic moment measurements. As we shall see, the optical spectrum of some of the crystals of insulating TMOs looks very similar to the spectrum obtained when the isolated transition metal ion is placed in an environment equivalent to that of the parent crystal except that there are no unpaired electrons on neighboring cation sites. This spectrum shows the relatively sharp line structure which has been successfully described by the local crystal or ligand field theory.[3] The similarity of the spectrum in the pure TMO to that in a dilute crystal suggests that the interaction of electrons on neighboring cations has only a small effect on the essentially localized wave functions of the d electrons.

The magnetic measurements suggest the same interpretation. In those crystals which are either antiferromagnetic, ferromagnetic or ferrimagnetic, it is usually found that the magnetic moment in the paramagnetic state obeys a Curie-Weiss law in which the spin moment is close to the integral spin-only value found in the isolated ion. For these materials, as in magnetic metals,[4] the ordered magnetic state necessarily requires an interaction among the d electrons on different cation sites. However the energy differences between ordered and non-ordered states is approximately given by the ordering temperature which is around .05 eV for many materials. This is an indication that the perturbing effect of the inter-cation interaction is small compared to the intra-cation electron-electron interactions which are a few eV, as observed in the optical spectrum.[5] This small magnetic interaction has been successfully

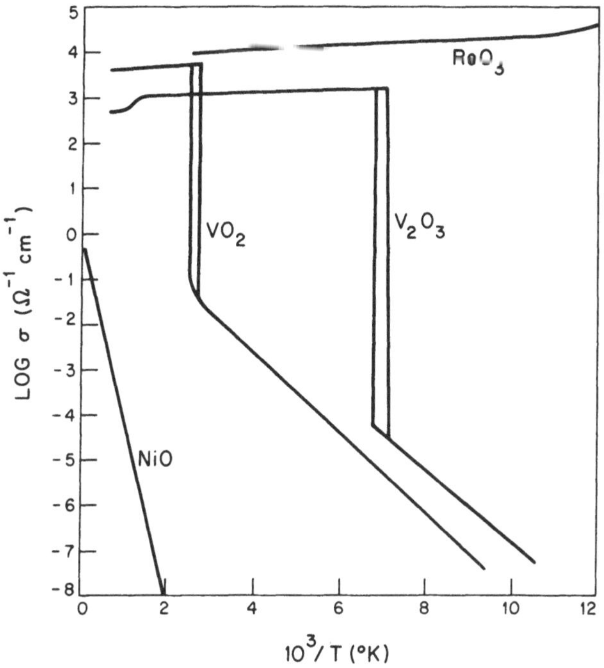

Fig. 1. Conductivity versus reciprocal temperature for different representative TMOs. Data compiled from references 7, 8, 9, 30.

described by the superexchange theory in which the d-electron wave functions are only slightly perturbed from the localized d-electron states.[6]

This strictly localized model runs into difficulty in explaining the wide range of conductivities found in apparently similar oxides. The variety of behavior is illustrated in Fig. 1 for four TMOs with different numbers of d electrons. We will characterize the oxides by the number of d electrons remaining on the cation if the material were purely ionic and the oxygen ion was in its usual O^{2-} valence state. Rhenium is a 5d transition metal and in this oxide, ReO_3, it has one d electron ($5d^1$). ReO_3 is seen to be a good metal[7] at all temperatures with a conductivity about 0.1 that of copper. Vanadium is a 3d transition metal and both oxides V_2O_3($3d^2$) and VO_2($3d^1$) are seen to have poorer metallic behavior at high temperatures. However, a first order transition occurs in these oxides in which the conductivity decreases by several orders of magnitude and then has a typical semiconductor temperature dependence at low temperatures.[8] In contrast, NiO($3d^8$) is typical of most of the TMOs in that it has a low conductivity at all temperatures with the characteristic semiconductor temperature dependence.[9] Certainly metallic conductivity cannot be ascribed to localized d electrons, and this experimental fact has given strong support to attempts to describe the d electrons in the TMOs by a one electron band model like those that have been so successfully used to explain the properties of most metals and other semiconductors.

The simplest band model calculation, in the tight binding approximation, also starts from localized, atomic wave functions for the electrons.[10] In this case the electron wave functions are again determined first by the interaction with the core potential, giving the localized like function, but then these are strongly perturbed by the periodic potential of the lattice. The interaction with the other electrons is taken into account by a correction to the average core potential. Thus, in general we may say that the difference between this band and localized models is in the importance of the intra-atomic electron-electron interaction relative to the periodic potential of the lattice.[11] The results differ greatly: the d electrons are localized on a particular site in one model while the wave function in the band model puts equal amplitude of the wave function for each electron on every lattice site. In the band model, therefore, if each band arising from d orbitals are not entirely filled or entirely empty (for the pure crystal at $T = 0°K$) an electron at the Fermi level in the partly filled band has states close in energy which are empty. With an electric field applied, the electron is free to move in these states, giving metallic conduction. More generally, if there are carriers in these bands, the mobility of the carrier in an electric field is limited by collision and has a relatively weak dependence on temperature. On the other hand, in the localized model the intra-atomic electron repulsion causes a large energy difference between the states having n and n + 1 electrons on a given site. Thus, even though the orbitals are not filled, the repulsion prevents the transfer of an electron from one d^n ion to another d^n ion, making the crystal an insulator at $T = 0°K$. In general, the mobility of a carrier in this model will require a thermal activation, $\mu \sim e^{-E_a/kT}$, in order to supply the energy required for the carrier transfer.[13]

What are the criteria that determine when the band model, in this simple form, fails to give an adequate description of the d electrons in the TMOs? As we shall show, conductivity and optical data for the four oxides above, as well as for many other TMOs, suggest that the simple band model is adequate when there are only one or two d electrons. When there are more than three d electrons, the interaction among the d electrons on each cation site causes a relatively large energy difference among the many-electron wave functions that describe this interaction. The simple one-electron band model is not adequate to describe this situation for the many d-electron TMOs.

As we noted, the temperature dependence of the mobility of the carriers is a natural indicator of whether the conduction occurs by band-like or localized carriers. The mobility is not easily measured directly but is determined indirectly from conductivity, thermoelectric, and Hall effect data. The conductivity alone is not a sufficient indicator, except in the case of metallic conductivity, since the mobility appears in conjunction with the carrier density n: $\sigma = ne\mu$. In the case of interest, the insulating TMOs, both band and localized models give zero conductivity at $T = 0°K$ since either the bands are necessarily filled, or, in the localized case, there is no thermal activation. For conduction it is necessary to ionize carriers or to produce a mixed valency[12] in order to have conducting carriers. Thus an activation process is also associated with n: $n \alpha e^{-E'/kT}$ or ℓn $n \alpha E'/kT$. Here E' could be either an impurity ionization energy or half the energy gap for an intrinsic band conductor. The thermoelectric power, $\alpha = k/e(E_F/kT + \beta)$, where β is a small constant, is related to the Fermi energy or the number of carriers. If it is plotted in the following units: $\alpha/(198\mu V/°K) = E''/kT + C''$, the slope of a 1/T plot will give the carrier density activation energy. Similarly, the Hall coefficient: $R_H \sim 1/n$, when plotted as: $\log R_H = E'''/kT + C'''$, it will also give the carrier density activation energy. Therefore if the conductivity, Hall effect and thermoelectric power are plotted in the units shown, the three lines obtained will be parallel, or have the same slope, if only the carrier density is the activated quantity and the mobility has only a small temperature dependence. Similar slopes indicate a band type of conduction process. If the slopes are not equal, some of the activation may be in the mobility, indicating a hopping type of mobility that is expected for localized carriers.

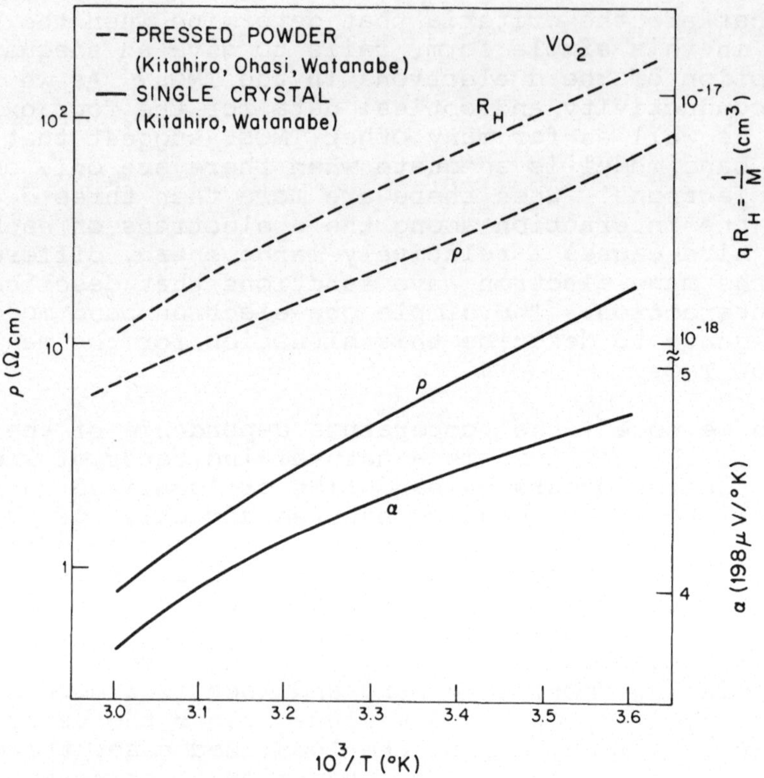

Fig. 2. Thermoelectric power (α), resistivity (ρ) and Hall constant (R_H) for VO_2 versus reciprocal temperature. Data plotted from references 14, 15.

Such data has been obtained for the semiconducting phase of the oxide VO_2 as shown in Fig. 2. The curves were plotted from the work of I. Kitahiro, T. Ohasi and A. Wantanabe,[14] and from I. Kitahiro and Wantanabe.[15] The metallic phase of the vanadium oxides, and also ReO_3, clearly shows the bandlike nature of the conduction, but this data shows that the semiconducting phase of VO_2 also has a bandlike conduction: the slopes of the three curves are very nearly equal indicating that the mobility is not thermally activated. The small mobility ($\sim$0.2 cm^2/V-sec) of the semiconducting phase and the low conductivity of the metallic phase, suggest that the bands involved are quite narrow. Nevertheless, the band model seems to be appropriate for describing the d electrons in these oxides, if indeed it is the d electrons that are responsible for the conduction.

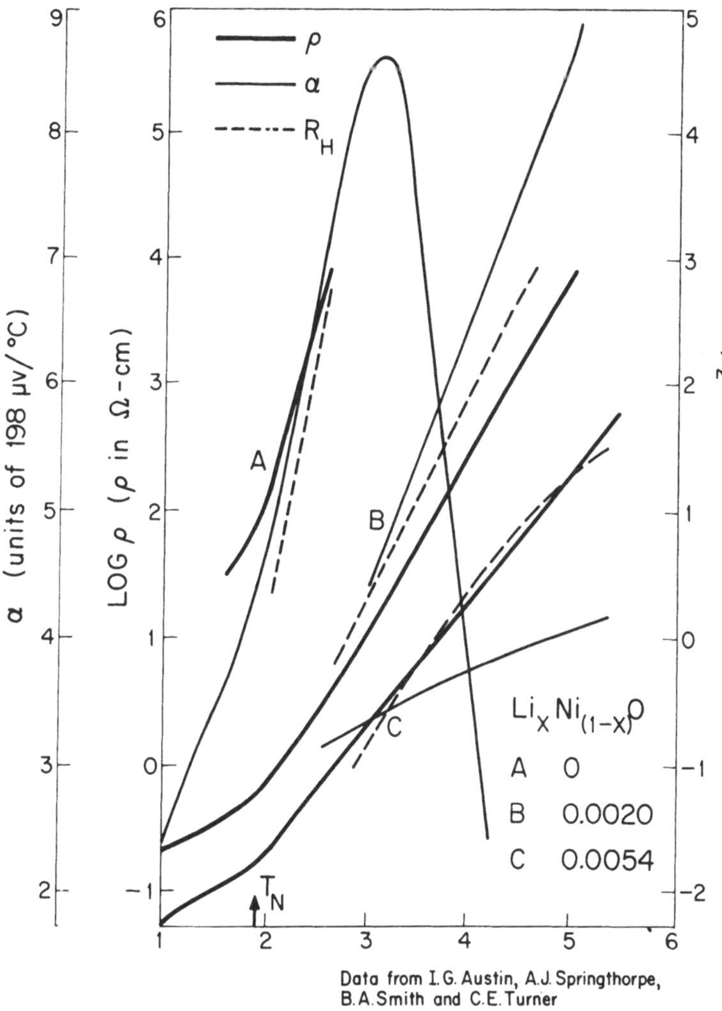

Fig. 3. α, ρ and R_H versus reciprocal temperature for NiO single crystals with different Li dopings. Data is compiled from reference 22.

We might add here that ReO_3 does not have any paramagnetic moment and is diamagnetic[16] (or Pauli paramagnetic) over the entire temperature range. VO_2 is also non-magnetic in both phases.[17,18] V_2O_3 appears to be antiferromagnetic in the semiconducting phase and becomes paramagnetic at the transition.[19,20] The moment on the cation appears to be much smaller than that expected for a d^2 ion[21] and this too would be consistent with a bandlike behavior of the d electrons.

Transport data have also been recently obtained for NiO[22] and are shown in Fig. 3. This oxide is an example of the typical insulating TMOs where the conductivity increases with the addition of impurities. In this case the Li goes into the lattice as a Li^{1+} ion is formed from one of the normal Ni^{2+} ions in order to maintain charge neutrality. Up until five years ago, before this data and similar data by Ksendsov[23] in the U.S.S.R., Bosman[24] in the Netherlands and others[25] became available, it was believed that the hole on the Ni^{3+} site moved through the lattice of Ni^{2+} ions by a thermally activated hopping mobility,[12,13] as was expected for the insulating TMOs. But as this data shows, in a limited temperature range around room temperature, the three curves of ρ, α and R_H are nearly parallel, as it was in VO_2. The mobilities were found to have only a low temperature dependence and to be much higher ($\sim$0.1 cm^2/V-sec) than had been previously calculated from the hopping interpretation ($\sim 10^{-5}$ cm^2/V-sec). This very strongly suggests that the conduction process does take place in narrow bands.

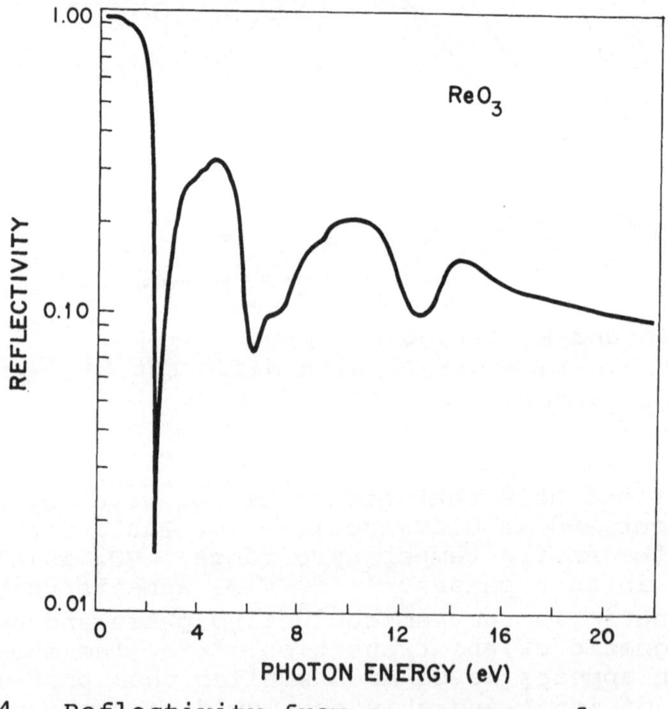

Fig. 4. Reflectivity from an as grown face of an ReO_3 single crystal. Data from reference 26.

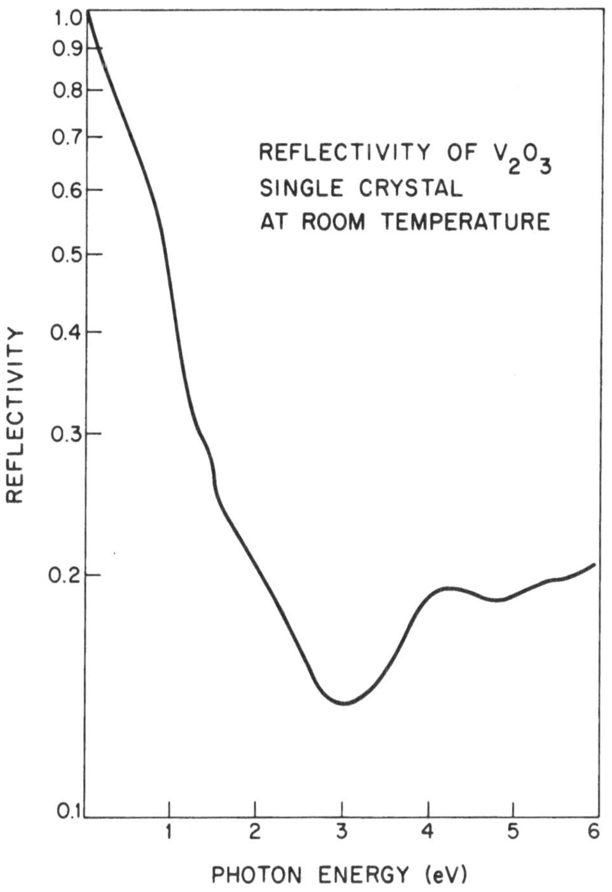

Fig. 5. Reflectivity of single crystal V_2O_3.

Optical data which have been obtained for these oxides also show characteristics for localized or band type behavior. Figure 4 shows our data for the reflectivity of ReO_3.[26] Near 100% reflectivity at low energy is the behavior expected of free electrons in a metal. The sharp drop in reflectivity is the free electron plasma edge modified by the onset of the first strong interband absorption. These data are fit very well by an effective mass of about 0.86 m_o and a carrier density of one electron per Re ion. This suggests that the $5d^1$ electron, which is the only one remaining on the Re ion after the oxygen band is filled, is the conduction electron. Recent de Haas-van Alphen data[27] confirm the optical data, and Knight shift measurements by Narath and

Barham[28] strongly support the suggestion that the conduction electron is in a d like conduction band.

For V_2O_3 in the metallic phase we have recently seen similar free electron behavior as shown in Fig. 5.[29] The free electron reflectivity falls off much faster with photon energy than it does in ReO_3. This indicates first, that the conduction band is much narrower, and secondly, that there are other d bands nearby in energy to cause low energy interband transitions. No transmission is observed in the metal phase as expected, but as the temperature is lowered through the transition to the semiconducting phase, transmission is observed below 0.4 eV as shown in Fig. 6.[30] A crystal

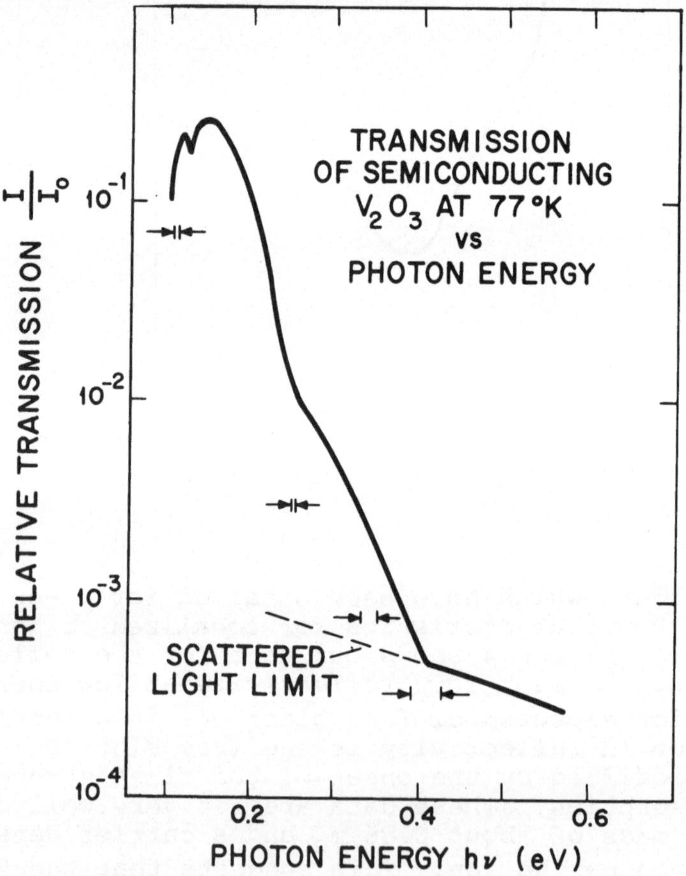

Fig. 6. Transmission of 25μ thick single crystal of V_2O_3 in semiconducting phase. Data from reference 30.

structure change occurs at the phase transition in V_2O_3[31] so that we could only obtain rather crude data because of the difficulty in keeping the thin transmission sample from cracking. The data, however, suggests that an energy gap of between 0.3 and 0.1 eV opens up in the partly filled d band of the metallic phase when the transition occurs. Very similar data to these have been obtained for VO_2 by Barker et al.[32] and Ladd and Paul.[33] Here too a structure change accompanies the metal to semiconductor transition.[34]

Unlike the conductivity data, the optical absorption data for NiO does not show only the absorption edge usually seen for a band-like semiconductor. Newman and Chrenko[35] have observed that a strong absorption edge occurs near 4 eV characteristic of the energy gap of an insulator. However, they also find a series of weaker, but sharp absorption lines with decreasing energy to about 0.24 eV. In Fig. 7 we show this set of lines for pure NiO and also the absorption in the series of crystals $Ni_xMg_{1-x}O$, the data was taken by D. Reinen.[36] What is so remarkable about this data is the clear evidence that the lines arise from the isolated d^8 ion. As the concentration of Ni^{2+} increases in the series of mixed oxides, the lines shift slightly and there is some change in relative intensity, but there is a clear one to one correspondence of the lines in the dilute mixture to that of pure NiO.

We mentioned earlier that these lines are the multiplet structure for the d^8 configuration of a single ion. The transitions occur from the many-electron ground state to the crystal field split, excited states of the multiplets. We need not look at the details of calculating the multiplet structure since this is a subject in itself and results for the d^8 ion in various crystal field symmetries and including spin orbit effects are thoroughly treated in the book by Griffith.[5] But let us recall the physical idea.

For two d electrons on an isolated ion, a wave function can be written which is a product of spin-orbitals for each electron: $\psi' = d_1(1)d_2(2)\alpha(1)\beta(2)$ where α and β refer to up and down spin electrons respectively, and d_1 and d_2 are two d orbitals. The correct antisymmetric and normalized wave function, in accordance with the Pauli exclusion principle, is given as the determinant:

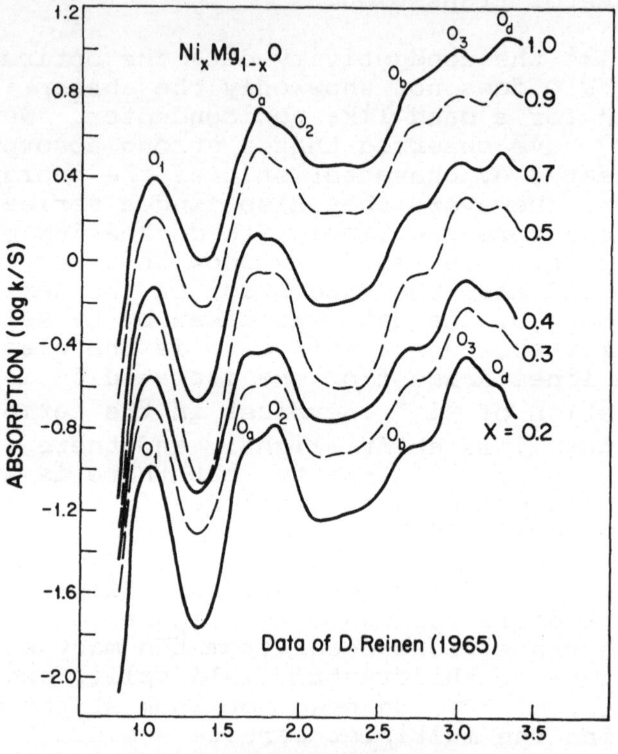

Fig. 7. Absorption calculated from diffused reflectance measurements of powdered samples of the mixed crystal $Ni_xMg_{1-x}O$. Data taken from reference 36.

$$\psi = \frac{1}{\sqrt{2}} \begin{vmatrix} d_1(1)\alpha(1) & d_1(2)\alpha(2) \\ d_2(1)\beta(1) & d_2(2)\beta(2) \end{vmatrix}$$

This many-electron wave function can be characterized by a total orbital momentum, L, and spin, S, which are the sum of the orbital and spin quantum numbers of the individual spin-orbitals. For a partly filled shell, L and S can take on several different values. Each LS state has a degeneracy of (2L + 1)(2S + 1), corresponding to the various values of the total quantum numbers M_L and M_S, and this multiplet state is labeled as ^{2S+1}L, e.g., ^3F for S = 1, L = 3. The energy of the eigenstates for different LS would have the same value if the electrons were non-interacting. However, because of the repulsion between electrons, $-e^2/r_{12}$, the different LS eigenstates have different energies and the wave function must usually be expressed as a sum of determinants. Under the influence of a crystal field which depends on the symmetry of the environment about the ion, these multiplet states interact and split up further. The result is a set of eigenstates with different energies which may be on the order of several electron volts for a $3d^8$ ion. The important point is that these are many-electron states, and even though they each come from the $3d^8$ configuration, they differ widely in energy. In this scheme, therefore, it is clear that if an electron is entirely removed from one ion and placed on another, leaving one $3d^7$ and one $3d^9$ ions, it will require a considerable change in energy, at least comparable to the several eV of the $3d^8$ multiplet splittings. The implication is that if this type of multiplet structure is the correct interpretation of the optical data, then there are no empty states in the d levels of pure NiO which can carry current. The insulating nature of the oxide is accounted for simply by the intra-atomic interaction of the d electrons rather than by special energy gaps in the d band arising from the periodicity of the lattice as would be required by the band model.

Many other types of measurements[37] have been made on NiO and I will mention some of those which have a direct bearing on the nature of the band structure. Several attempts have been made to observe photoconductivity,[38] but only Drabkin and Ksendsov[39] have reported a positive result. No photoconduction is associated with the line structure, we ascribed to crystal field states of the $3d^8$ ion. This result would be expected since these transitions involve transitions from the ground, $3d^8$, state to other excited, $3d^{8*}$, states. No valency changes, e.g., $3d^7$ or $3d^9$ states, are produced to provide carriers for conduction. However, in their experiment, photoconductivity was observed at photon energies above the 4 eV absorption edge seen in the optical spectrum. They also have measured directly a drift mobility[40] for these photo-excited carriers. Above 4 eV therefore, free carriers are produced which give band-type of conduction.

Another measurement of direct interest is the ac conductivity of NiO doped with Li. The high frequency conductivity measurements by Kabashima and Kawakubo[41] have indicated that below 250 K,° the carriers produced by the doping are trapped near the impurity or vacancy, and that they contribute to the ac conductivity by hopping among the sites surrounding the impurity. It is found that this hopping process requires no activation energy. The staying time, or the time between hops is measured as $\sim 10^{-10}$ sec. This time is much longer than a typical lattice vibration frequency and in terms of a bandwidth, it would suggest that the bands are less than .001 eV wide. The carriers are thought to be the hole produced when a Ni^{3+} site is created by the doping and can be labeled as a $3d^7$ state. These data suggest that the mobility of such a hole when excited into the $3d^8$ band is far too small to account for the band like conduction observed at higher temperatures.

The data I have presented demonstrate several aspects of the behavior of d electrons in the TMOs. In the following section I shall try to describe models of the band structure in these oxides which we believe can describe the d electrons.

The most straightforward behavior to explain is the metallic conductivity and band-type behavior observed for ReO_3. A tight binding calculation[42] has shown that the band structure of ReO_3 is compatible with a simpler quantitative calculation by Kahn and Leyendecker[43] for $SrTiO_3$ and discussed by Frederickse in these lectures. A complete APW band structure calculation has just recently been done for ReO_3 by Mattheiss[44] and appears to be in good agreement with the deHaas van Alphen measurements of the Fermi surface[27] and the optical data. However, we shall use a sketch of these results as derived from the $SrTiO_3$ calculation since the latter more directly illuminates the qualitative features of the band structure. $SrTiO_3$ has the same cubic crystal structure as ReO_3, except that the Sr ion occupies the body center position which is vacant in ReO_3. However the tight binding calculation ignores the effect of the Sr orbitals. The qualitative results are seen in Fig. 8.

For $SrTiO_3$, it is found that there are two sets of non-overlapping bands. The lower set are composed primarily of oxygen 2p wave functions and are the filled valence bands. The upper set are predominantly 3d bands and are the empty conduction bands, the energy gap of this semiconductor is about 3 eV and is the separation of these two sets of bands. We deduced from the optical data that this band structure will qualitatively describe ReO_3, and this has now been confirmed by the APW calculation. In ReO_3 the oxygen bands are also filled. However, the remaining $5d^1$ electron on the Re goes into the d conduction band, and this partly filled band gives the observed metallic behavior. Evidence that this conduction band is a d band comes from several sources, aside from the band calculation. The metal ion in the oxide WO_3, has one electron fewer than the Re ion in ReO_3. WO_3 is found to be an insulator, corresponding to the insulating structure of $SrTiO_3$. When Na is put in WO_3, this doped oxide has metallic properties[45] like ReO_3. Apparently the additional electron from the Na ion goes into the otherwise empty conduction band of WO_3. Knight shift measurements by Narath and Fromhold[46] have shown that the conduction electron in this compound is d-like, and recently, Narath and Barham[28] have verified the same d-like character for the conduction electron in ReO_3.

The optical data verify other features of this band structure and the quantitative results of the APW calculation. The free electron like region below 2 eV, as seen in Fig. 4, suggests that the 5d bands are quite wide so that no transitions between d bands and the Fermi level occur below 2 eV. The optical effective mass was calculated to be $\sim 0.8\ m_o$. At 2.1 eV, transitions between d bands occur which give rise to the plasma edge and the following increase in reflectivity. Above 2.5 eV, transitions from the oxygen 2p valence bands to the d band begin and reach a maximum in intensity near 4.5 eV. This is in good agreement with the joint density of states computed from the band structure.

The surprising thing here is that the band structure calculations give such good agreement with the data, and confirm the normal band type behavior of the d electrons. In ReO_3, the crystal structure is such that the Re ions are well separated from each other and are surrounded by

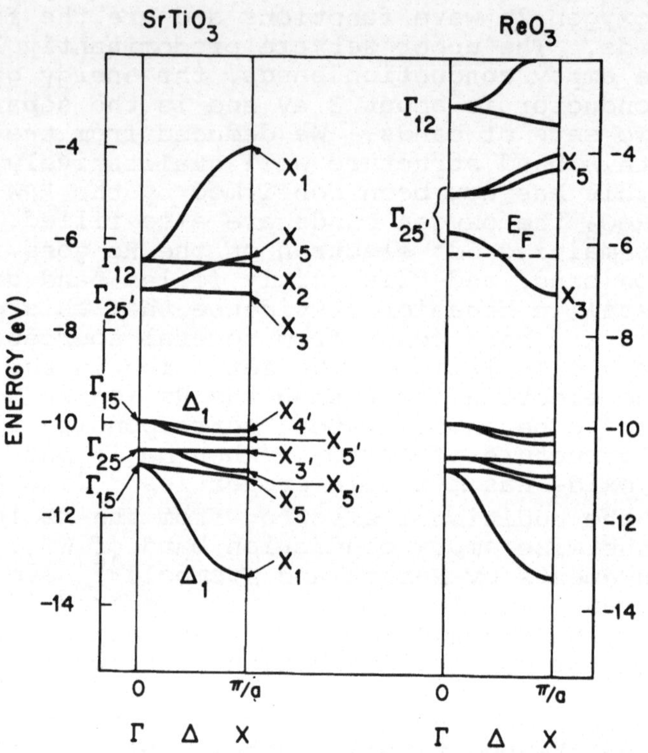

Fig. 8. Band structure along the Δ axis of the Brillouin zone. For $SrTiO_3$ (reference 43), the Fermi level lies in the gap between Γ_{15} and Γ_{25}. For ReO_3 (reference 26), the Fermi level

oxygens. Since the wide d bands could not come from the direct overlap of d wave functions on cation sites, it is clear that the overlap must come about by a relatively strong covalent mixing with the oxygen wave functions. The band calculations confirm this. For $SrTiO_3$ it was found[43] that a large covalency was necessary to match the observed valence band to conduction band energy gap of 3 eV. For ReO_3 it was found[44] that this oxygen 2p and Re 5d admixture widened both the 2p and 5d bands. However, it was surprising to find that while this effect made the effective mass of the d conduction band edge small, the oxygen valence band edge was not greatly perturbed and the masses at the top of the oxygen 2p band were large. This latter result has not been experimentally confirmed in these oxides, but as we shall see, it provides indirect evidence that 2p band conduction may have low mobilities and this is pertinent to our discussion of NiO. Since the oxygen covalency with the d electrons of the cation is so important in giving the band like properties to the d electrons, one might expect a similar result for the other oxides, at least those with only a few d electrons as in ReO_3.

With this motivation we have tried to construct a similar band structure for the vanadium oxides[47] as shown in Fig. 9. For VO_2 we again assume that the oxygen 2p bands are filled and separated from the V 3d bands. In the metallic phase, the crystal has tetragonal symmetry so that we get the band splittings indicated by the solid curves. The VO_2 unit cell contains two V^{4+} ions, so that the d bands have orbital degeneracies. These are indicated by the numbers near each solid curve. The lowest d band is depicted as the one in which the d orbitals are oriented along the c axis of the crystal, and the 2 fold degenerate band is labeled t_c. This band ordering was based on the chemical bonding arguments put forward by Goodenough,[2] since no band calculations are available.[56]

In this scheme the $3d^1$ electron from each of the two V^{4+} ions per unit cell will only half fill the t_c band which can hold four electrons. This band will then give the low mobility (since it is narrow), but metallic properties, observed. When the temperature is decreased below 338°K, a crystal structure change takes place in VO_2 accompanied by the transition to a semiconductor. This transition can also be described by the band structure of Fig. 9 using the model of Adler and Brooks.[42] The idea is as follows: The crystal structure change

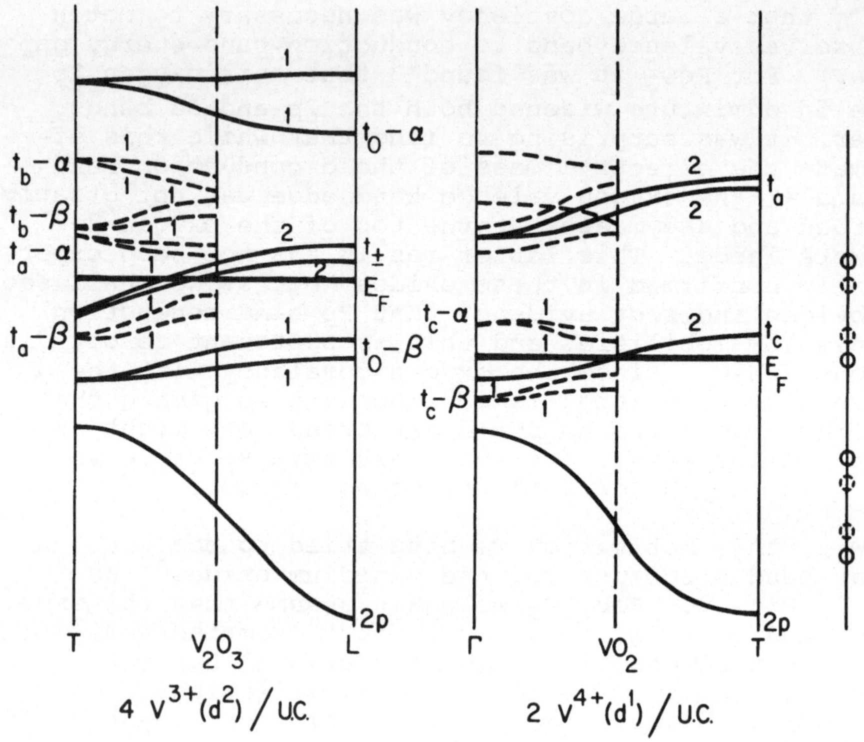

Fig. 9. Schematic band structure of the two vanadium oxides showing an insulator to metal transition. Bands shown along direction which must clearly show distortion from metal (solid curves) to semiconductor (dashed curves). B.Z. change is indicated by dashed vertical lines. Data based on analysis of B.Z. in references 56 and 47.

has the feature that the equally spaced V ions along the c axis in the metallic phase (shown by the solid circles on the extreme right of the figure) move together in such a way that pairs of ions are formed along this axis. This distortion causes the unit cell size to be doubled, and in the reciprocal lattice scheme on which the band structure is depicted, the Brillouin zone size between Γ and T is cut in half. Since energy gaps between bands are formed at the zone edges, there is now the possibility that an energy gap opens up at this new edge where there was no gap before. This is shown in Fig. 9 by the dashed curves. Because the unit cell size is increased, new d band degeneracies arise as shown by the numbers near the dashed curves. The new splitting occurs between the bands labeled $t_c - \beta$ and $t_c - \alpha$. Now if each $3d^1$ electron per V^{4+} ion is put in this modified band structure, it is found that the $t_c - \beta$ band is just filled, leaving the $t_c - \alpha$ band empty. Thus we have the observed semiconducting state.

The Adler and Brooks model showed that such a distortion in the crystal structure will occur as a first order transition if the bands are sufficiently narrow. They argue that the energy gap is caused by the distortion, causing the $t_c - \beta$ bands to be stabilized or lowered in energy while the $t_c - \alpha$ bands are raised in energy. Since the electrons go into the lower band, the total electron energy of the crystal is lowered, compared to the undistorted phase. The energy gap depends on the distortion. Thus when an electron is excited into the conduction band it is in an antibonding state which tends to destabilize the distortion and reduce it slightly. When the carrier density in the excited state reaches a critical value near T_c, the model shows that the energy gap drops catastrophically, inducing the transition to the metallic phase. The free energy of the metal phase above the critical temperature T_c is lower than that in the semiconducting phase because of the increase in entropy by having the electrons free to move at the Fermi surface, and also because the strain energy is removed by the disappearance of the distortion. The model holds equally well for V_2O_3 and quantitative predictions of the ratio of the energy gap to the transition temperature have been borne out by our pressure and stress experiments,[30] as well as the optical measurement of the energy gap.[30]

In VO_2 there is no magnetic ordering above or below the transition.[17,18] In V_2O_3 there is a change from the low temperature antiferromagnetic state with a small moment, to a paramagnetic state.[19-21] Mossbauer data show that the sublattice magnetization remains nearly saturated up to the transition temperature.[20] The Neel temperature extrapolated from this data is much higher than the phase transition temperature. This is good evidence that, as in VO_2, a magnetic ordering is not driving the transition. However, the magnetic ordering is still compatible with the one electron band model used to describe the transition. By having a periodic ordering of the spins, it is possible to double the apparent size of the unit cell as in the case of a distortion. Again the ordered state can lower the energy of the crystal by introducing new energy gaps. We don't think this is the main reason for the energy gap in V_2O_3 and it is not applicable to VO_2. However, we find that while the numerical value for the conductivity discontinuity in VO_2 as seen in Fig. 1, agrees[47] with the Adler and Brooks model, the measured discontinuity is much larger than the predicted for V_2O_3. It may be that the antiferromagnetic ordering helps to increase the energy gap beyond the prediction of the model for the distortion alone. We feel[49] that the one electron band model is sufficiently exact so that the inclusion of the magnetic ordering will bring the predicted conductivity discontinuity into better agreement with experiment. The hope is that this band model can explain the experimental results without resorting to the inclusion of the many-electron effects apparently needed to describe the electronic structure of the oxides with many d electrons, like NiO.

It is clear from the data we have presented here that the simple band models used to describe the vanadium and rhenium oxides cannot easily be applied to NiO. Even though NiO has a simple rocksalt structure, it may be possible, in principle, to construct a crystal potential including antiferromagnetic ordering, inter-atomic exchange (i.e., a separation into up spin and down spin bands) and crystal field splitting, and somehow come up with a band structure that makes pure NiO and insulator at 0°K. That is, it may be possible to split up the d band sufficiently to have a filled band separated from the next empty band by a band gap. Aside from the practical difficulties, there are conceptual difficulties to have such a scheme stay within the one electron band model. First, these different possible interactions would have to be applied differently to each of the many insulating TMOs in order to achieve insulating behavior. And secondly, even in principle, the resulting band

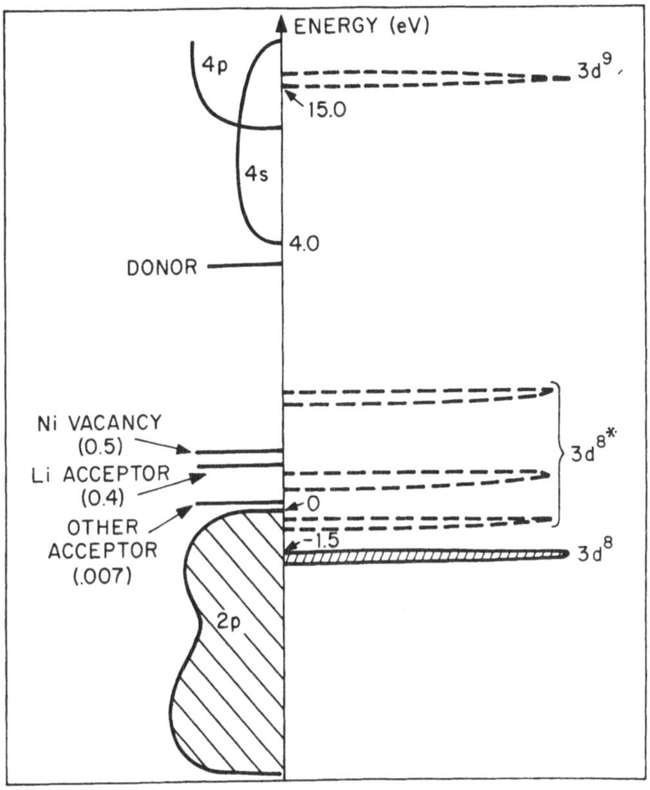

Fig. 10. Schematic density of states versus single particle energy relative to the top of the 2p band in NiO. Left side shows one-electron bands, right side is single particle representation of many electron d states. Taken from reference 53.

structure may not reproduce the optical spectrum observed, since there seems to be no way of obtaining the many-electron states from a single determinant, one electron band model.[50] A difficulty with any model arises when trying to describe the localized optical spectrum with the same electronic states used to describe the apparent band-like conduction. All models that take into account the localized optical spectrum will equally well describe pure NiO at 0°K since there is no conduction because the carriers are frozen out. The interesting question is to decide what electronic states are involved in the conduction that takes place at moderate temperature in doped NiO. As we noted, a Li impurity is thought to produce a $3d^7$ hole state. The current carrying ability of the hole, when it is ionized from the acceptor impurity center, has been variously ascribed to motion in a band formed from $3d^8$ states,[1,22-24,51] or in a polaron band[52] composed from $3d^8$ electronic states. The absence of a hopping conductivity, and the evidence of a localized optical spectrum, suggest that a model with a simple, one electron $3d^8$ band cannot explain the band-like conduction of these holes. Instead of trying to adapt a rather complicated polaron theory to explain the conduction in a d band, we have proposed[53] a model which requires only simple bands and localized states. In this model the band-like conduction takes place in the oxygen 2p valence band, while the d electrons are essentially localized, making only a small contribution to the conduction. The proposed model is displayed in a density of states sketch in Fig. 10.

Consider first the left-hand side of Fig. 10. Here we show the oxygen 2p and Ni 4s states as normal one-electron bands. The 2p band is filled and the 4s band empty at 0°K. The 2p-4s energy gap is put at 4 eV so as to describe the measured optical absorption[35,38] and photoconductivity edges.[39] Since the $3d^8$ states in pure NiO are filled, localized states, NiO will be a semiconductor with an intrinsic energy gap of 4 eV. If the crystal is doped with Li^{1+}, a Ni^{3+} hole state is produced which acts as an acceptor for the 2p band. These holes are trapped at the Li^{1+} impurity at low temperatures and contribute only to the ac conductivity[41] by hopping to the Ni^{2+} sites surrounding the impurity. At higher temperatures, the hole is ionized into the 2p valence band, as in the usual acceptor models. The hole mobility in the 2p band may be expected to be small, for as we mentioned earlier, the effective masses at the top of the 2p band are likely to be large. The band-type conduction is thus simply accounted for in this scheme.

We must also consider how the d states enter this scheme.

On the right side of the figure we have drawn in the ground state of the Ni^{2+} ion and labeled it $3d^8$. This is the many-electron wave function for the eight strongly interacting d electrons localized on a Ni. The excited, crystal field states for the $3d^8$ configuration are grouped under the label $3d^{8*}$. Transitions between the ground and excited states with energy differences less than 4 eV are then seen as line structure in the optical absorption spectrum. Crystal field transitions with larger energy separations are masked by the strong 2p-4s absorption edge. Other fine structure transitions, such as a sharp line at 0.24 eV which is associated with the antiferromagnetic ordering, have not been included for simplicity, but these also do not contribute to photoconduction. One transition which produces a potentially current carrying state is the transition from the $3d^8$ ground state to the state labeled $3d^9$. In this transition an electron is ionized from one $3d^8$, making it a $3d^7$, and goes to another Ni^{2+} ion making it a Ni^{1+} ($3d^9$). In this localized scheme, however, this transition energy is given as the ionization energy minus the electron affinity, and is thought to be near 16 eV.[38,54]

We have said nothing about transitions which cross from one side of the diagram to the other, and indeed the relative energies of the states on the right are not related to the zero of energy which was taken as the top of the 2p valence band. The difficulty is that the states on the left are one electron states, while those on the right are the energies for the many-electron states of the d electrons. To describe a transition from say the $3d^8$ state to a 4s band, we must use a consistent one-electron energy scheme. We have done this by the following approximation. We assume that the "core" states for NiO are the configurations O^{4+} and $Ni^{3+}(3d^7)$. The six electrons filling the 2p band are the valence electrons, and the eighth electron on a Ni^{2+} ion is considered to take on the energy values equal to the difference between the $3d^8$ many-electron energy states and the core state energy, $3d^7$. We thus construct a one-electron diagram by assigning the true many-electron energy levels to a single electron of the $3d^8$ configuration. The right side of the figure remains the same, but now we can also portray transitions from the $3d^8$ states to the band states by the energy differences shown on the diagram. We must be careful to follow the rules required by this approximation. For example, only transitions that leave a $3d^7$ core as a final state are allowed to cross the diagram. Also, the sum of occupied $3d^8$ and $3d^{8*}$ states must equal the number of Ni^{2+} ions in the crystal - this prevents a transition from, say the 2p band to a $3d^{8*}$ state. To summarize these rules on the diagram, we must not allow any transitions from the solid curves on the left to the dashed states on the right.

The relative energies are set empirically from experimental data. We believe that the optical data indicate that the $3d^8$ to 4s band transition occurs at an energy of 5.5 eV.[38,55] If this is correct, then the energies on the right can be set relative to the energies on the left by placing the $3d^8$ states 5.5 eV below the 4s band, as we have done in the figure. The diagram then gives the energies for optical transitions as the differences in energy of the labeled states. At the same time, the impurity levels relative to the valence band is all so easily portrayed to describe the transport data. Another virtue of this band scheme for NiO is that it can easily be applied to all other insulating TMOs that have more than a few d electrons per cation, because it can describe both localized and band excitations in a simple way. However, we must keep in mind that no experiment has yet distinguished between the prediction of this model and those of the more complex polaron band models.

The experimental data and the theoretical models presented here are by no means the final work for the TMOs. Significant progress has just recently been made in the materials preparation of many of these oxides, and only very recently have samples been made which can be used for the detailed investigations, e.g., the de Haas van Alphen measurements, which are needed to answer the many theoretical questions raised. We may consider the recent APW calculation of ReO_3 as a breakthrough on the theoretical side for the oxides with only a few d electrons, but the insulating TMOs have yet to be explained successfully by anything like an *ab initio* band calculation. At the present, there is little formal theoretical support for the idea of combining the localized and band states, as we have done, because of the difficulty in approximating the strong effect of the intra-atomic electron interaction and the effect of the periodic potential in a one electron calculation. However, it seems that within the concepts of the band models and crystal field models, it is possible to describe the behavior of the d electrons in the TMOs.

REFERENCES

1. D. Adler, Solid State Phys. 21, 1 (1968); D. Adler, Rev. Mod. Phys. (to be published).

2. J. B. Goodenough, Magnetism and the Chemical Bond, Interscience Publishers, New York, (1963).

3. C. J. Ballhausen, Ligand Field Theory, McGraw-Hill Book Company, Inc., New York (1962).

4. C. Herring, "Exchange Interactions Among Itinerant Electrons," Magnetism, Vol. IV, edit. G. T. Rado and H. Suhl, Academic Press, New York (1966).

5. J. S. Griffith, The Theory of the Transition Metal Ions, Cambridge University Press, Cambridge (1961).

6. P. M. Anderson, Magnetism, Vol. 1, ed. G. T. Rado and H. Suhl, Academic Press, New York, p. 25, (1963).

7. A. Ferretti, D. B. Rogers, and J. B. Goodenough, J. Phys. Chem. Solids 26, 2007 (1965).

8. F. J. Morin, Phys. Rev. Letters 3, 34 (1959).

9. J. H. DeBoer and E. J. W. Verwey, Proc. Phys. Soc. (London) A49, Suppl. 59 (1937).

10. J. C. Slater and G. F. Koster, Phys. Rev. 94, 1498 (1954).

11. N. F. Mott, Proc. Phys. Soc. (London) A62, 416 (1949); N. F. Mott, Phil. Mag. 6, 287 (1961).

12. E. J. W. Verwey, Semiconducting Materials, Butterworths Scientific Publishers, London (1951).

13. R. R. Heikes and W. D. Johnston, J. Chem. Phys. 26, 582 (1957); S. van Houten, J. Chem. Phys. Solids 17, 7 (1960).

14. I. Kitahiro, T. Ohashi and A. Wantanabe, J. Phys. Soc. Japan 21, 2422 (1966).

15. I. Kitahiro and A. Wantanabe, J. Phys. Soc. Japan 21, 2423 (1966).

16. J. H. E. Griffiths, J. Owen and I. M. Ward, Proc. Roy. Soc. A219, 526 (1953).

17. J. Umeda, H. Kusumoto, K. Narita and E. Yamada, J. Chem. Phys. 42, 1458 (1965).

18. K. Kosuge, J. Phys. Soc. Japan 22, 551 (1967).

19. E. D. Jones, Phys. Rev. 137, A978 (1965); E. D. Jones, J. Phys. Soc. Japan 20, 1292 (1965).

20. T. Shinjo, K. Kosuge, S. Kachi, H. Takaki, M. Shiga and Y. NaKamura, J. Phys. Soc. Japan 21, 193 (1966); T. Shinjo and K. Kosuge, J. Phys. Soc. Japan 21, 2622 (1966).

21. A. Paloetti and S. J. Pickart, J. Chem. Phys. 32, 308 (1960); H. Kendrick, A. Arrott, and S. A. Werner, J. Appl. Phys. (to be published).

22. I. G. Augsin, A. J. Springthorpe, B. A. Smith and C. E. Turner, Proc. Phys. Soc. (London) 90, 157 (1967).

23. Ya. M. Ksendzov, L. N. Ansel'm, L. L. Vasil'eva, and V. M. Latysheva, Soviet Phys. Solid State 5, 1116 (1963).

24. A. J. Bosman and C. Crevecour, Phys. Rev. 144, 763 (1966).

25. V. P. Zhuze and A. I. Shelykh, Soviet Phys. Solid State 5, 1278 (1963); M. Roilos and P. Nagels, Solid State Commun. 2, 285 (1964).

26. J. Feinleib, W. J. Scouler and A. Ferretti, Phys. Rev. 165, 765 (1968).

27. S. M. Marcus (to be published); J. Graebnor (to be published).

28. A. Narath and D. C. Barham (to be published).

29. J. Feinleib (unpublished data).

30. J. Feinleib and W. Paul, Phys. Rev. 155, 841 (1967).

31. E. P. Warekois, J. Appl. Phys. Suppl. 31, 3465 (1960).

32. A. S. Barker, Jr., H. W. Verleur, and H. J. Guggenheim, Phys. Rev. Letters 17, 1286 (1967).

33. L. Ladd and W. Paul (to be published).

34. K. Kosuge, J. Phys. Soc. Japan 22, 551 (1967); S. Minomura and H. Nagasaki, J. Phys. Soc. Japan 19, 131 (1964).

35. R. Newman and R. M. Chrenko, Phys. Rev. 114, 1507 (1959).

36. D. Reinen, Ber. Bunsenges 69, 82 (1965).

37. See references in 1.

38. R. J. Powell, Tech. Rept. No. 5220-1, Stanford Electronics Laboratory, (1967) (unpublished).

39. Ya. M. Ksendzov and I. A. Drabkin, Soviet Phys. - Solid State 7, 1519 (1965).

40. Ya. M. Ksendzov, B. K. Avdeenko and V. V. Makarov, Soviet Phys. - Solid State 9, 828 (1967).

41. S. Kabashima and T. Kawakubo, J. Phys. Soc. Japan 24, 493 (1968).

42. J. M. Hanig, J. O. Dimmock and W. H. Kleiner (to be published).

43. A. H. Kahan and A. J. Leyendecker, Phys. Rev. 135, A1321 (1964).

44. L. Mattheiss (to be published).

45. B. W. Brown and E. Banks, Phys. Rev. 84, 609 (1951); H. R. Shanks, P. H. Sidles and G. C. Danielson, Advances in Chemistry, R. Ward, ed., (Am. Chem. Soc., Washington, D.C. [1963], Series 39.

46. A. T. Fromhold, Jr. and A. Narath, Phys. Rev. 152, 585 (1966).

47. D. Adler, J. Feinleib, H. Brooks and W. Paul, Phys. Rev. 155, 851 (1967).

48. D. Adler and H. Brooks, Phys. Rev. 155, 826 (1967).

49. D. Adler and J. Feinleib (unpublished).

50. J. C. Slater, J. Appl. Phys. 39, 761 (1968).

51. J. Hubbard, Proc. Roy. Soc. A277, 237 (1964).

52. T. Holstein, Ann. Phys. (New York) 8, 343 (1959).

53. J. Feinleib and D. Adler, Phys. Rev. Letters 21, 1010 (1968).

54. C. Bonnelle and C. K. Jorgensen, J. Chim. Phys. 61, 826 (1964).

55. E. Rossi and W. Paul (to be published).

56. W. H. Kleiner, Lincoln Laboratory Solid State Research Report (1967:3) p. 44.

THE ELECTRONIC BAND STRUCTURE OF STRONTIUM TITANATE: THEORY AND EXPERIMENT*

H. P. R. Frederikse

National Bureau of Standards, Washington, D. C.

The appropriate description of the 3d-valence electrons in transition metal oxides has been a subject of discussion for many years. Considering the heavy mass and low mobility of 3d-electrons (as compared to the values for these parameters in the "conventional" semiconductors), the question arises whether the band model is a legitimate representation of the energy states. Several features of the transition metal compounds have slowed down progress towards solution of this problem: magnetic ordering (NiO, CoO, MnO), complicated crystal structure (Fe_2O_3, Ti_2O_3), and low quality monocrystalline or polycrystalline samples. However, during the last few years considerable advances have been made. Better crystals have opened the door to more and better experiments. Although the nature of the transitions in many of these solids is extremely complex, some understanding of the mechanisms involved has been achieved (see the lecture by Dr. J. Feinleib). It has been established that the band model is more widely applicable than was previously assumed. In this area the theoretical and experimental work on $SrTiO_3$ (and related compounds) has been of major importance. The material has proven to be an ideal representative of the oxide family. It lends itself very well to simplified theoretical calculations; at the same time nearly the entire spectrum of electron, phonon and structural properties can be attacked successfully by experiment. Single crystals of reasonable perfection can be prepared. By doping or reduction the compound

*Major parts of the research discussed in this review have been supported by the U. S. National Aeronautics and Space Administration.

can be transformed from an insulator to a semiconductor (and superconductor). The material has the simple O_h^1 structure (characteristic for the very large perovskite family, LMX_3) over most of the temperature range; below 110 K it transforms to a very slightly tetragonal phase. In the following we will discuss the energy band structure calculated for cubic $SrTiO_3$ and $BaTiO_3$ and touch upon a number of experimental investigations nearly all of which support the theoretical predictions.

ENERGY BAND CALCULATION

The study of the energy band structure by Kahn and Leyendecker (KL)[1] is based on the LCAO method. A first estimate of the energy states of the different ions in the crystal is obtained as the sum of the Madelung potential and the ionization potential. For the full ionic charge (Ti^{4+} and O^{2-}) the result is an energy gap $E_g \sim 15$ eV between the Ti-3d and the O-2p states. By reducing the ionic charge on the Ti and O ion this gap decreases rapidly until the experimental value ($E_g \approx 3.25$ eV) is reached for a charge of -1.7 e on the oxygen ion (ionicity of 85%).

Using the atomic wave functions (3d and 2p orbitals), one forms Bloch sums for all lines of symmetry in the Brillouin zone. Subsequently diagonal elements $\langle Ti|H|Ti\rangle$ and $\langle O|H|O\rangle$ as well as off-diagonal elements $\langle Ti|H|O\rangle$ are composed from these Bloch sums and the energy matrix is constructed. Magnitudes of the overlap integrals are estimated or calculated using analogous situations in other solids. Solving the secular equation, one obtains a plot of E vs k as shown in Fig. 1 [along the 100 direction (Δ) and on the 100 face (Z and T)].

The lowest conduction band $\Delta_{2'}$ is completely flat due to vanishing of the Ti-O overlap integrals in this direction. The effect of second-nearest-neighbor overlap (Ti-Ti) causes this band to curve downward; the X_3 point at the edge is expected to be 0.02 - 0.05 eV below $\Gamma_{25'}$. Hence the lowest conduction band structure consists of 6 (or 3) ⟨100⟩ ellipsoids similar to that of silicon. Spin-orbit interaction will split the bands at the center by an amount of $\sim$ 0.03 eV. The total widths of all the conduction bands is predicted to be about 3.5 - 4.0 eV, while the valence bands span an energy range of 4.5 - 5.0 eV.

The above energy band calculation has been criticized by Šimánek and Šroubek.[2] They point out that a lowering of the titanium and oxygen charges (the latter from -2.0 e to -1.7 e) not only changes the Madelung potential, but also the ionization potentials. Taking this change into account, a transfer of -1.8 e is required which would make the compound predominantly covalent.

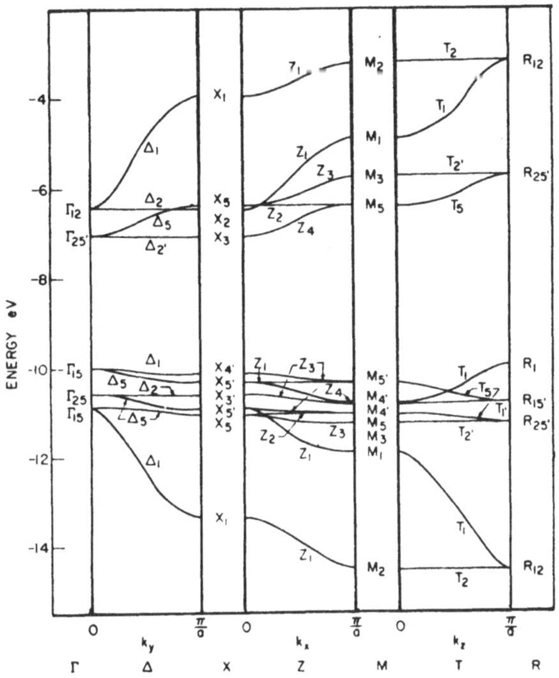

Fig. 1. Calculated energy band structure of strontium titanate (ref. 1). The point X_3 is calculated to be several hundredths of one eV below $\Gamma_{25'}$.

The above authors point out, however, that several other corrections should be considered: the electron-hole interaction and the polarization energy. It appears that, as a result of the latter two contributions, the original ionicity (85-90%) of the Kahn and Leyendecker calculation is restored, and hence the validity of this energy scheme re-established.

An augmented-plane-wave calculation has been attempted by Mattheis.[3] The result of this investigation suggests that the lowest minimum is at the center of the zone, but that the Fermi surface has also 6 narrow "arms" extending along the <100> directions. However, experimental verification of such a model would be difficult.

EXPERIMENTAL EVIDENCE

The reflectivity spectrum of $SrTiO_3$ and $BaTiO_3$ has been measured by Cardona.[4] Figure 2 shows the spectrum for $SrTiO_3$; the one for $BaTiO_3$ is very similar. Most of the structure occurs in an energy range of about 13-15 eV, in good agreement with the KL-calculation (bottom of valence band to top of conduction band $\approx$ 12 eV). Cardona suggests that the A_1 and A_2 peaks result from the $X_4' \rightarrow X_5$ and $X_5' \rightarrow X_5$ transitions, respectively, while the B peaks are associated with transitions to the X_1 level.

Other corroboration concerning the width of the valence band comes from soft x-ray emission data.[5] The breadths of the Ti-$K_{\beta 5}$ bands were 3.55, 2.86, and 3.63 eV, respectively. The calculation indicated a valence band width of 4.0 - 4.5 eV.

A number of experimental investigations yield information concerning the density-of-states in the conduction band. (So far no p-type $SrTiO_3$ has been produced; hence our knowledge of the valence band is still mostly theoretical.) Determination of the

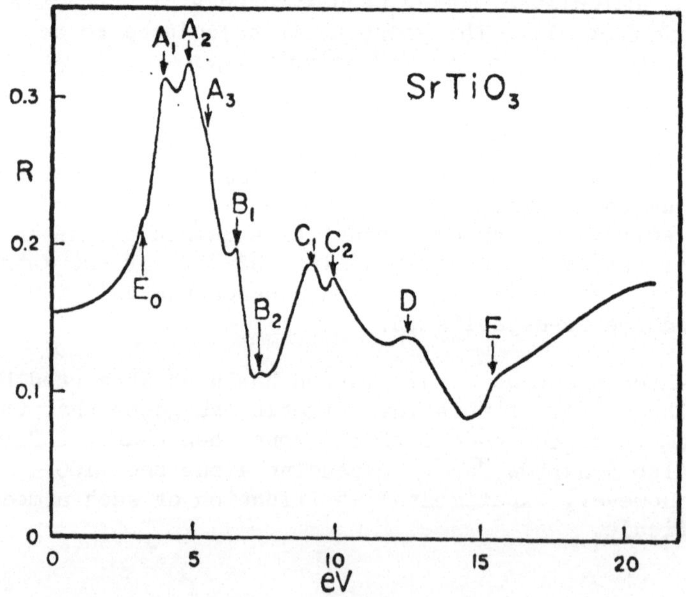

Fig. 2. Reflectivity spectrum of $SrTiO_3$ at room temperature (ref. 4).

Hall- and Seebeck- coefficient enables us to calculate the density-of-states effective mass m^*_D [$= \nu^{2/3} (m_t^2 m_\ell)^{1/3}$, where ν = number of ellipsoids of the Fermi surface]. Such experiments[6] have been carried out on a series of reduced and doped samples over a wide temperature range (4-300 K). The low-temperature Seebeck effect contains only a negligible phonon drag component due to the very small phonon mean-free-path. Hence a straightforward analysis can be made yielding a value $m_D^* = 6\ m_o$ at 78 K. At room temperature one finds a value of 16 m_o which is not confirmed by other measurements.

Another interesting method to find m_D^* is a measurement of the magnetic susceptibility χ.[7] In the case of $SrTiO_3$ the electronic contribution χ_e can be determined as the difference of χ (insulator) and χ (reduced semiconducting $SrTiO_3$). The susceptibility of the conduction electrons contains both a diamagnetic and a paramagnetic contribution:

$$\chi_e = \frac{n\mu^2}{kT} \frac{F'_{1/2}(\zeta)}{F_{1/2}(\zeta)} \left(1 - \frac{m^2}{3m_D^{*2}}\right) , \qquad (1)$$

where n = number of charge carriers, $\mu = g\ \mu_B/2$, μ_B = Bohr magneton, $F_{1/2}$ and $F'_{1/2}$ = integral functions[8] of $\zeta = E_F/kT$, E_F = Fermi energy. For complete degeneracy χ_e becomes independent of temperature (Pauli paramagnetism). Table I shows the values of the susceptibilities and of the resulting effective mass m_D^* for a sample of $SrTiO_3$ before and after several hydrogen reductions.

Table I. Susceptibilities of a sample of $SrTiO_3$ before and after several hydrogen reductions.

Sample	n(4.2 K) cm^{-3}	$T_{degen.}$ °K	χ or $\Delta\chi$ x 10^7 cm^3 g^{-1} 300 K	78 K	4.2 K	$\left(\frac{m_D^*}{m_o}\right)_{4.2}$
Pure	--	--	-1.016	-1.012	-0.919	--
R_1	6×10^{18}	28	+ .037	+ .073	--	--
R_2	7.5×10^{19}	148	+ .270	+ .523	+ .928	5.1
R_3	5.3×10^{20}	550	+1.704	+1.763	+1.719	4.9

Heat capacity measurements[9] also yield information concerning the density of states in the conduction band. A sample containing $1.4 \pm 0.3 \times 10^{20}$ electrons was investigated between 0.3 and 4.0 K. The result could be represented by terms proportional to T, T^3, and T^5. From the linear term one calculates $m_D^* = 5\ m_0$.

Still another approach is a study of the infrared reflectivity of pure and semiconducting $SrTiO_3$. Barker[10] measured both pure and strongly reduced samples. For high concentrations ($n > 3 \times 10^{20}$ carriers/cm^3) polaron effects can be neglected and the plasma edge frequency is given by: $\omega_n^2 = 4\pi ne^2/\epsilon_\infty m^*$. Barker finds values of (2.5 - 2.8) m_0 at room temperature and 1.1 m_0 at 90°K. However, this m^* is a "mobility-mass" $[(m^*)^{-1} = \frac{1}{3}(2/m_t + 1/m_\ell)]$. Assuming a ratio $m_\ell/m_t = 4$ (see below) and $m_D^* = 5\ m_0$, one calculates $m_t = 1.6\ m_0$, $m_\ell = 6.4\ m_0$, and $m^* = 2.1\ m_0$. This agrees reasonably well with Barker's room temperature values, but not with his results at 90 K. The temperature dependence, however, is somewhat questionable considering the magnetic susceptibility values of Table I and Baer's experiments[11] on free carrier absorption (who finds no temperature shift).

SYMMETRY OF THE FERMI SURFACE

Among the methods for probing the symmetry of the energy bands in semiconductors, the magnetoresistive effects (both low and high magnetic field) and the piezoresistive phenomena have been particularly successful. All three of these methods have been brought to bear on semiconducting $SrTiO_3$.

In the weak-magnetic-field case[12] the magnetoresistance is proportional to H^2. Given a current $\underline{J}$ and a magnetic field $\underline{H}$, the magnetoresistance has the form:

$$M_J^H = \Delta\rho/\rho H^2 = b + c\,\frac{(\underline{J}\cdot\underline{H})^2}{J^2H^2} + d\,\frac{J_1H_1^{\,2} + J_2H_2^{\,2} + J_3H_3^{\,2}}{J^2H^2}. \qquad (2)$$

Cubic symmetry suggests four possible energy surfaces in $\underline{k}$ space: a sphere centered at $k = 0$, or sets of ellipsoids along the three major crystallographic directions, respectively. The four models are distinguished from each other by specific relations between the coefficients b, c, and d.

In general :	$b + c + xd = 0,$		(3)
for spherical symmetry :	$x = 0,$	$d = 0,$	(3A)
for <100> type spheroids:	$x = 1,$	$d < 0,$	(3B)
for <111> type spheroids:	$x = 0,$	$d > 0,$	(3C)
for <110> type spheroids:	$x = -1,$	$d > 0.$	(3D)

Transverse ($\underline{H} \perp \underline{J}$) and longitudinal ($\underline{H} \parallel \underline{J}$) magnetoresistive effects were measured at 4.2 K on a number of rectangular samples cut along the <100> or <110> directions (J $\parallel$ <100> or $\parallel$ <110>, respectively) with different carrier concentrations. A rotation diagram is shown in Fig. 3 for a <100> sample: The longitudinal effect appears to be much smaller than the transverse effect. Values of b, c, d, and x were calculated from the experimental data. The results indicate a negative d and x = 1.20 ± 0.04; these values strongly support the case of <100> type spheroids in agreement with the KL-calculation.

The magnitude of the magnetoresistive effect is related to the ellipticity K (= m_ℓ/m_t) of the spheroids. For degenerate statistics the coefficient b can be written as follows:

$$b = \mu^2 \left[\frac{8}{9} \frac{(K^2 + K + 1)(2K + 1)}{K(K + 2)^2} - 1\right]. \tag{4}$$

The experimental value for b leads to K = 4.0 ± 1 or K = 0.34 ± 0.07. (The theoretical band calculation indicates a preference for the former value.)

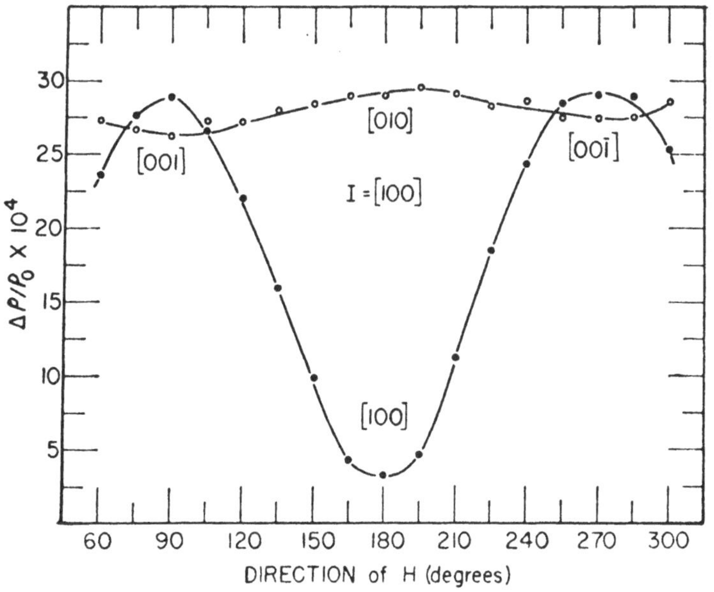

Fig. 3. Angular dependence of the magnetoresistance for sample HR-51 (n = 10^{19} electrons/cm^3) at 7370 Oe and 4.2 K. The open circles refer to a rotation of H about the <100> axis, the solid points to a rotation around an <010> axis.

A more direct measurement of the tensor elements of the effective mass can be obtained from high-field magnetoresistance experiments.[13] The magnetic field H should be large enough to observe the cyclotron orbits of the charge carriers. The conditions necessary for this to occur are:

$$\omega_c\tau > 1, \; \hbar\omega_c > kT, \text{ and } E_F > kT, \qquad (5)$$

where: $\omega_c = eH/m_c c$ [m_c = cyclotron mass (in a plane ⊥ H)], τ = collision time, and E_F = Fermi energy. Considering the large masses in $SrTiO_3$, it is obvious that large magnetic fields are required. As a matter of fact, the experiments were performed in the large solenoid at the U. S. Naval Research Laboratory where fields up to 150 kOe could be achieved. Nb-doped and hydrogen-reduced samples with carrier concentrations in the range 4 to 8 x $10^{18}/cm^3$ were investigated at 1.4 to 4.2 K. At these temperatures the mobilities ranged from 1370 to 10,500 cm^2/volt-sec and the condition $\omega_c\tau > 1$ was fulfilled. Above 50 kOe oscillations in the magnetoresistance were observed (see Fig. 4). These oscillations are periodic in 1/H with a period $\Delta(1/H) = (e\hbar/m_c)(1/E_F)$.

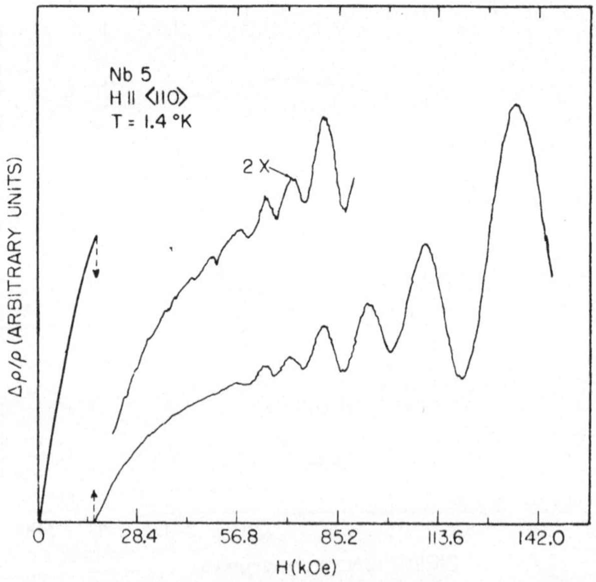

Fig. 4. Oscillatory magnetoresistance of Nb-doped $SrTiO_3$ (n = 6 x 10^{18} electrons/cm^3) at 1.4 K. I and H ∥ <110>.

The measurements were performed with H parallel to the <100>, <110>, and <111> directions. The magnitude of the different periods (and hence m_c's) indicates that the case of <100> spheroids applies. Although the crystal is very slightly tetragonal below 110 K, we assume that the cubic structure is a good approximation at 1-4 K. This statement is based on the fact that the crystal consists of a large number of domains, which will distribute the directions of the c-axes evenly among the three cubic edges throughout the sample. In this case the different m_c's are simply related to the mass tensor elements m_ℓ and m_t. Combining the results for the cyclotron masses with the figure for the density-of-states mass ($m_D^* = 5\ m_0$), one finds that 3 spheroids yield considerably more consistent values than a 6-spheroid model. Furthermore one calculates: $m_t = 1.5 \pm 15\%$ and $m_\ell = 6.0 \pm 40\%$ and K ranging from 2.8 to 6.9. While the error margin is somewhat larger, the results are consistent with the values quoted above. A few experiments concerning the saturation of $\Delta\rho/\rho$ as well as the temperature and magnetic field damping of the oscillations also yield data which confirm the magnitude of m_t, m_ℓ, and K.

The piezoresistive effect has been measured by Tufte and Stelzer[14] and by Frederikse et al.[15] The former group arrived at the conclusion that two sets of conduction band minima played an essential role: 3 minima at the edge of the Brillouin zone in the <100> directions and a slightly lower minimum at the center of the zone. It is difficult to visualize how such a bandstructure could produce Hall coefficients independent of temperature from 1 to 1000 K.

The piezoresistance of semiconductors is usually measured on long thin samples cut along the <100>, <110>, and <111> directions. The longitudinal configuration is easily realized; compressive stresses up to 1 kbar are being applied. In order to extract the three piezoresistive coefficients (for a cubic material), at least one transverse measurement or the hydrostatic pressure experiment has to be performed. In the case of $SrTiO_3$ most information can be deduced from low temperature measurements. However, the crystal undergoes a very small transformation from the cubic to the tetragonal phase at 110 K (c/a = 1.0003 at 78 K). The piezoresistance of <100> and <110> samples saturates at very low stresses, while <111> samples show a continuous linear effect. Hence piezoresistive coefficients are not meaningful. However, the difference between the behavior of the <100> and <110> specimens on the one hand and the <111> specimens on the other is particularly important. It tends to confirm a Si-like bandstructure: A compression along the <111> direction does not produce any line-up of the tetragonal domains or a shift of carriers from one valley to the others; only hydrostatic pressure changes the resistance. In contrast, compression along the <100> direction will line up the tetragonal domains with their c-axes in a plane perpendicular to the stress. It can be shown that one valley (parallel to the c-axis) will move upwards in energy and the other two valleys downwards. A much larger effect results due to a transfer of carriers from the higher to the lower valleys. This is the cause of the early saturation in both the <100> and <110> samples.

SUMMARY

The host of experiments discussed in this paper together make a strong case for the KL-bandstructure of $SrTiO_3$. Some of the strongest support comes from the experimental and theoretical work on superconductivity in this compound. Koonce et al.[16] have had impressive success in calculating the superconducting transition temperature as a function of carrier concentration. In their attempt to reproduce the experimental curve they made use of the electron band structure in all its detail. They conclude that "the normal-state properties are probably close to the values in Table I" (presented in their paper) ". . . since large variations from these values produce T_c vs n_c curves which lie outside the limits of the experimental data."

References

(1) A. H. Kahn and A. J. Leyendecker, Phys. Rev. 135, A1321 (1964).
(2) E. Šimanek and Z. Šroubek, Phys. Stat. Sol. 8, k47 (1965).
(3) Discussed by C. N. Berglund and W. S. Baer, Phys. Rev. 157, 358 (1967).
(4) M. Cardona, Phys. Rev. 140, A651 (1965).
(5) M. A. Blokhin and A. T. Shuvaev, Bull. Acad. Sci. U. S. S. R., Physical Series 26, 429 (1962), Columbia Technical Translations.
(6) H. P. R. Frederikse, W. R. Thurber, and W. R. Hosler, Phys. Rev. 134, A442 (1964).
(7) H. P. R. Frederikse and George A. Candela, Phys. Rev. 147, 583 (1966).
(8) O. Madelung, in Handbuch der Physik, edited by S. Flügge (Springer Verlag, Berlin, 1957), Vol. 20, p. 78.
(9) E. Ambler, J. H. Colwell, W. R. Hosler, and J. F. Schooley, Phys. Rev. 148, 280 (1966).
(10) A. S. Barker, Proceedings of the International Colloquium on Optical Properties and Electronic Structure of Metals and Alloys, Paris, 1965 (North Holland Publishing Company, Amsterdam, 1965).
(11) W. S. Baer, Phys. Rev. 144, 734 (1966).
(12) H. P. R. Frederikse, W. R. Hosler, and W. R. Thurber, Phys. Rev. 143, 648 (1966).

(13) H. P. R. Frederikse, W. R. Hosler, W. R. Thurber, J. Babiskin, and P. G. Siebenmann, Phys. Rev. 158, 775 (1967).
(14) O. N. Tufte and E. L. Stelzer, Phys. Rev. 141, 675 (1966).
(15) H. P. R. Frederikse, W. R. Hosler, and R. C. Casella (to be published: International Conference on the Physics of Semiconductors, Moscow, 1968).
(16) C. S. Koonce et al., Phys. Rev. 163, 380 (1967).

SUPERCONDUCTING SEMICONDUCTORS*

H. P. R. Frederikse

National Bureau of Standards, Washington, D. C.

For many years solid state scientists have been intrigued by the possibility of superconductivity in semiconductors. The whole question received new impetus as a result of the understanding of superconductivity achieved by Bardeen, Cooper and Schrieffer (BCS)[1] in 1957. In the early sixties Gurevich, Larkin and Firsov[2] discussed the possibility of superconductivity in polar semiconductors. At about the same time, Marvin L. Cohen[3] took on the challenge and investigated the likelihood of germanium becoming a superconductor by applying the BCS theory to this semiconductor. His calculation indicated that germanium with 10^{20} electrons/cm^3 should become a superconductor at temperatures of the order of 10^{-4} K. Although this result is hard to test -- considering that T_c is in a rather inaccessible temperature range -- the calculation was nevertheless very useful. It clearly indicated the conditions that had to be optimized in order to obtain a more easily reachable transition temperature. On the basis of this theoretical study, a number of solids were explored for superconductivity. The first semiconducting compound to display superconducting behavior was degenerate GeTe[4] with carrier concentrations around 10^{21} cm^{-3}. (This case is particularly interesting because the carriers are holes.) Shortly afterwards superconductivity was found in reduced $SrTiO_3$.[5] Since then a few more compounds were reported to be superconductors. In most cases (except $SrTiO_3$) the concentration of charge carriers is in the range 10^{21} - 10^{23} cm^{-3}; the transition temperatures of the three "real" semiconductors GeTe, SnTe, and $SrTiO_3$, all are between 0.1 and 0.4 K.

*Parts of the research discussed in this review have been supported by the U.S. National Aeronautics and Space Administration.

In this lecture we will discuss briefly the major points of Cohen's theory. Following this exposé a number of experiments on semiconducting $SrTiO_3$ will be described establishing its superconducting behavior and, in particular, a series of phenomena which show it to be a type-II superconductor.[6] Experimental results concerning the dependence of the transition temperature T_c on carrier concentration n will be presented.[7] One of the highlights of this whole area of research is the success of Cohen's theory in reproducing this curve of T_c vs n.[8]

Finally we will touch upon the occurrence of superconductivity in mixed titanates[9] (such as Sr-Ba and Sr-Ca titanates) and other compounds that can be classified as strongly degenerate semiconductors or semimetals (energy gap ≈ 0).

MAJOR POINTS OF COHEN'S THEORY

One of the most important results of the BCS-theory[1] is the expression for the superconducting transition temperature:

$$kT_c = 1.14\ k\ \theta_D \exp\left(\frac{-1}{N(0)V}\right) \qquad (1)$$

where: $N(0)$ = electronic density of states ($\sim m_D^* n^{1/3}$),
m_D^* = density-of-states effective mass,
n = number of charge carriers,
θ_D = Debye temperature, and
V = effective attractive electron-electron interaction.

It is immediately clear that a measurable T_c (let's say, $T_c >$ 0.05 K) requires a large V to compensate for the small value of N(0) in semiconductors (as compared with metals).

The salient point of Cohen's contribution[3] is his recognition of the advantages of the many-valley band structure over the single-valley case. In such a structure both intravalley and intervalley electron-phonon scattering processes are possible. The latter will increase the attractive electron-electron interaction for the following reasons:

1. Intervalley processes involve large momentum-transfer; hence the screening will be diminished.
2. More valleys in a degenerate semiconductor provide more states into which an electron can be scattered.

With this bandstructure in mind, Cohen goes back to the "gap-equation" of the BCS theory:

$$\Delta_k = -\frac{1}{2}\sum_{k'} \frac{V_{kk'}\,\Delta_{k'}}{E_{k'}} \tanh\left(\frac{E_{k'}}{2\,k_B\,T}\right) \qquad (2)$$

where: $E_k = (\epsilon_k^2 + \Delta_k^2)^{1/2}$,
ϵ_k = normal state one-electron energy (measured from the Fermi level E_F),
Δ_k = superconducting energy gap,
T = temperature,
k_B = Boltzmann constant, and
$V_{kk'}$ = the electron-electron scattering matrix element (between states k and k').

Equation (2) can be transformed into an integral equation. Writing $D_k = (k/k_F)\Delta_k$ and a similar expression for $\Delta_{k'}$, one finds:

$$D_k = -\int \frac{D_{k'}\, K(c,\delta)}{E_{k'}} \tanh\left(\frac{E_k}{2\, k_B\, T}\right) d\epsilon_{k'} \qquad (3)$$

where: $c = k/k_F$ and $\delta = \hbar\omega/E_F$. The electron-electron interaction is now contained in the kernel $K(c,\delta)$, which corresponds to the "N(0)V" parameter in Eq. (1). This kernel comprises two parts associated with intravalley processes ("ra") and intervalley processes ("er"), respectively; each of these include an attractive (phonon-induced) interaction and a repulsive (screened-coulomb) interaction. Hence

$$K(c,\delta) = K_c^{ra} + K_{ph}^{ra} + K_c^{er} + K_{ph}^{er}. \qquad (4)$$

The evaluation of these kernels is the most difficult and complicated task of the whole problem and requires a knowledge of the entire electron and phonon energy band structures of the material involved. We will not enter into the details of this calculation, but only point out some of the major parameters.

The first term includes the Lindhard dielectric function[10]:

$$\epsilon(\beta,\delta) = \epsilon_1(\beta,\delta) + i\epsilon_2(\beta,\delta) \qquad (5)$$

where: $\beta = q/2k_F$, and $q = |k' - k|$. The static dielectric constant enters through both ϵ_1 and ϵ_2. Similarly, the intervalley coulomb interaction (3rd term) contains ϵ and ϵ_0; however, both are modified to take care of screening. One can show that $K_c^{ra} \gg K_c^{er}$.

The second and fourth terms involve electron coupling to both acoustical and optical phonons. As far as intravalley processes are concerned, the coupling to acoustical phonons can be expressed in terms of the deformation potentials Ξ_u and Ξ_d. The latter are easily deduced from piezoresistance measurements.[11] Coupling to the optical phonons is described by means of an optical deformation potential, which can be derived from hot-electron experiments.[12,13] The intervalley processes that contribute to the fourth term bring the phonons at the edge of the zone into the picture. The interaction of electrons with these phonons is lumped together in an average coupling constant or intervalley deformation potential η. This constant η is used as the only adjustable parameter in the calculation of the transition temperature T_c. In turn, η may be calculated from the value of T_c at a particular carrier concentration n_0, and subsequently, the dependence of T_c on n predicted using this value for $\eta(n_0)$.

The theory which we have sketched here very briefly leads to some important conclusions with respect to conditions favorable for superconductivity in semiconductors at a reasonable temperature. These conditions can be summarized as follows:

1. large density of states,
2. many degenerate valleys,
3. large static dielectric constant,
4. maximum phonon degeneracy,
5. large coupling constant η.

Cohen tested his theory first on Ge and concluded that the transition temperature of heavily doped Ge (10^{20} carriers/cm^3) would be around 10^{-3} - 10^{-4} K assuming reasonable values (7 to 8 eV) for the intervalley deformation potential η.

The next step was Si-Ge alloys. For the appropriate ratio of the two elements, one finds ten degenerate minima in the electron energy band structure. Although the case looks promising, no experiments have been performed yet.

As mentioned before, two tellurides as well as semiconducting $SrTiO_3$ have actually been proven to be superconductors.

SUPERCONDUCTIVITY IN $SrTiO_3$

In 1964, the resistance of reduced and doped samples of $SrTiO_3$ was investigated down to 0.05 K using adiabatic demagnetization. It appeared that the resistance of a specimen with 3×10^{19} carriers/cm^3 suddenly dropped into the noise at about 0.25 K. This result was confirmed by a measurement of the magnetic susceptibility χ_0. Below about 0.25 K, χ increased sharply to values of the order of $-(1/4\pi)$ (Meissner effect). Subsequently some 27 specimens were studied; the carrier concentration n ranged from 7×10^{18} to 5.5×10^{20} cm^{-3} as deduced from Hall measurements. The relation between n and the transition temperature T_c is shown in Fig. 1.[7,8] T_c increases with n from 0.05 K at $n = 8 \times 10^{18}$ cm^{-3} up to about 0.3 K for $n = 10^{20}$ cm^{-3}. For larger n the transition temperature <u>decreases</u>; no superconductivity can be detected when n exceeds 3×10^{20} cm^{-3}. Calculation of T_c as outlined in the previous section reproduces the experimental results very well. The adjustable parameter η was derived from a

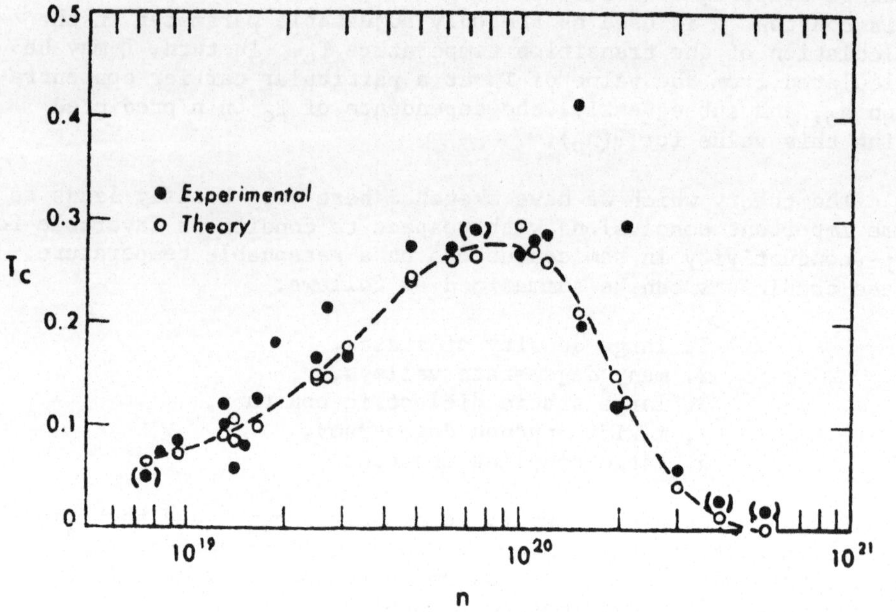

Fig. 1. Transition temperature T_c vs charge carrier concentration n for semiconducting $SrTiO_3$.
● experimental O theory

fit of T_c vs n near the maximum. Some of the normal state parameters used in this calculation have been discussed in the earlier lecture; the best figure for $\eta \equiv 15$ eV. The decrease of T_c for large values of n is the result of screening of the interactions by the charge carriers, preventing formation of Cooper pairs.

Besides a study of the carrier concentration dependence of the transition temperature, $SrTiO_3$ lends itself also to an exploration of the effect of lattice changes on T_c.[9] One can easily substitute ions like Ba^{2+} and Ca^{2+} for Sr^{2+}. A series of sintered specimens of $Ba_xSr_{1-x}TiO_3$ and $Ca_ySr_{1-y}TiO_3$ were prepared and reduced to various degrees. In the Ba-Sr titanate with x = 0.075, superconductivity persisted down to electron concentrations just below 10^{18} cm^{-3}; at the same time the maximum value of T_c -- still for n = 10^{20} cm^3 -- was shifted upwards to about 0.5 K. The titanates with Ba in them remained superconductors up to x = 0.11. The Ca-Sr mixed titanates showed a similar behavior: The maximum T_c increased again to 0.5 K for n = 2 x 10^{19} cm^{-3}, but superconductivity disappeared below n = 3 x 10^{18} cm^{-3}. X-ray data seem to indicate that the mixed titanates (with Ba and Ca) possess an increasingly non-cubic structure. Therefore the degeneracy of the three valleys

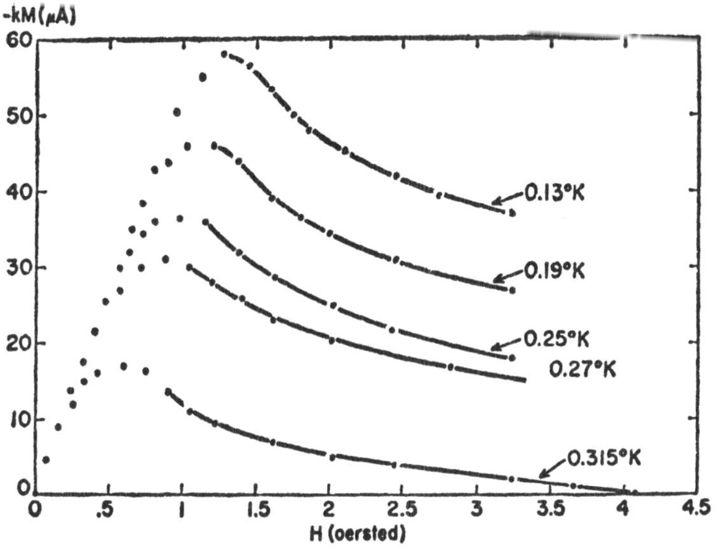

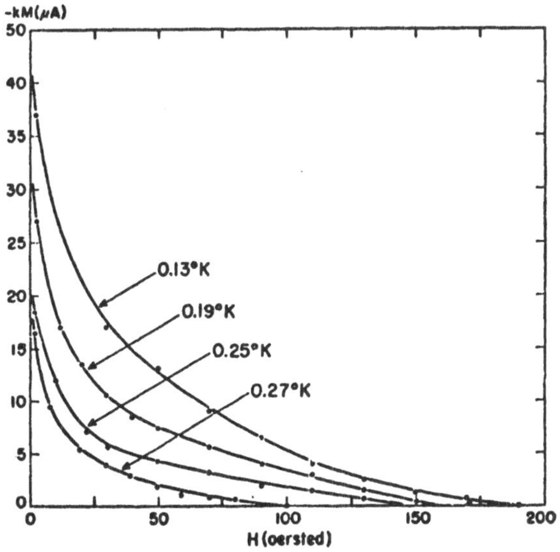

Figs. 2a and 2b. Magnetization of superconducting $SrTiO_3$ (sample HR-24, $n = 1.0 \times 10^{20}$ carriers/cm^3) as a function of magnetic field for different temperatures.

at the <100> zone edge disappears and one would expect a decrease of T_c. However the opposite is observed. Hence the question arises if the dielectric properties do not play a more important role than originally envisaged.

TYPE-II SUPERCONDUCTIVITY

Because of the small concentration of electrons and hence the large penetration depth, one expects $SrTiO_3$ to be a type-II superconductor.[14] It is therefore not surprising that one of the early experiments performed on $SrTiO_3$ was the measurement of the magnetization[6] as a function of magnetic field and of temperature. Results of such an investigation are shown in Figs. 2a and 2b. For this sample (containing 1.0×10^{20} carriers/cm^3), the extrapolated H_{c1} and H_{c2} (at T = 0 K) are 1.95 and 420 Oe, respectively. Using the relation: $H_{c2}/H_{c1} = \sqrt{2}\,\kappa^2 (\ln \kappa + 0.24)^{-1}$, one finds $\kappa = 13$. Other samples yield values for κ of the order 8 to 10. (κ = order parameter.)

In this case of a "dilute" superconductor it is possible to measure the penetration depth λ absolutely.[15] The magnetic susceptibility χ of thin slabs (0.25 - 1 mm) and a small sphere (diameter = 3 mm) was measured as a function of temperature. This susceptibility is less than that calculated for the geometric volume of the sample because of the absence of superconductivity over a distance λ from the surface. The temperature dependence of the penetration depth is given by: $\lambda(T) = \lambda_0(1 - t^4)^{-1/2}$, where $t = T/T_c$. Plotting the normalized susceptibility $\chi(t)/\chi(0)$ versus t, for different values of λ, one can find which λ best fits the experimental data. This comparison is shown in Fig. 3 for a 1-mm slab containing 1.1×10^{20} carriers/cm^3. The penetration depth λ_0 appears to be 3.2×10^{-3} cm. The London-Pippard expression for this parameter is:

$$\lambda_0 = \left[\frac{m^*c^2}{4\pi n e^2} \cdot \frac{\xi_0}{\xi(\ell)}\right]^{1/2} \qquad (6)$$

where: $\xi(\ell) = \ell\xi_0/(\ell + \xi_0)$,
ξ_0 = coherence length, and
ℓ = mean free path.

The theoretical value for λ_0 [Eq. (6)] is about 3 to 10 times smaller than the experimental result depending on the value chosen for m^*.

Another experiment that provides important information on superconductors is a determination of the heat capacity. Such

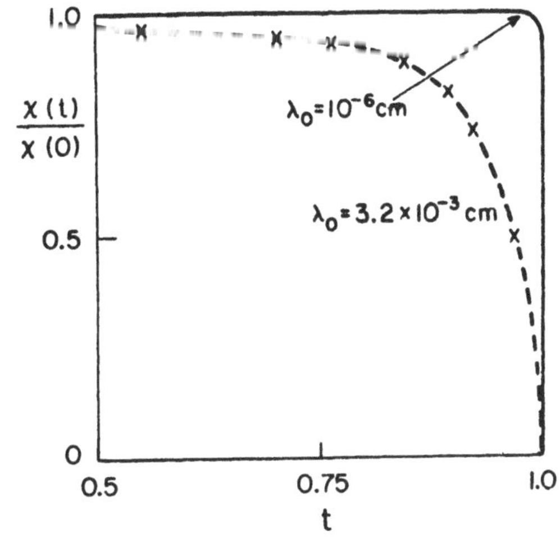

Fig. 3. Normalized susceptibility vs normalized temperature for a single crystal slab ($n = 1.1 \times 10^{20}$ carriers/cm^3).

measurements have been made both on Nb-doped $SrTiO_3$ and on reduced $Ba_{0.08}Sr_{0.92}TiO_3$ between 0.3 and 4.0 K.[6,16] Results on the latter sample in zero field and in fields of 365 Oe and 410 Oe are shown in Fig. 4. This graph shows that the heat capacity C of the lattice is negligible compared to the electronic contribution. The curve begins to bend upward at about 0.55 K, in reasonable agreement with the transition temperature determined by susceptibility measurements ($T_c \approx 0.45$ K). The gradual rather than discontinuous rise of C is probably due to an inhomogeneous electron distribution or to strains in the sample. The data show that magnetic fields decrease the contribution of the superconducting electrons to the heat capacity. No fields high enough to bring the whole sample back into the normal state were applied.

OTHER SUPERCONDUCTING SEMICONDUCTORS

Two other semiconductors, GeTe[4] and SnTe,[17] have been proven to become superconductors below 1.0 K. These compounds are slightly nonstoichiometric; their transition temperatures as a function of carrier concentration are shown in Fig. 5. Specific heat measurements[18] have been made on GeTe without magnetic field and in a field of 270 Oe, confirming the results of other superconductivity tests (Meissner effect). Experimental values for the critical fields H_{c_1} ($\sim$ 15 Oe) and H_{c_2} ($\sim$ 130 Oe) led to an order parameter $\kappa \approx 5$-10. In the case of GeTe it has been possible to measure the superconducting energy gap using the tunneling

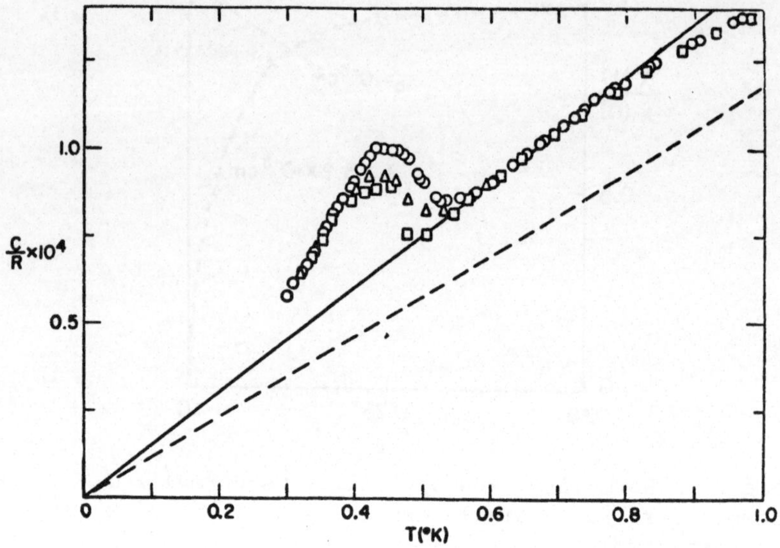

Fig. 4. Heat capacity of $Sr_{0.925}Ba_{0.075}TiO_3$ below 1 K.

O earth field; △ 410 Oe applied field at 0.3 K; □ 365 Oe applied at 4 K.

The solid curve corresponds to an "electronic coefficient" $\gamma/R = 1.50 \times 10^{-4}$, and the dashed line to $\gamma/R = 1.15 \times 10^{-4}$.

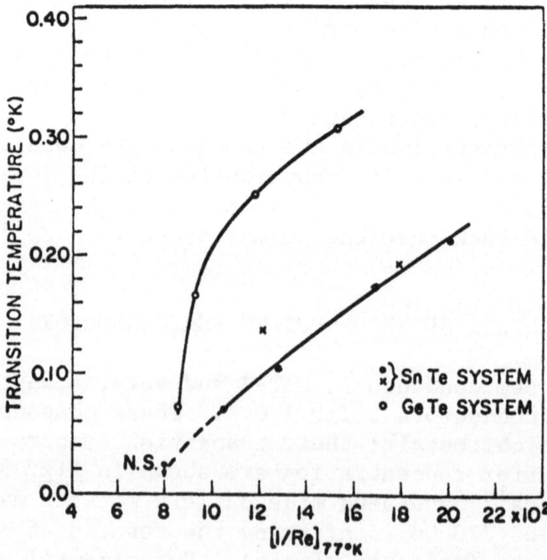

Fig. 5. Transition temperature T_c vs carrier concentration n for GeTe and SnTe.

technique.[19] An Al-Al_2O_3-GeTe tunnel junction was cooled to about 0.09 K. The derivative curve (dI/dV vs V) showed the double maximum characteristic of tunneling betwen two superconductors. The GeTe gap appeared to have the value 4.3 kT_c, slightly higher than the BCS prediction of 3.5 kT_c. A magnetic field of 20 Oe perpendicular to the plane of the junction destroyed the double peak; a single peak remained corresponding to 3.5 kT_c (Al). This peak disappeared with application of a field of 1500 Oe (the critical field of Al).

Two other obvious candidates for superconductivity tests (closely related to the above mentioned compounds) are PbS (or PbTe) and $KTaO_3$. Although PbTe can be doped with alkali metals to about 5 x 10^{20} carriers/cm^3, no superconductivity has been detected in such specimens.[20]

All three lead salts (PbS, PbSe, and PbTe) have been reported superconducting around 5-7 K at some time during the last twenty years. The most recent report came from Lalevic[21]; however, his conclusions were not confirmed by other workers.[20] Actually, the rather small effective masses in the lead salts do not favor superconductivity at such small carrier concentrations.

$KTaO_3$ containing 1 to 3 x 10^{19} electrons/cm^3 was tested for superconductivity, but no sign of such behavior was found.[22] This is the highest concentration of charge carriers that has been achieved in this compound so far. The effective masses are a factor 5 to 6 smaller than those of $SrTiO_3$.[23] Hence the absence of superconductivity (above 0.05 K) is not entirely surprising.

There are a few groups of compounds that show semiconduction in certain ranges of composition and have proven to be superconductors, usually at or near the point where the semiconductor-to-metal transition takes place. An example is the family of tungsten bronzes. Na_xWO_3 is a semiconductor for $x < 0.25$; at about $x = 0.25$ the compound becomes a metal, and for $x = 0.28$ it appears to be superconducting.[24] Another group is that of the carbon halides. These compounds have a layer-like structure like graphite and are prepared by heating pyrolytic graphite together with very pure alkali metal. Depending on the amount of metal, a range of superconducting transition temperatures was observed (see Table I).

The energy gap of several semiconductors can be decreased under high pressure. Once the solid has reached the metallic state ($E_g \approx 0$) the material becomes a superconductor.

The behavior of all these compounds is in rather good agreement with Cohen's idea of many degenerate valleys favoring a transition to superconductivity. A list of the materials discussed in this lecture is presented in Table I.

Table I. Superconductive Semiconducting and Semimetallic Compounds

Compound	Composition	T_c(°K)	n(cm^{-3})	Structure	Ref.
Ge_xTe	x = .976	.08	(8 - 16). 10^{20}	f.c.c.	a
	.963	.17			
	.950	.26			
	.937	.30			
Sn_xTe	.97	.16	(8→ 22). 10^{20}	f.c.c.	b
$SrTiO_{3-x}$	x=10^{-2}-10^{-4}	0.3 (max)	~ 10^{20}	perovskite	c
		0.05→0.3→0.05	5 x 10^{18}-10^{21}		
$Sr_{1-y}Ba_y TiO_{3-x}$	y= 0 - 0.12	0.5 (max)	5 x 10^{17}-10^{21}	pseudocubic	d
$Sr_{1-z}Ca_z TiO_{3-x}$	z= 0 - 0.30	0.5 (max)	5 x 10^{18}-10^{21}	pseudocubic	d
C_8K		0.55		hex.	e
C_8Rb		0.023 - 0.151		hex.	e
C_8Cs		0.020 - 0.135		hex.	e
Na_xWO_3	x= 0.28	0.55 ± 0.02		tetr. I	f,g
	x< 0.28, x> 0.35	no superc.		tetr. II	g
K_xWO_3	0.27 < x < 0.31	0.5 ± 0.2		perovskite	g
	.40 < x < 0.57	1.5 ± 0.5		tetr.	g
Rb_xWO_3	.27 < x < 0.29	1.98 ± 0.29		hex.	g
Cs_xWO_3	x = 0.32	1.12 ± 0.32		hex.	g
Ca_xWO_3	x = 0.10	1.4 - 3.4		tetr.	h
Sr_xWO_3	x = 0.08	2.0 - 4.0		tetr.	h
Ba_xWO_3	x = 0.14	? - 2.2		tetr.	h,j
In_xWO_3	x = 0.11	? - 2.8		tetr.	h
Tl_xWO_3	x = 0.30	2.0 - 2.14		tetr.	h
Under Pressure (P)					
	P (atm)				
Te	56,000	3.3	under pressure	hex	k
InTe	~ 30,000	3.2 - 3.45	under pressure	f.c.c.	ℓ
InSb	~ 25,000	1.6 - 2.1	under pressure	tetr. (white Sn)	m

References, Table I.

(a) R. A. Hein, J. W. Gibson, R. Mazelsky, R. C. Miller and J. K. Hulm, Phys. Rev. Letters 12, 320 (1964).
(b) R. A. Hein, J. W. Gibson, R. S. Allgaier, B. B. Houston, R. Mazelsky, and R. C. Miller, Proceedings of the Ninth International Low Temperature Conference (Plenum Press, New York, 1965), part A, p. 604.
(c) J. F. Schooley, W. R. Hosler, E. Ambler, J. H. Becker, M. L. Cohen, and C. S. Koonce, Phys. Rev. Letters 14, 305 (1965).
(d) H. P. R. Frederikse, J. F. Schooley, W. R. Thurber, E. Pfeiffer, and W. R. Hosler, Phys. Rev. Letters 16, 579 (1966).
(e) N. B. Hannay, T. H. Geballe, B. T. Matthias, et al., Phys. Rev. Letters 14, 225 (1965).
(f) C. J. Raub, A. R. Sweedler, M. A. Jensen, S. Broadston, and B. T. Matthias, Phys. Rev. Letters 13, 746 (1964).
(g) A. R. Sweedler, C. J. Raub, and B. T. Matthias, Phys. Rev. Letters 15, 108 (1965).
(h) P. A. Bierstedt, T. A. Bither, and F. J. Darnell, Solid State Comm. 4, 25 (1966).
(j) A. R. Sweedler, J. K. Hulm, B. T. Matthias, T. H. Geballe, Phys. Letters 19, 82 (1965).
(k) B. T. Matthias and J. L. Olsen, Phys. Letters 13, 202 (1964).
(ℓ) B. R. Tittman, A. J. Darnell, H. E. Bömmel, and W. F. Libby, Phys. Rev. 135, A1453 and A1460 (1964).
(m) M. D. Banns, et al., Science 142, 662 (1963). Also, S. Geller, A. Jayaraman, and W. G. Hull, Appl. Phys. Letters 12, 474 (1964).

References

1. J. Bardeen, L. N. Cooper and J. R. Schrieffer, Phys. Rev. 108, 1175 (1957).
2. V. L. Gurevich, A. I. Larkin, and Y. A. Firsov, Soviet Phys. -- Solid State 4, 131 (1962).
3. M. L. Cohen, Phys. Rev. 134, A442 (1964).
4. R. A. Hein, J. W. Gibson, R. Mazelsky, R. C. Miller, and J. K. Hulm, Phys. Rev. Letters 12, 320 (1964).
5. J. F. Schooley, W. R. Hosler, and M. L. Cohen, Phys. Rev. Letters 12, 474 (1964).
6. E. Ambler, J. H. Colwell, W. R. Hosler, and J. F. Schooley, Phys. Rev. 148, 280 (1966).
7. J. F. Schooley, W. R. Hosler, E. Ambler, J. H. Becker, M. L. Cohen, and C. S. Koonce, Phys. Rev. Letters 14, 305 (1965).
8. C. S. Koonce, M. L. Cohen, J. F. Schooley, W. R. Hosler, and E. R. Pfeiffer, Phys. Rev. 163, 380 (1967).
9. J. F. Schooley, H. P. R. Frederikse, W. R. Hosler, and E. R. Pfeiffer, Phys. Rev. 159, 301 (1967).

10. J. Lindhard, Danske Videnskab. Sęlskab., Mat. -- Fys. Medd. 28, 8 (1954).
11. R. W. Keyes, Solid State Phys. 11, 149 (1960).
12. H. J. G. Meyer, Phys. Rev. 112, 298 (1958).
13. H. G. Reik and H. Risken, Phys. Rev. 126, 1737 (1962).
14. V. L. Ginzburg and L. D. Landau, Zh. Eksperim. i Teor. Fiz. 20, 1064 (1950); A. A. Abrikosov, Soviet Phys. -- JETP (English transl.) 5, 1174 (1957); L. P. Gorkov, Soviet Phys. -- JETP (English transl.) 10, 998 (1960).
15. J. F. Schooley and W. R. Thurber, Proceedings of the International Conference on the Physics of Semiconductors, Kyoto, 1966, J. Phys. Soc. Japan Suppl. 21, 639 (1966).
16. J. H. Colwell, Phys. Letters 25A, 623 (1967).
17. R. A. Hein, J. W. Gibson, R. S. Allgaier, B. B. Houston, R. Mazelsky, and R. C. Miller, Proceedings of the Ninth International Low Temperature Conference (Plenum Press, New York, 1965), part A, p. 604.
18. L. Finegold, Phys. Rev. Letters 13, 233 (1964).
19. P. J. Stiles, L. Esaki, and J. F. Schooley, Phys. Letters 23, 206 (1966).
20. J. K. Hulm, C. K. Jones, R. C. Miller, and T. Y. Tien, Proceedings of the Tenth International Low Temperature Conference, Moscow, 1966 (to be published).
21. B. Lalevic, Phys. Letters 16, 206 (1965).
22. J. F. Schooley, private communication.
23. L. S. Senhouse, G. E. Smith, and M. V. DePaolis, Phys. Rev. Letters 15, 776 (1965).
24. C. J. Raub, A. R. Sweedler, M. A. Jensen, S. Broadston, and B. T. Matthias, Phys. Rev. Letters 13, 746 (1964).

BEHAVIOR OF TYPE II SUPERCONDUCTORS IN VERY HIGH FIELDS

K. HECHLER and E. SAUR

Institute of Applied Physics

University of Giessen, Giessen, W. Germany

I. EXPERIMENTAL RESULTS

In some preliminary papers[1] we have reported on measurements carried out on some type II superconductors up to very high fields. In the meantime more measurements have been done, and the results on NbN have been published in detail[2]. The details on critical data of A 15 (ß-W) compounds are to be published[3,4]. From these data the critical field curves for some compounds are given in Fig. 1. The critical fields up to 150 kOe have been measured directly, but the higher critical fields at 4.2 °K for the ß-W compounds have been determined by extrapolation of the quenching curves to currents as low as 5 mA. The quenching curves for V_3Si and V_3Ga with various conditions of preparation are given as examples in Fig. 2 and 3 respectively. The experimental values of the upper critical field $H_{c2}(4.2)$ at 4.2 °K and the extrapolated values of the upper critical field $H_{c2}(0)$ at zero temperature are listed in Table 1. All measurements have been done at the MIT Francis Bitter National Magnet Laboratory, Cambridge, Mass., USA, using Bitter type solenoids for the generation of the high fields.

II. THE SHAPE OF THE CRITICAL FIELD CURVES

To check the critical field curves on parabolic shape the critical field values have to be plotted versus T^2 or $t^2 = (T/T_c)^2$. For the values given in Fig. 1 this is shown in Fig. 4. The deviation from the parabolic shape

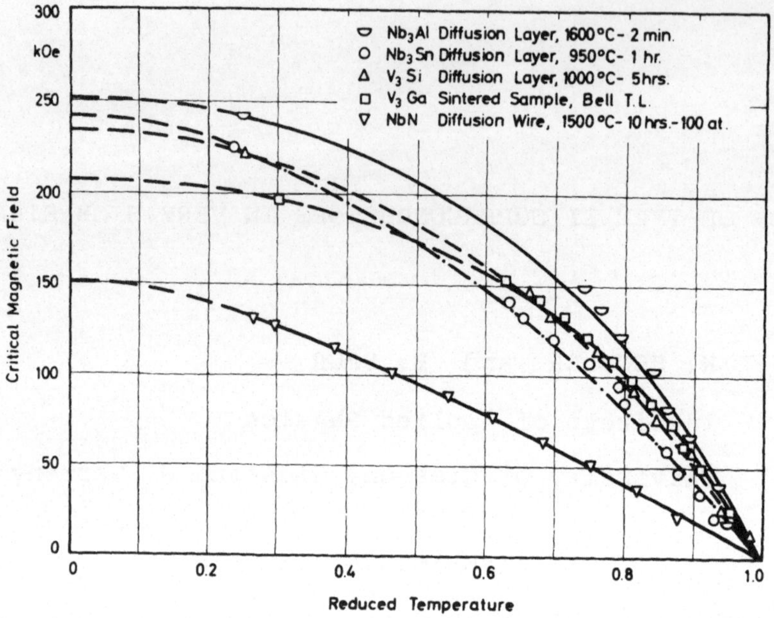

Fig. 1. Critical field curves for Nb_3Al, Nb_3Sn, V_3Si, V_3Ga and NbN.

Table 1: Upper Critical Fields in kOe at 4.2 °K and Zero Temperature

	Nb_3Al	Nb_3Sn	V_3Si	V_3Ga	NbN
$H_{c2}(4.2)$	242	225	220	196	132
$H_{c2}(0)$	252	245	235	208	153

is clearly to be seen. Only the critical field values for Nb_3Sn and V_3Si are represented nearly by straight lines.

The real shape of the critical field curves can be obtained from a formula given by WERTHAMER et al.[5] taking into account Pauli-paramagnetism as well as spin-orbit scattering in the superconductor. By using the experimental values of Maki-parameters α and fitting the curves by selected values of λ_{so}, which measures the spin-orbit effect, the theoretical curves for the ß-W compounds in Fig. 5 have been drawn. The matching of experimental values and theoretical curves is fairly good and gives a method of determining λ_{so}. For more details compare the paper of HECHLER et al.[4]

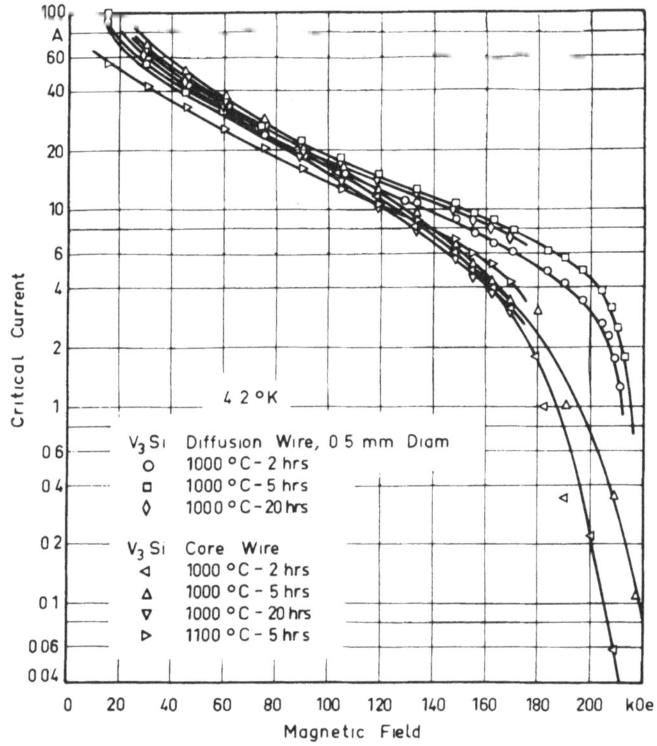

Fig. 2. Quenching curves for different samples of V_3Si.

III. COMPARISON OF EXPERIMENTAL RESULTS WITH RECENT THEORY

In a recent theory HELFAND and WERTHAMER[6] give an equation for the reduced initial slope $(-dh/dt)_{t=1}$ of the critical field curve as a function of the reduced mean number of collisions ρ, which is defined by:

$$\rho = 8.85 \times 10^{3} \times \gamma^{1/2} \times \rho_n / K_0 \tag{1}$$

where γ_3 is the electron specific heat coefficient in $erg/cm^3 (^{\circ}K)^2$, ρ_n the normal state resistivity just above T_c in Ohm·cm, and K_0 the Ginzburg-Landau parameter in the pure limit. The reduced slope of the upper critical field curves at T_c or $t = 1$ may be calculated from experimental data by the formula[7]:

$$(-dh/dt)_{t=0} = \frac{0.5807}{K_0} \cdot \frac{T_c}{H_c(0)} \cdot (-dH_{c2}/dT)_{T=T_c} \tag{2}$$

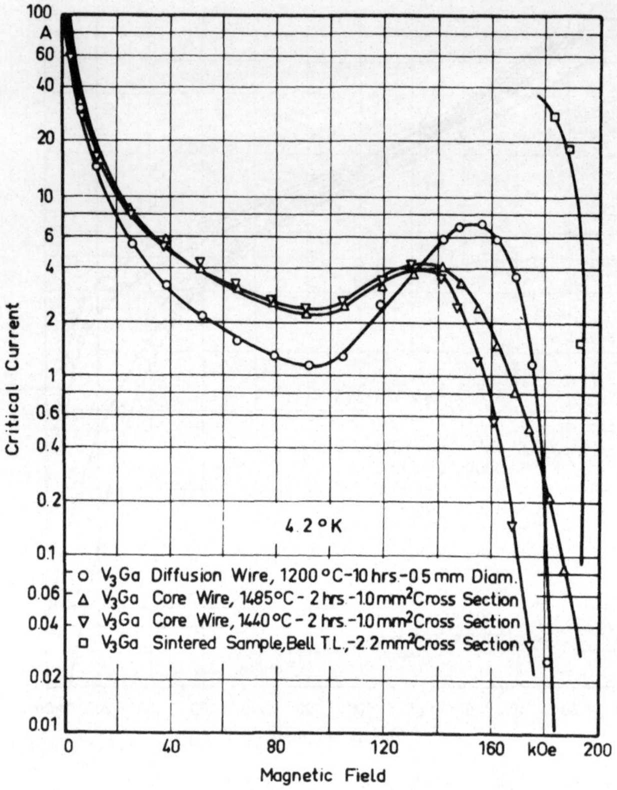

Fig. 3. Quenching curves for different samples of V_3Ga.

where $H_c(0)$ is the thermodynamical critical field at zero temperature:

$$H_c(0) = 2.24 \cdot \gamma^{1/2} \cdot T_c \tag{3}$$

and $(-dH_{c2}/dT)_{T=T_c}$ the initial slope of the critical field curves, which may be taken from Fig. 1. In Fig. 5 the experimental data of the reduced initial slope $(-dh/dt)_{t=1}$ for a variety of type II superconductors are plotted versus ρ. They fit well the theoretical curve after HELFAND and WERTHAMER[6], which is also shown. FIETZ and WEBB[7] have presented results for other type II superconductors in the range of ρ from 0.001 to 20. So our results give a continuation up to ρ = 300. More details will be given in the paper by HECHLER et al.[4]

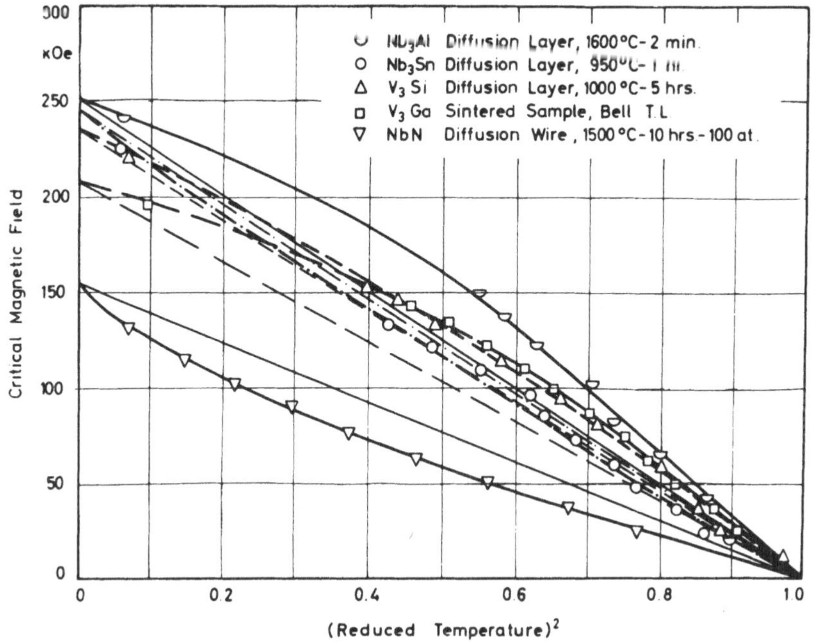

Fig. 4. Critical fields as a function of the square of reduced temperature.

ACKNOWLEDGMENTS

We wish to thank the Deutsche Forschungsgemeinschaft, the Stiftung Volkswagenwerk and the Fraunhofer-Gesellschaft for financial support.

REFERENCES

1 SAUR, E. and H. WIZGALL: Les Champs Magnetiques Intenses, Colloque International, Grenoble 1966, p. 223.

SAUR, E. and H. WIZGALL: Proc. 10th Intern. Conf. Low Temp. Phys., Moscow 1966, p. 125.

SAUR, E. and H. WIZGALL: Proc. 1st Intern.Cryog. Eng. Conf., Tokyo and Kyoto 1967, p. 156.

2 HECHLER, K., E. SAUR and H. WIZGALL: Z. Physik 205, 400 (1967).

3 OTTO, G., E. SAUR and H. WIZGALL: J. Low Temp. Physics (in press).

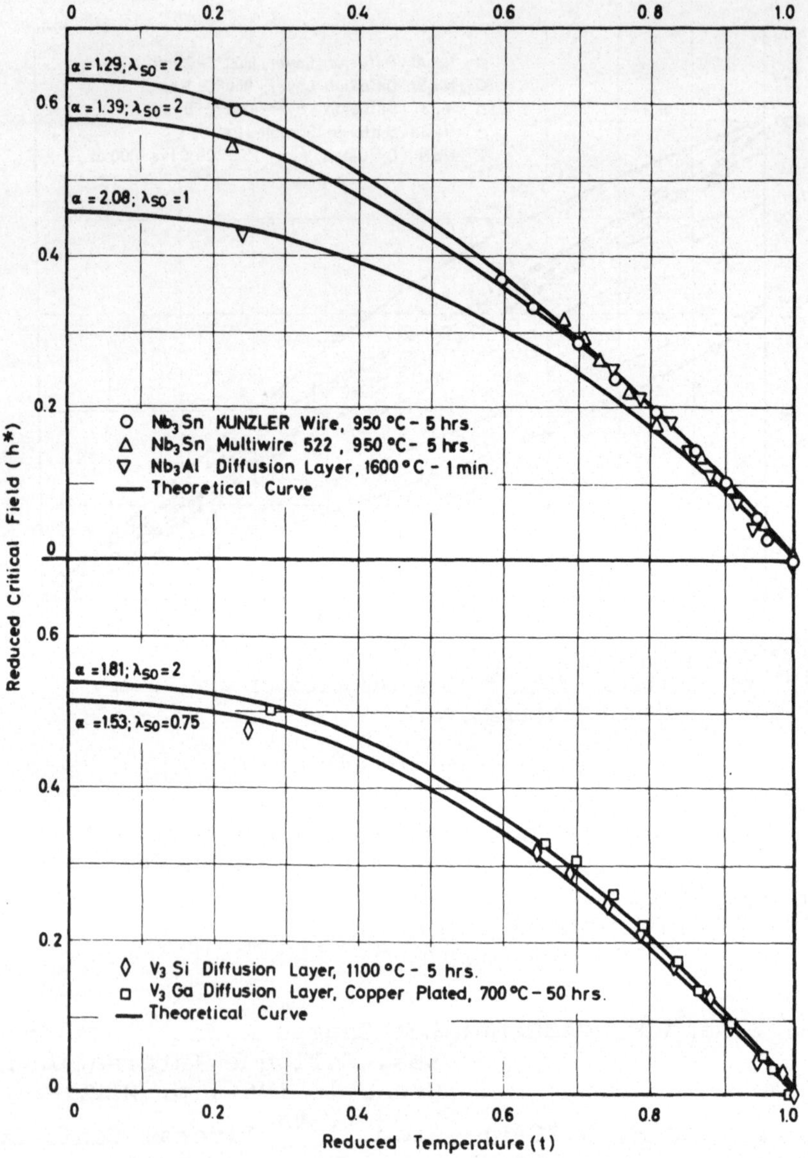

Fig. 5. Reduced critical field curves for some β-W compounds, experimental and theoretical.

References (cont.):

4 HECHLER, K., G. HORN, G. OTTO and E. SAUR: J. Low Temp. Physics (in press).

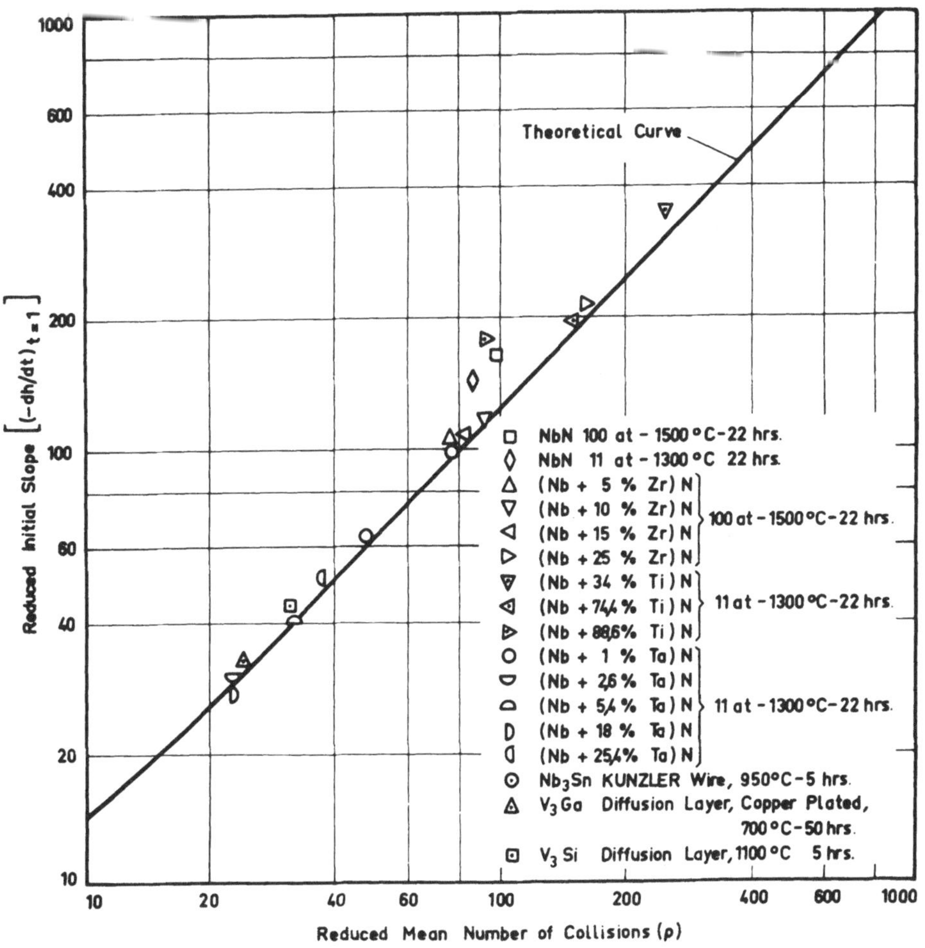

Fig. 6. Reduced initial slope of critical field curves versus reduced mean number of collisions.

References (cont.):

[5] WERTHAMER, N.R., E. HELFAND and P.C. HOHENBERG: Phys. Rev. 147, 295 (1966).

[6] HELFAND, E. and N.R. WERTHAMER: Phys. Rev. 147, 288 (1966)

[7] FIETZ, W.A. and W.W. WEBB: Phys. Rev. 161, 423 (1967)

VORTEX MOTION IN TYPE II SUPERCONDUCTORS

W. L. McLean

Rutgers University

Analogies have been drawn between the motion of vortex lines in the classical perfect fluid and some of the non-equilibrium phenomena in type II superconductors. We consider here a few of the difficulties encountered in such an approach.

The flow of electric currents in superconductors, like the flow of currents in the classical perfect non-viscous fluid, is usually free from dissipation - the motion of the electrons being similar to the motion of the atoms of the fluid, except that in the first case electromagnetic forces arise because of the unbalanced charge on the electron [1]. In the mixed state of a type II superconductor vortices exist in the electronic superfluid [2]. There are important differences - which we discuss below - between these vortices and those in the perfect classical fluid.

EQUATION OF MOTION OF THE CLASSICAL FLUID

The basic equation of motion of the classical uncharged non-viscous fluid is the Euler equation [3], which is equivalent to Newton's second law: acceleration = force density/mass density. In the simplest case, the force density arises from the pressure of adjacent fluid on an element of unit volume and is $-\underline{\nabla P}$, where $\underline{\nabla P}$ is the pressure gradient. The earth's gravitational field contributes a term $\rho\underline{g}$ to the force density where ρ is the mass density and $\underline{g}$ the acceleration due to gravity. The equation of motion of the fluid is then

$$\rho \frac{d\underline{v}}{dt} = -\nabla P + \rho \underline{g} , \qquad (1)$$

where $\underline{v} = \underline{v}(\underline{r})$ is the velocity of the fluid particles (averaged over a small volume element at the point with position vector $\underline{r}$). $\underline{v}(\underline{r})$ is in general a function of time so that

$$\frac{d\underline{v}}{dt} = \frac{\partial \underline{v}}{\partial t} + \underline{v} \cdot \nabla \underline{v} .$$

When the velocity field $\underline{v}(\underline{r})$ is irrotational, i.e. when $\text{curl}\ \underline{v} = 0$, then, $\underline{v}\cdot\nabla\underline{v} = \frac{1}{2}\nabla v^2$ - since $\underline{v} \times \text{curl}\ \underline{v} \equiv \frac{1}{2}\nabla v^2 - \underline{v}\cdot\nabla\underline{v}$. If the fluid is incompressible so that $\nabla\rho = 0$, the equation of motion can be written

$$\rho\frac{\partial \underline{v}}{\partial t} = -\nabla(P + \tfrac{1}{2}\rho v^2 - \rho \underline{g}\cdot\underline{r}) , \qquad (2)$$

so that in the steady state, when $\frac{\partial \underline{v}}{\partial t} = 0$, we arrive at the well-known Bernoulli equation

$$P + \tfrac{1}{2}\rho v^2 - \rho\underline{g}\cdot\underline{r} = \text{constant}. \qquad (3)$$

EQUATION OF MOTION OF THE ELECTRONIC SUPERFLUID

The analogue of equation (1) for the superconductor at low temperatures, where all the electrons can be regarded as being in the superfluid phase, is [4]

$$m \frac{d\underline{v}}{dt} = -\nabla\mu_o + e\underline{E} + \frac{1}{c}e\underline{v} \times \underline{B} .$$

We have neglected the effect of gravity here but included the forces due to electric and magnetic fields. μ_o is the chemical potential and the term in its gradient is analogous to the force in (1) arising from the pressure gradient. In a superconductor, we assume that the velocity field $\underline{v}(\underline{r})$ is related to the magnetic induction $\underline{B}(\underline{r})$ by the London equation [1] $\text{curl}\ \underline{v} = -e\underline{B}(\underline{r})/mc$. The analogue of equation (2) is thus

$$\frac{\partial}{\partial t}(m\underline{v} + \frac{e}{c}\underline{A}) = -\nabla(\mu_o + \tfrac{1}{2}mv^2 + eV)$$

where $\underline{A}$ is the magnetic vector potential and V the scalar electric potential, which are related to the magnetic induction and the electric field by

$$\underline{B} = \text{curl}\ \underline{A},\ \underline{E} = -\frac{1}{c}\frac{\partial \underline{A}}{\partial t} - \nabla V .$$

Under equilibrium conditions $\mu_o + \frac{1}{2}mv^2 + eV = \text{constant}$.

The left hand side of this equation can be identified with the total chemical potential (electrochemical potential) of thermodynamics. Alternatively, $\underline{E}=-\underline{\nabla}V=\underline{\nabla}(\mu_0+\frac{1}{2}mv^2)$. The relative sizes of the contributions from the two terms on the right hand side are approximately in the ratio of the Fermi energy to the rest mass energy of the electron, i.e. 1 : 10^5, so that the electric field is $\underline{E} = \underline{\nabla}(\frac{1}{2}mv^2)$.

The existence of such a field has been recently demonstrated by Bok and Klein [5] by an elegant measurement of the electrical potential difference between two points on the surface of a cylinder placed in a transverse magnetic field. Earlier attempts in many laboratories to observe this analogue of the Bernoulli effect in superconductors failed because contact had been made with the surface by the probes of a conventional (current measuring) voltmeter, which is sensitive to the total chemical potential difference $\Delta\mu$ rather than to the electric potential difference ΔV. But the former is zero since μ is uniform. Interesting effects in ΔV arising from the band structure of the superconductor are discussed by Adkins and Waldram [6].

THE VORTEX IN THE CLASSICAL NON-VISCOUS FLUID

In deriving equations (2) and (3) above it was assumed that the flow was irrotational i.e. that $\underline{\text{curl } v} = 0$. The simplest case of rotational flow is in a vortex line with the velocity field $\underline{v}(\underline{r}) = (\underline{K}\mathrm{x}\underline{r})/2\pi r^2$, where $\underline{K}$ is the circulation or "strength" of the vortex, a vector parallel to the line which forms the vortex axis - the line $\underline{r}_\perp = (\underline{K} \times \underline{r}) \times \underline{K}/K^2 = 0$. The curl of the velocity field is still zero, except on the axis where $\underline{v}(\underline{r}) \to \infty$ as $r_\perp \to 0$. In a vortex in water or in a whirlwind this limit is never achieved because of turbulence or other effects.

THE VORTEX IN TYPE II SUPERCONDUCTORS

When a long needle-like cylinder of a superconductor is introduced into an initially uniform magnetic field of strength H_e parallel to the axis of the cylinder, the magnetic flux, Φ, passing through a section of the cylinder perpendicular to its axis will have one of the types of dependence on H_e shown in figure 1 - the first if it is a type I, the second if it is a type II superconductor. In regions (1), apart from a small effect

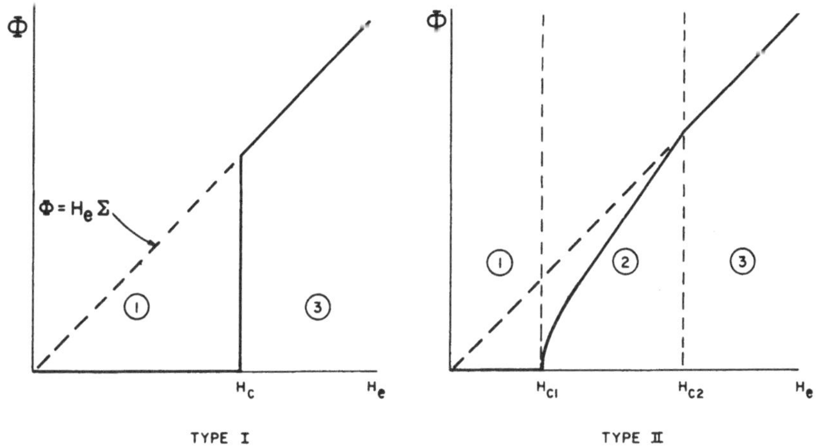

Figure 1. The dependence of magnetic flux passing through the cross-section of long cylindrical superconductors on magnetic field intensity.

near the surface, there is complete diamagnetism - the so-called Meissner effect, embodied in the London equation referred to above. In regions (3), again neglecting surface effects, the metal is in the normal state and $\Phi = \Sigma H_e$, where Σ is the area of the cross-section (we neglect the small Landau diamagnetism [7] of the normal state). Region (2) is peculiar to the type II superconductor and is known as the mixed state. Magnetic flux enters at H_{c1} in the form of bundles of flux lines with the density falling off monotonically with distance away from the centre of the bundle as shown in figure 2. But from Maxwell's equation $\underline{J} = c\ \underline{\text{curl}\ B}/4\pi$, it can be seen that there is a velocity field in the electronic superfluid qualitatively similar to that in the classical vortex line. It is shown in appendix 1, using the single-valuedness of the many-particle wave-

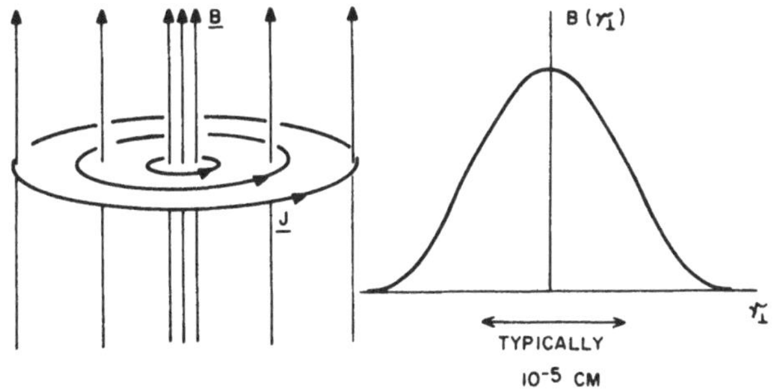

Figure 2. The magnetic flux distribution near the core of a vortex in a superconductor.

function of the electrons, that the total magnetic flux Φ_L of the flux bundle is quantized [8]: i.e. $\Phi_L = hc/2e$. Most of this flux is confined to an area which typically is about 10^{-5} cm in diameter.

Most of the equilibrium properties of the mixed state have been satisfactorily described by the Ginzburg-Landau theory [9], which was initially derived from the thermodynamic treatment of second-order phase transitions [10] but later from the quantum mechanics of the interacting electrons in a superconductor [11]. Near to H_{c2} the vortex lines are close together and form a regular array [12]. Neutron diffraction [13] and other [14] experiments have shown this to be a triangular lattice. Professor Saur will deal in his lecture with the detailed comparison of some of the equilibrium properties with theory. Other results are discussed in the review by Serin [15].

VORTEX MOTION IN THE CLASSICAL FLUID

Without for the moment inquiring how such a motion can be brought about, suppose that there is a uniform movement in the y direction of well-separated vortex lines, each line having its circulation $\underline{K}$ parallel to the z axis. If the vortex lines cross a unit length of the x axis at the rate ν, it can be shown from Euler's equation that the pressure gradient along the x axis is proportional to νK. This pressure gradient can be regarded as a reaction to the Magnus force that a vortex line experiences as it moves through the fluid. If the velocity of the vortex line relative to the fluid at a large distance from the line is $\underline{v}_L$, the Magnus force per unit length of the line is given by $\rho \underline{K} \times \underline{v}_L$. Physically this is the same phenomenon as the Bernoulli effect described by equation (3). The fluid velocity around a moving vortex line is no longer the same at equal distances from its axis. Consequently the pressure is no longer the same at equal distances from the axis.

VORTEX MOTION IN SUPERCONDUCTORS

Although zero resistance was thought to be a necessary property of superconductors, it has been found that when a current is passed through a type II superconductor in its mixed state a potential difference appears, which approximately obeys Ohm's law [16]. It should be noted that a conventional voltmeter records the difference in total chemical potential (electrochemical potential) and not just electric potential. By considering the electric field induced by the uniform drift of a pattern of magnetic flux, it has been shown [17] that the flux-flow voltage gradient - the gradient in electrochemical potential - is $|\underline{\nabla\mu}| = 2ev_{Lt}B/c$, where v_{Lt} is the component of $\underline{v_L}$ perpendicular to $\underline{\nabla\mu}$. But $v_{Lt}(B/\Phi_L) = \nu$, where ν is the rate at which the vortex lines cross unit length of a line parallel to $\nabla\mu$. Thus $|\underline{\nabla\mu}| = 2e\nu\Phi_L/c = \nu h$ - a result which should be compared with the result above for the pressure gradient in the classical fluid. (We note that Φ_L corresponds to the vortex strength K. Also, in an incompressible classical fluid, $\underline{\nabla\mu_o} = (\underline{\nabla P}/\rho)$.

The quantum mechanical description of this phenomenon has been discussed by Anderson [18]. As is shown in appendix 2, the electro-chemical potential is proportional to the rate of change of the contribution of a single particle to the phase of the many-particle wavefunction of the fluid: $\mu = \hbar\, ds/dt$. The average value of the electrochemical potential difference between two points, $(\mu_1 - \mu_2)$, is thus $(\mu_1 - \mu_2) = \hbar d(s_1 - s_2)/dt$. Figure 3 shows how $(s_1 - s_2)$ changes as a vortex line moves across the line joining the points 1 and 2. Because every vortex in a type II superconductor contains only

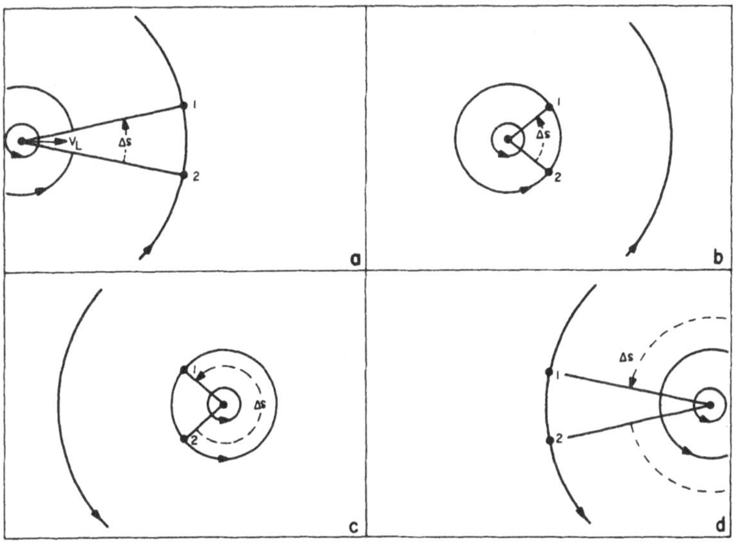

Figure 3. Illustrating how the difference between the phases at the points 1 and 2 varies as a vortex line with velocity v_L passes by.

one unit of flux, the phase of the many body wavefunction changes by 2π in one complete circuit around the vortex. Thus as the vortex moves from a to d, the phase difference $(s_1 - s_2)$ changes by almost 2π. It follows that if the line joining the points 1 and 2 is of unit length

$$(\mu_1 - \mu_2) = \hbar\ \nu\ 2\pi = \nu h.$$

Note that the electronic charge does not appear in this result which would be the same for an uncharged fluid. Thus although the flux-flow voltage may appear to arise from electromagnetic induction, it also has an analogue in the uncharged fluid. The many-particle wavefunction approach can also be used to derive Euler's equation in pure superconductors at absolute zero [19].

So far, we have assumed the vortex to be in motion, without considering how such motion is brought about. As in the classical fluid, the motion of the background relative to the vortex gives rise to the Magnus effect - a net force on the vortex which, in the steady state, is balanced by other forces that we consider below. For a stationary vortex the Magnus force is the same as the so called "Lorentz force", the force on the electron fluid due to its motion in the local magnetic field. (Care has to be taken to consider whether or not the field lines should be considered rigidly attached to the fluid). The "Lorentz force" due to a transport current $\underline{J}_t$ (as opposed to the current flow in the vortex itself) is $\underline{J}_t \times \underline{\Phi}_L$. Thus, if the vortex array is initally at rest, it can be put into motion by passing a current $\underline{J}_t$ through the array.

Having noted some similarities between the perfect classical fluid and superconductors, we now pass on to the differences which limit the usefulness of analogies between the two systems.

THE CORE OF THE VORTEX IN A SUPERCONDUCTOR

Just as the velocity in the classical vortex increases as r is reduced, the velocity rises near the centre of a vortex in the electronic superfluid. At a distance r_c from the axis, the velocity reaches the critical value at which excitations can form in the superfluid at the expense of its kinetic energy. This happens when the superfluid velocity satisfies the condition $v(r_c)p_F = \Delta$, where p_F is the Fermi momentum and Δ the energy gap - the energy per electron required to destroy the electron pairing, which is one of the main characteristics of the superconducting ground state [20]. The excitations are like normal electrons in many of their properties. Thus the central part of the vortex can be thought of as consisting not of superfluid but of normal electrons, which can give rise to electrical resistance and in alloys are capable of carrying heat. The theoretical work of Caroli, de Gennes and Matricon [21] and of Bardeen and Kümmel [21] has shown that the electrons exist in bound states which, although discrete, form almost a continuum without any appreciable energy gap.

INFLUENCE OF THE CORE ON THE FLUX-FLOW RESISTIVITY AND HALL EFFECT

When a vortex in a superconductor moves, current passes through its central core causing a dissipation of energy. Although this dissipation was ignored in our earlier discussion of the flux-flow voltage, this is where most of the power loss implied by the presence of a voltage actually occurs. In fact, such a loss is necessary if the vortex is to have any motion perpendicular to the main current in the wire. If the core region were completely negligible so that no dissipative scattering could occur, the vortex-line would move with the same velocity as the fluid (this can be seen from the equation for the Magnus force. If the only force is the Magnus force, it must be zero if the net force is zero). We saw above that the voltage difference between two points 1 and 2 is connected with the component v_{Lt} of the vortex line velocity perpendicular to the line joining 1 and 2. Consequently, the transverse (or Hall) voltage in a wire of a type II superconductor is related to the component of $\underline{v}_L$ parallel to the direction of current flow. Thus the Hall angle - the angle whose tangent is equal to the ratio of the transverse voltage to the longitudinal voltage - is determined by the behaviour of the core of the vortex. There has been much controversy over this matter. Two different results have been predicted from models of vortex motion based on Euler's equation but modified to allow for the effect of the core of the vortex. Nozières and Vinen [4] have predicted that the tangent of the Hall angle should not depend on external field as the field is reduced below H_{c2} - in contrast to the linear field dependence expected in the normal state of a free-electron-like metal with negligible magnetoresistance. On the other hand the model of Bardeen and Stephen [22] predicts a continuation of the normal state behaviour below H_{c2}. The experimental work has produced even wider disagreement. In contrast to most of the early direct measurements, the tangent of the Hall angle found by Maxfield [23] from a study of the propagation of helicon-like waves in the mixed state of pure niobium agreed with the model of Nozières and Vinen. The cause of the discrepancy has recently been illuminated by the work of Fiory and Serin [23], who have shown that the vortex lines are subject to an additional force due to inhomogeneities in the superconductor. Providing this force is small in comparison with the forces of electromagnetic origin, the tangent of the Hall angle that would be measured in a defect-free specimen is equal to the ratio of the differential voltages (dV/dJ) - rather than the actual voltages, as would be the case in the defect-free specimen. Further complications in apparently defect-free alloys due to surface effects have just been found by Serin[24].

INTERACTIONS BETWEEN VORTICES IN SUPERCONDUCTORS

When the external field in figure 1(b) passes through H_{c1}, it becomes energetically favourable for a phase change to take place - for vortices to form in the superconductor. The free energy is lowered each time a vortex enters. Vortices continue to enter until they move close enough together that they begin to interact, repelling each other. Near to H_{c1} their interactions are weak enough that they can still be thought of as separate entities. But near to H_{c2} this is no longer true. The collective modes that occur in that case have been discussed from the hydrodynamical point of view by a number of authors [25]. However, in this case, it is better to think of the mixed state as a perturbation of the normal state. This has been done by Kulik [26], using a phenomenologically derived time-dependent form of the Ginzburg-Landau equations, and by Schmid [27], using a similar equation obtained from the microscopic theory approach of Gor'kov. A more direct quantum mechanical approach has been used by Caroli and Maki [28], who have carried their work into great detail and have not only derived the flux-flow resistivity and Hall coefficient for direct currents but have also dealt with the thermoelectric properties of the mixed state and also the high frequency surface impedance (see appendix 3). One phenomenon emphasized in some of these papers is that when a direct current is passed through the superconductor causing the array of flux lines to move steadily with a velocity v_L, there should be an induced oscillating electric field (generated by induction) with the frequency $\sim v_L/d$, where d is the spacing of the vortex lines or the repetition distance of the vortex lattice. Attempts have been made to observe this oscillating field but so far without success. Most methods used, or envisaged as being feasible, turn out to involve spatial averages over such large regions of the superconductor that the measured effect is zero. Vinen [29] has suggested looking for electric quadrupole resonances induced by the motion of the magnetic field pattern. At the moment, no direct observation of the oscillating field has yet been made.

Appealing though the analogy between the classical vortex and its counterpart in type II superconductors may be, the complications introduced by the core and by the interactions between vortices destroy the simplicity of the picture. However, there is no doubt that the work done so far on the hydrodynamic models has been extremely revealing, even though the most general point that has been made clear is that the phenomena are very complicated.

APPENDIX 1. QUANTIZATION OF MAGNETIC FLUX IN SUPERCONDUCTORS

We have already referred to the theory of superconductivity proposed by Bardeen, Cooper and Schrieffer[20]. According to this theory, the electrons in the superfluid phase are correlated in pairs.

Some of the properties of the superconductor can be discussed in terms of the coordinates of the centres of mass of the pairs while others require a consideration of the relative motion. In the former case, it will be assumed that the phase S of the many-particle wave-function of the N pairs of electrons,

$$\Psi(\underline{r}_1\underline{r}_2---\underline{r}_N) = e^{iS(\underline{r}_1--\underline{r}_N)}|\Psi(\underline{r}_1\underline{r}_2---\underline{r}_N)|,$$

where $\underline{r}_\alpha$ is the coordinate of the centre of mass of the αth pair of electrons, has the special form $S(\underline{r}_1---\underline{r}_N) = s(\underline{r}_1) + s(\underline{r}_2) + --- + s(\underline{r}_N)$. This special form of the phase is the same as that assumed by Feynman [29] for the superfluid phase of helium and is the simplest form that allows non-uniform flow to occur.

In the presence of a magnetic field of induction $\underline{B}(\underline{r}) = \underline{\text{curl}}\ \underline{A}(\underline{r})$, where $\underline{A}(\underline{r})$ is the magnetic vector potential, the velocity $\underline{v}_\alpha$ and generalized momentum $\underline{p}_\alpha$ of the centre of mass motion of an electron pair are related by $\underline{v}_\alpha = [\underline{p}_\alpha - 2e\,\underline{A}(\underline{r}_\alpha)/c]/m$ where $e = -4.8 \times 10^{-10}$ e.s.u. is the electronic charge. The expectation value of the current density at the point $\underline{r}$ is the expectation value of the operator

$$2e \sum_{\alpha=1}^{N} [(\underline{p}_\alpha - 2e\,\underline{A}(\underline{r}_\alpha)/c)\delta(\underline{r}-\underline{r}_\alpha) + \delta(\underline{r}-\underline{r}_\alpha)(\underline{p}_\alpha - 2e\,\underline{A}(\underline{r}_\alpha)/c)]/2m$$

where $\underline{p}_\alpha = -i\hbar\,\underline{\nabla}_\alpha = -i\hbar\,\partial/\partial\underline{r}_\alpha$. We thus find for the current density at $\underline{r}$, $\underline{J}(\underline{r}) = [\hbar\,\underline{\nabla}s - 2e\,\underline{A}(\underline{r})/c]\rho(\underline{r})/m$ where $\rho(\underline{r}) = \int \Psi(\underline{r}_1 --- \underline{r}_{N-1}\underline{r})^* \Psi(\underline{r}_1 --- \underline{r}_{N-1}\underline{r})\,d^3\underline{r}_1 --- d^3\underline{r}_{N-1}$, the density at $\underline{r}$.

Now consider an isolated vortex line in a superconductor and, in a plane perpendicular to its axis, take a contour surrounding the vortex but so far away from its axis that the current density $\underline{J}(\underline{r}) = c\,\mathrm{curl}\,\underline{B}(\underline{r})/4\pi$ is negligible. Then $J(r) \sim 0 \therefore \hbar\underline{\nabla}s = 2e\,\underline{A}(\underline{r})/c$ since $\rho(\underline{r}) \neq 0$. The line integral of $\underline{\nabla s}$ taken completely around the contour gives the change in $s(\underline{r})$ when $\underline{r}$ is taken once around the contour. If $\Psi(\underline{r}_1 --- \underline{r}_N)$ is to be single-valued in each of the $\underline{r}_\alpha$ then $\oint \underline{\nabla}s \cdot \underline{d\ell} = \text{integer} \times 2\pi$. Thus

$$\oint \underline{A}(\underline{r}) \cdot \underline{d\ell} = \oint \hbar c\,\underline{\nabla s} \cdot \underline{d\ell}/2e = \text{integer} \times hc/2e$$

But $\oint \underline{A} \cdot \underline{d\ell} = \int \underline{\mathrm{curl}\,\underline{A}(\underline{r})} \cdot \underline{n}\,d\Sigma = \int \underline{B}(\underline{r}) \cdot \underline{n}\,d\Sigma$ is the total magnetic flux Φ_L of the vortex. Thus $\Phi_L = \text{integer} \times hc/2e$. Consideration of the kinetic energy of the electrons in the supercurrent shows that, given the amount of flux $2 \times (hc/2e)$, the total energy of the system is smaller if the flux is shared equally between two vortices than if it is all contained by one vortex. It can also be shown that the energy of the two vortices is least when they are farthest apart, but that even when they are close enough together to interact, it is always most favourable - from the point of view of minimizing the free energy - to share the flux between the vortices.

APPENDIX 2. THE EQUATION OF MOTION OF $s(\underline{r})$

Anderson has drawn attention to the connection between the phase $s(\underline{r})$ (not to be confused with $S(\underline{r}_1\underline{r}_2 --- \underline{r}_N)$) and the number operator for the electron pairs and has used this to relate the motion of the phase to the total chemical potential. The following treatment is based on his discussion [18].

In the many-particle wavefunction, we suppose that the phase be changed uniformly by δs, where $N\delta s \ll 1$. The new wavefunction becomes

$$\Psi'(\underline{r}_1\underline{r}_2---\underline{r}_N) = \exp\{i \sum_{\alpha=1}^{N} [s(\underline{r}_\alpha)+\delta s]\} \, |\Psi(\underline{r}_1---\underline{r}_N)|$$

$$= \exp\{iN\delta s\} \, \Psi(\underline{r}_1---\underline{r}_N)$$

$$\sim (1+iN\delta s) \, \Psi(\underline{r}_1---\underline{r}_N)$$

Thus $\delta\Psi=\Psi'-\Psi = iN\delta s\Psi$, or

$$-i \, \partial\Psi/\partial s = N\Psi.$$

The operator $-i \, \partial/\partial s$ is equivalent to the number operator. That is, N and s are conjugate variables just like $p_x/\hbar$ and x. Similarly we have $i \, \partial/\partial N \equiv s$.

The Heisenberg equations of motion for the phase and number operators are:

$$\hbar \, \dot{s} = i[H,s] = + \, \partial H/\partial N \quad ; \quad \hbar \, \dot{N} = i[H,N] = - \, \partial H/\partial s$$

Taking the expectation value of the first equation and using the definition of the total chemical potential μ, we get

$$\hbar ds/dt = \mu.$$

APPENDIX 3. THE SURFACE IMPEDANCE OF TYPE II SUPERCONDUCTORS

Measurements of the surface impedance of superconductors have been important in the past in suggesting how the current density and fields are related in superconductors [30], in checking on the electrodynamic aspects of the Bardeen, Cooper, Schrieffer theory [31] and in confirming the surprising suggestion from the flux-flow resistance measurements [15] that the transport currents in the mixed state are not short-circuited by the superconducting regions but apparently have a uniform density [32].

The surface impedance, Z, is defined as

$$Z \equiv R + iX \equiv R + i\omega L = E_0 \Big/ \int_0^\infty J \, dz$$

where E_0 is the complex electric field at the surface produced by an incident beam of radio - or micro-waves, J is the current density at a depth z below the surface. R, X and L are called the surface resistance, reactance and inductance respectively. ω is the angular frequency of the waves. The quantity $L/4\pi = \lambda$ is called the penetration depth. From Maxwell's equations it can be seen that $\lambda = \text{Re} \int_0^\infty B\, dz/B_0$, where B is the magnetic induction of the oscillating field at a depth z and B_0 the value of B at the surface.

Measurements of λ at 5 mc/s by Carlson [33] also confirm the interpretation mentioned above in terms of the flux-flow theory. On the other hand, the theory of the electromagnetic oscillations of Caroli and Maki [26] predicts quite different results - as might be expected from the fact that in pure materials, the limit as the frequency is allowed to tend to zero does not give the behaviour observed in the direct current flux-flow experiments. Some aspects of the theory do agree with the penetration depth measurements; but this is when the oscillating currents induced by the radio-waves are parallel to the vortex lines and thus produce no driving force - a configuration where the surface impedance depends on the static properties of the mixed state. Systematic errors which arise and make a comparison difficult are discussed in a forth-coming paper by Carlson and the author [34].

REFERENCES

*Work supported by the National Science Foundation and the Rutgers Research Council.

1. F. London, "Superfluids" Volume I, Dover Publications, Inc., New York (1961).
2. See, for instance: P. G. de Gennes, "Superconductivity of Metals and Alloys", W. A. Benjamin, Inc., New York (1966); E. A. Lynton and W. L. McLean, "Type II Superconductors", Advances in Electronics and Electron Physics, Vol. 23, Academic Press (1967).
3. See, for instance: L. M. Milne-Thomson, Theoretical Hydrodynamics", 3rd Edition, Macmillan Ltd., London (1955).
4. See, for instance: P. Nozières and W. F. Vinen, Phil. Mag. 14, 667 (1966).
5. J. Bok and J. Klein, Phys. Rev. Letters 20, 660 (1968).
6. C. J. Adkins and J. R. Waldram, Phys. Rev. Letters 21, 76 (1968).
7. C. Kittel, "Introduction to Solid State Physics", 3rd Edition, John Wiley and Sons, Inc., New York (1966).

8. See, for instance: M. Tinkham, Low Temperature Physics, edited by C. deWitt, B. Dreyfus and P. G. deGennes, Gordon and Breach, New York (1962).
9. V. L. Ginzburg and L. D. Landau, Soviet Phys. JETP 20, 1064 (1950).
10. See, for instance: L. D. Landau and E. M. Lifshitz "Statistical Physics", Pergamon Press, Oxford (1958).
11. L. P. Gor'kov, Soviet Phys. JETP 9, 1364 (1959).
12. A. A. Abrikosov, Soviet Phys. JETP 5, 1174 (1957).
13. D. Cribier, B. Jacrot, L. M. Rao and B. Farnoux, Phys. Letters 9, 106 (1964).
14. Field mapping using nuclear magnetic resonance, A. G. Redfield, Phys. Rev. 162, 367 (1967); Electron microscope examination of template of precipitated ferromagnetic material, U. Essmann and H. Träuble, Phys. Letters 10, 526 (1967).
15. B. Serin "Type II Superconductors" in "Superconductivity", edited by R. D. Parks, to be published by Marcel Dekker, Inc., New York.
16. Y. B. Kim, C. F. Hempstead and A. R. Strnad, Phys. Rev. 139, A1163 (1965).
17. B. D. Josephson, Physics Letters 16, 242 (1965).
18. P. W. Anderson, Quantum Fluids, edited by D. F. Brewer, North Holland Publishing Company, Amsterdam (1966).
19. This is discussed in many of the papers in the same volume as ref. 18.
20. J. Bardeen, L. N. Cooper and J. R. Schrieffer, Phys. Rev. 108, 1175 (1957).
21. C. Caroli, and J. Matricon, Phys. Condensed Matter 3, 380 (1965); J. Bardeen and R. Kümmel, unpublished.
22. J. Bardeen and M. J. Stephen, Phys. Rev. 140, A1197 (1965); A. G. van Vijfeijken and A. K. Niessen, Philips Res. Rep. 20, 505 (1965).
23. B. W. Maxfield, Solid State Commun. 5, 585 (1967); A. T. Fiory and B. Serin, Phys. Rev. Letters 21, 359 (1968).
24. B. Serin, private communication.
25. P. G. de Gennes and J. Matricon, Revs. Mod. Phys. 36, 45 (1964); A. A. Abrikosov, M. P. Kemoklidze and I. M. Khalatnikov, JETP 21, 506 (1965); A. L. Fetter, P. C. Hohenberg and P. Pincus, Phys. Rev. 147, 140 (1966).
26. I. O. Kulik, Soviet Physics JETP 23, 1077 (1966).
27. A. Schmid, Phys. Condensed Matter 5, 302 (1966).
28. C. Caroli and K. Maki, Phys. Rev. 159, 306 (1967).
29. W. F. Vinen, private communication.
30. A. B. Pippard, Proc. Roy. Soc. A216, 547 (1953).

31. J. R. Waldram, Advances in Physics 13, 1 (1964).
32. B. Rosenblum and M. Cardona, Phys. Rev. Letters 12, 657 (1964).
33. D. E. Carlson, private communication.
34. D. E. Carlson and W. L. McLean, to be published, Phys. Rev., Dec. 10, 1968.

References 1, 2 and 8 and the volume referred to in 18 contain a good coverage of the subject of superconductivity. For an introduction to superconductivity, see E. A. Lynton "Superconductivity", 3rd Edition, Methuen.

QUANTUM TRANSPORT THEORY

J. Hajdu

Institut für Theoretische Physik

Universität zu Köln, Köln, Deutschland

1. Transport coefficients

1.1. Definitions and interrelations

Processes in the neighbourhood of local thermodynamic equilibrium are phenomenologically described by the following linear relations between currents ($\underline{J}_p$) and forces ($\underline{X}_p$)

$$\underline{J}_p = \sum L_{pq} \underline{X}_q \;, \quad p,q = 1,2,\dots \tag{1}$$

These relations define the transport coefficients L_{pq}. In the case of electronic transport processes in solids we have in particular

$$\begin{aligned} \underline{J}_1 &= L_{11}\left[-\frac{1}{e}\left(\nabla\zeta - \frac{\zeta}{T}\nabla T\right) + E\right] + L_{12}\left[-\frac{\nabla T}{T}\right], \\ \underline{J}_2 &= L_{21}\left[-\frac{1}{e}\left(\nabla\zeta - \frac{\zeta}{T}\nabla T\right) + E\right] + L_{22}\left[-\frac{\nabla T}{T}\right]. \end{aligned} \tag{2}$$

Here $\underline{J}_1$ is the electric and $\underline{J}_2$ the energy current density, ζ the chemical potential, T the absolute temperature and $\underline{E}$ the electric field. The gradients $\nabla\zeta$ and ∇T are assumed to be constant in space and time. The transport coefficients L_{pq} are second rank tensors ($L_{pq}^{\mu\nu}$, μ,ν = x,y,z); L_{11} is the tensor of electric conductivity, L_{12} is the Peltier tensor,

L_{21} is the thermal e.m.f. tensor and L_{22} is the heat conductivity tensor.

The tensor components are connected by the Onsager relations

$$L_{pq}^{\mu\nu}(\underline{B}) = L_{qp}^{\nu\mu}(-\underline{B}), \tag{3}$$

where $\underline{B}$ is the magnetic field. Furthermore the so-called electrical-thermal symmetry, which is also experimentally established, requires

$$L_{12} = L_{21} \,. \tag{4}$$

Notice that in (2) $\nabla\xi$ and $\underline{E}$ appear in the combination

$$\nabla\xi/e + \underline{E} = \nabla\xi/e + \nabla\varphi = \nabla(\xi/e + \varphi) = \nabla\xi^{0}/e$$

only, where ξ^{0} is the electrochemical potential. Thus the electrical conductivity tensor and the diffusion tensor are identical. This is the content of the so-called Einstein relation.

1.2. Methods of calculation

The transport coefficients can be calculated by means of transport equations or using linear response theory, i.e. Kubo's formulae. These two methods are actually not as different as they might appear at first sight. I shall consider both methods in some detail. For sake of simplicity I restrict myself to the most simple model system considered in the theory of electronic transport phenomena in solids. This system consists of free electrons interacting with randomly distributed static impurities, and is, therefore, dynamically a single electron system. The total single electron Hamiltonian is

$$\begin{gathered} H_T = H_0 + V + H_F = H + H_F, \\ H_0 = \frac{1}{2m}\left(\underline{p} - \frac{e}{c}\underline{A}(\underline{r})\right)^2, \quad V = \sum_{j=1}^{N_S} v(\underline{r} - \underline{R}_j), \\ H_F = -e\underline{E}\cdot\underline{r}\,. \end{gathered} \tag{5}$$

Being interested in the volume averages of the currents only, we calculate the statistical average of the quantities

$$\underline{J}_1 = e\underline{v}\ ,\ \underline{J}_2 = \frac{1}{2}(H\underline{v} + \underline{v}H). \tag{6}$$

Here $\underline{v}$ is the electron velocity. Using an appropriately defined distribution function, these averages are given by

$$\underline{\mathcal{J}}_q = \int d^3r \int d^3p\, f(\underline{r},\underline{p},t)\,\underline{J}_q(\underline{r},\underline{p})\ ,\ q = 1,2, \tag{7}$$

or using gauge invariant quantities

$$\underline{\mathcal{J}}_q = \int d^3r \int d^3w\, f(\underline{r},\underline{w},t)\,\underline{J}_q(\underline{r},\underline{w}), \tag{8}$$

where $\underline{w} = m\underline{v} = (\underline{p} - \frac{e}{c}\underline{A}(\underline{r}))$ is the gauge invariant momentum.

2. Transport equations

2.1. Validity of the Boltzmann-Bloch equation

In the traditional kinetic transport theory $f(\underline{r},\underline{w},t)$ obeys the well known Boltzmann-Bloch equation. For the system considered here this equation has the simple form

$$\frac{\partial f}{\partial t} + \frac{\partial f}{\partial \underline{r}}\cdot\frac{\underline{w}}{m} + \frac{\partial f}{\partial \underline{w}}\cdot\left(e\underline{E} + \frac{\underline{w}}{mc}\times\underline{B}\right) = \Gamma f, \tag{9}$$

$$\Gamma f = 2\pi n_s \int \frac{d^3w}{(2\pi)^3}\,|v(\underline{w}-\underline{w}')|^2\,\delta(\varepsilon_w - \varepsilon_{w'})\left[f(\underline{r},\underline{w}',t) - f(\underline{r},\underline{w},t)\right].$$

Here is $n_s = N_s/\Omega$ the density of impurities, $v(\underline{q})$ the Fourier transform of the interaction potential and $\varepsilon_w = \frac{w^2}{2m}$.

Rigorous derivation of equ. (9) shows that it is only valid for a large system under the restrictions

$$(\hbar)\omega < kT,\ \tau_a \ll \tau,\ d \ll R,\ edE \ll \varepsilon_F,\ \omega\tau_a \ll 1. \tag{10}$$

Here $\omega = \frac{eB}{mc}$ is the cyclotron frequency, k the Boltzmann constant, τ_a the atomic time or duration of a collision, τ the relaxation time or duration between subsequent collisions, d the diameter of an impurity, R the range of inhomogeneity, and ε_F the Fermi energy. Notice that no restriction must be imposed on $\omega\tau$. The condition $(\hbar)\omega \ll kT$ means that the magnetic quantum effects are completely smeared out. Being interested in these quantum effects, we have to seek a more general quantum transport equation.

2.2. Simple derivation of a quantum transport equation

By Liouville's theorem the total time derivative of the canonical single electron distribution function $f(\underline{r},\underline{p},t)$ vanishes

$$0 = \frac{df}{dt} = \frac{\partial f}{\partial t} + \frac{\partial f}{\partial \underline{r}}\cdot\dot{\underline{r}} + \frac{\partial f}{\partial \underline{p}}\cdot\dot{\underline{p}} = \frac{\partial f}{\partial t} + \frac{\partial f}{\partial \underline{r}}\cdot\frac{\partial H_T}{\partial \underline{p}} - \frac{\partial f}{\partial \underline{p}}\cdot\frac{\partial H_T}{\partial \underline{r}}, \tag{11}$$

$$\frac{\partial f}{\partial t} + \{H_T, f\} = 0, \tag{12}$$

$$\{A,B\} = \frac{dA}{d\underline{p}}\cdot\frac{dB}{d\underline{r}} - \frac{dB}{d\underline{p}}\cdot\frac{dA}{d\underline{r}}. \tag{13}$$

Now instead of f we introduce a new function f^1 by

$$f = f^{\ell} + f^1 \tag{14}$$

where f^{ℓ} is the local equilibrium distribution function

$$f^{\ell} = g(u^{\ell}), \quad g(u) = \begin{cases} e^{-u} \\ \dfrac{1}{e^u \pm 1} \end{cases} \tag{15}$$

$$u^{\ell} = \frac{H - \zeta(\underline{r})}{kT(\underline{r})}. \tag{16}$$

We now assume that the local chemical potential and the local temperature are slowly varying functions,

$$T(\underline{r}) = T + \nabla T\cdot\underline{r}, \quad \zeta(\underline{r}) = \zeta + \nabla\zeta\cdot\underline{r} \tag{17}$$

the gradients ∇T and $\nabla\zeta$ being sufficiently small. Expanding f^{ℓ} up to linear terms in ∇T and $\nabla\zeta$ we obtain

$$f^{\ell} = f^a + \frac{df^a}{dH}\left\{\left(\frac{\zeta}{T}\nabla T - \nabla\zeta\right) - H\frac{\nabla T}{T}\right\}\cdot\underline{r}. \tag{18}$$

Here $f^a = g(u^a)$ with $u^a = (H-\zeta)/kT$ is the absolute equilibrium distribution function. Neglecting also $\nabla T\cdot\underline{E}$, $\nabla\zeta\cdot\underline{E}$ and higher order terms, we end up with the linearised equation

$$\frac{\partial f^1}{\partial t} + \{F, f^a\} + \{H, f^1\} = 0 \tag{19}$$

with

$$F = -\underline{X}_1 \cdot \underline{r}_1 - \underline{X}_2 \cdot \underline{r}_2 , \tag{20}$$

$$\underline{r}_1 = e\underline{r} , \quad \underline{r}_2 = \tfrac{1}{2}(H\underline{r} + \underline{r}H) , \quad \underline{J}_q = \frac{d\underline{r}_q}{dt} . \tag{21}$$

By means of the well known correspondence principle

$$\{A,B\} \rightarrow i[A,B] \tag{22}$$

we now go over to quantum theory

$$i\frac{\partial f^1}{\partial t} = [F, f^a] + [H, f^1] . \tag{23}$$

As initial condition we choose for $t = -\infty$, $\underline{E} = \nabla\zeta = \nabla T = 0$, $f = f^a$ and transform equ. (23) into the integral equation

$$f^1 = -i\int_{-\infty}^{t} e^{-iH(t-t')} \{[F, f^a] + [V, f^1(t')]\} e^{iH(t-t')} dt' . \tag{24}$$

Substituting equ. (24) back into equ. (23) we obtain

$$\frac{df^1}{dt} + i[H_0, f^1] + i[F, f^a] \tag{25}$$
$$+ \int_0^\infty [V, e^{-iH_0t'} \{[F, f^a] + [V, f^1(t-t')]\} e^{iH_0t'}] dt' = 0 .$$

The integral term in equ. (25) is already proportional V^2. Restricting ourselves to the Born approximation, we substitute into equ. (25)

$$f^1(t-t') = e^{iHt'} f^1(t) e^{-iHt'} \approx e^{iH_0t'} f^1(t) e^{-iH_0t'} . \tag{26}$$

After averaging over all possible impurity distributions we obtain finally the required transport equation

$$\frac{df^1}{dt} + i[H_0, f^1] + [F, f^a] + \Gamma\gamma = \Gamma f^1 , \tag{27}$$
$$\Gamma f^1 = -\int_0^\infty dt \langle [V, [e^{-iH_0t} V e^{iH_0t}, f^1]] \rangle_s ,$$

$$\langle A(\underline{R}_1,\underline{R}_2,\cdots,R_{N_s})\rangle_s = \int \frac{d^3R_1}{\Omega}\cdots\int \frac{d^3R_{N_s}}{\Omega} A(\underline{R}_1,\cdots,\underline{R}_{N_s}) . \tag{28}$$

γ is the solution of the equation

$$[\gamma, H_0] = -[F, f^0], \tag{29}$$

$$f^0 = g(u^0),\ u^0 = \frac{H_0-\zeta}{kT} . \tag{30}$$

2.3. Interference effect

Equ. (27) reduces to the Boltzmann-Bloch equation (9) in the case of vanishing magnetic field and for finite magnetic fields in the classical limit. The unusual term $\Gamma\gamma$ describes the action of the electric field and other driving forces during collisions (see later). This so-called interference effect is (in the Born approximation considered here) a pure quantum effect. Indeed, it vanishes (i) for zero magnetic field, (ii) for longitudinal magnetic field ($\underline{B} \parallel \underline{E}$), since the motion parallel to the magnetic field is not quantized, and (iii) for transverse magnetic field ($\underline{B}\perp\underline{E}$) in the classical limit (large quantum numbers). To see this notice that the quantum collision integral of equ. (27) is a straightforward generalization of the "classical" collision integral of equ. (9)+). The latter vanishes for any arbitrary function $f(\varepsilon_{\underline{k}}, \underline{r})$ and in particular for the local equilibrium distribution (because of energy conservation). In the quantum case, however, $\Gamma f(H,\underline{r})$ is in general different from zero, since the operators H and $\underline{r}$ do not commute and only the additional term $\Gamma\gamma$ causes the approach to local equilibrium. In fact, it can easily be shown that the local equilibrium distribution function is a stationary solution of the transport equation (27). In the limit $(\hbar)\omega > kT$, $\omega\tau \gg 1$ the relative

+) Apart of the δ-function the quantum collision integral is a symmetrized product of VVf^1 which must vanish if f^1 is a c-number. Thus, it must have the form $VVf^1 + f^1VV - 2Vf^1V = [[V,f^1],V]$. Energy conservation can be taken into account by means of

$$2\pi A_{\alpha\alpha'}\delta_-(\omega_{\alpha\alpha'}) = \int_0^\infty e^{-\varepsilon t}\langle\alpha|e^{iH_0t}Ae^{-iH_0t}|\alpha'\rangle dt$$
$$= \int_0^\infty e^{-\varepsilon t}(A(t))_{\alpha\alpha'}dt,\ \omega_{\alpha\alpha'} = \varepsilon_\alpha - \varepsilon_{\alpha'} .$$

contribution of the term $\Gamma\gamma$ (which describes the action of the electric field etc. during collisions) compared to the contribution of the term $i[F,f]$ (which describes the action of the electric field etc. between collisions) to the transverse current parallel to the electric field etc., is of the order of $\omega\tau\cdot(1/\omega\tau)=1$.

The first factor arises because in the limit $\omega\tau\to\infty$ the term $i[F,f^a]$ gives rise to the Hall effect only. In the absence of collisions for perpendicular electric and magnetic fields, a stationary state exists in which a constant Hall current flows. Setting $\underline{B} = (0,0,B)$, $\underline{E} = (E,0,0)$ and $\underline{A} = (0, -Bx,0)$ the single electron energy eigenvalues are

$$\tilde{\varepsilon}_\alpha = (\hbar)\omega(n+1/2) + \frac{(\hbar^2)k_z^2}{2m} - eEX . \tag{31}$$

Here X is the x-coordinate of the cyclotron orbit of the electron. Transforming to these stationary states the collision integral Γf^1 (equ. (27)) and the interference term fuse to a new collision integral which differs from the original one by replacing H_0 by $H_0 + H_F$. This means that the electrostatic potential energy of the electron will in general change in a collision process. Further remarks concerning the interference effect will be given in section 3.4.

3. Linear response theory

3.1. Kubo's formulae

About ten years ago Kubo and others have established general formulae expressing the transport coefficients in terms of microscopic quantities. These Kubo's formulae are

$$L_{pq} = \int_0^\infty \int_0^\beta \langle \underline{J}_p(t+i\lambda)\, \underline{J}_q(0) \rangle \, d\lambda\, dt . \tag{32}$$

Here $\underline{J}_p$ and $\underline{J}_q$ are the quantum mechanical many-particle operators associated with the macroscopic quantities $\underline{\mathfrak{J}}_p$ and $\mathfrak{J}_q$,

$$\langle \cdots \rangle = \mathrm{Tr}\left\{ \frac{e^{-\beta \mathbb{H}}}{Z} \cdots \right\}, \quad \underline{J}(t) = e^{i\mathbb{H}t}\, \underline{J}(0)\, e^{-i\mathbb{H}t}, \quad \beta = 1/kT .$$

$\mathbb{H} = \mathbb{H}_0 + \mathbb{V}$ is the many particle Hamiltonian of the system under consideration. Using Kubo's identity [+)]

$$e^{-\beta H}\int_0^\beta \dot{A}(-i\lambda)\,d\lambda = i\left[A, e^{-\beta H}\right] \tag{33}$$

for $\dot{A} = \dot{\underline{R}}_q = \underline{J}_q$, equ. (32) can be transformed into the form

$$L_{pq} = -i\int_0^\infty \mathrm{Tr}\left\{\underline{J}_p(t)\left[\frac{e^{-\beta H}}{Z}, \underline{R}_q\right]\right\} dt. \tag{34}$$

It can easily be shown that formulae (32) and (34) satisfy the relations (3) and (4).

3.2. Simple derivation of Kubo's formulae

Using again our initial conditions, the formal solution of equ. (23) is

$$f^1(t) = -i\int_{-\infty}^{t} e^{iH(t'-t)}\left[F, f^a\right] e^{-H(t'-t)}\,dt'. \tag{35}$$

Calculating the averages

$$\underline{\mathcal{J}}_p = \mathrm{Tr}\left\{f^1 \underline{j}_p\right\} \tag{36}$$

we obtain

$$\underline{\mathcal{J}}_p = \sum L_{pq} \underline{X}_q, \tag{37}$$

$$L_{pq} = -i\int_0^\infty \mathrm{Tr}\left\{\underline{j}_p(t)\left[f^a, \underline{r}_q\right]\right\} dt \tag{38}$$

These formulae are obviously the single electron versions of the formulae (34).

+) To prove this identity notice that both sides of equ. (33) vanish for $\beta \to 0$ and obey the same first order differential equation in β . Thus, they are equal altogether.

3.3. Electric conductivity as an example

Using the eigenstates of H_0, $H_0|\alpha\rangle = \varepsilon_\alpha|\alpha\rangle$, we obtain

$$\sigma_{\mu\nu} = L^{\mu\nu}_{11}$$

$$\sigma_{\mu\nu} = -ie^2\int_0^\infty dt \sum_{\substack{\alpha\beta\\ \gamma\delta}} (e^{iHt})_{\alpha\beta}(v_\mu)_{\beta\gamma}(e^{-iHt})_{\gamma\delta}([f^a, r_\nu])_{\delta\alpha} . \tag{39}$$

For sufficiently weak interaction V we can use the well known Wigner-Weisskopf damping theoretical approximation for calculating exp (iHt). $|\alpha\rangle$. Neglecting energy shifts, this gives

$$e^{\pm iHt}|\alpha\rangle = e^{\pm i\varepsilon_\alpha t - \Gamma_\alpha t}|\alpha\rangle, \tag{40}$$

where

$$\Gamma_\alpha = \pi\left\langle \sum_\beta |\langle\alpha|V|\beta\rangle|^2\right\rangle_s \delta(\varepsilon_\alpha - \varepsilon_\beta) \tag{41}$$

is one half of the total transition probability. This gives

$$\sigma_{\mu\nu} = -ie^2\int_0^\infty dt \sum_{\alpha\beta} e^{i(\varepsilon_\alpha-\varepsilon_\beta)t-(\Gamma_\alpha+\Gamma_\beta)t}(v_\mu)_{\alpha\beta}([f^a, r_\nu])_{\beta\alpha} . \tag{42}$$

Assuming - as a further simplification -

$$\langle|\langle\alpha|V|\alpha'\rangle|^2\rangle_s \quad = C = \text{constant}$$

we obtain

$$\Gamma_\alpha = \pi C \sum_\beta \delta(\varepsilon_\alpha - \varepsilon_\beta) = \pi C N(\varepsilon_\alpha), \tag{43}$$

where N(ε) is the density of states on the energy scale.

Using the equation of motion

$$-i\langle\alpha|\underline{v}|\beta\rangle = \langle\alpha|[H_0, \underline{r}]|\beta\rangle = (\varepsilon_\alpha - \varepsilon_\beta)\langle\alpha|\underline{r}|\beta\rangle, \tag{44}$$

we can easily varify that

$$\langle\beta|[f, \underline{r}]|\alpha\rangle = (f^a_\beta - f^a_\alpha)\langle\beta|\underline{r}|\alpha\rangle = -\frac{f^a_\beta - f^a_\alpha}{\varepsilon_\beta - \varepsilon_\alpha}\langle\beta|\underline{v}|\alpha\rangle \tag{45}$$

holds true. Now we consider three special cases.

(i) No magnetic field, B = 0. In this case we have

$$\alpha = \underline{k}, \; \varepsilon_k = \frac{k^2}{2m}, \; \langle \underline{k} | \underline{v} | \underline{k}' \rangle = \frac{\underline{k}}{m} \delta_{kk'} .$$

Writing $\tau_k = 1/(2\Gamma_k)$ we obtain from equ. (42), using equ. (45) for $\beta \to \alpha$

$$\sigma_{\mu\nu} = \sigma \delta_{\mu\nu} \tag{46}$$

$$\sigma = -\frac{e^2}{m} \sum_k k_x^2 \left(\frac{df^a}{d\varepsilon_k} \right) \int_0^\infty e^{-t/\tau_k} dt \tag{47}$$

$$= \frac{e^2}{m} \sum_k \tau_k k_x^2 \left(-\frac{df^a}{d\varepsilon_k} \right) . \tag{48}$$

Since $\tau_k = \tau_0(\varepsilon_k)$ depends only smoothly on ε, we can substitute in equ. (48) for low temperatures

$$-\frac{df^a}{d\varepsilon} \approx \delta(\varepsilon - \zeta) \tag{49}$$

and obtain then finally

$$\sigma = \frac{e^2 n_e \tau_0(\zeta)}{m} , \tag{50}$$

where n_e is density of electrons, $n_e = N_e/\Omega$. [Here we again used equ. (45):

$$\sum_k \frac{k_x^2}{m} \left(\frac{df^a}{d\varepsilon_k} \right) = \sum_k k_x \langle \underline{k} | \frac{p_x}{m} | \underline{k}' \rangle \frac{f_k^a - f_{k'}^a}{\varepsilon_k - \varepsilon_{k'}}$$

$$= -i \sum_k k_x \langle \underline{k} | [f^a, x] | \underline{k} \rangle = -i \mathrm{Tr} \{ p_x [f^a, x] \}$$

$$= -i \mathrm{Tr} \{ f^a [x, p_x] \} = \mathrm{Tr} f^a = n_e] .$$

(ii) Longitudinal magnetic field (B ∥ E). The situation is much the same as before. Choosing $\underline{B} = (0,0,B)$ and $\underline{A} = (0,Bx,0)$ the quantum numbers are $\alpha = (n, k_z, X)$. The energy eigenvalues and the density of states (for fixed n) are now respectively

$$\varepsilon_\alpha = \omega(n+1/2) + \frac{k_z^2}{2m} , \tag{51}$$

$$N(\varepsilon,n) = \frac{\Omega}{(2\pi)^2}\,\frac{1}{2}\,(2m)^{3/2}\,\frac{\omega}{\sqrt{\varepsilon-\omega(n+1/2)}} .$$

From equ. (43) using the Poisson summation formula we obtain

$$\tau(\varepsilon) = \tau_0\left[1-\frac{\sqrt{2}}{2}\left(\frac{\omega}{\xi}\right)^{1/2}\sum_{r=1}\frac{(-1)^r}{r^{1/2}}\cos\left(2\pi r\frac{\varepsilon}{\omega}-\frac{\pi}{4}\right)\right] , \tag{52}$$

where τ_0 is again the relaxation time for $\underline{B} = 0$. Substitution of equ. (49) yields

$$\sigma = \frac{e^2 n_e \tau(\xi)}{m} . \tag{53}$$

But in the present case this substitution is not permissible since $\tau(\varepsilon)$, given by equ. (52), is a rapidly varying function. Rigorous integration gives

$$\sigma = \frac{e^2 n_e \tau_0(\xi)}{m}\left[1-\sqrt{2}\pi^2\frac{kT}{\omega}\left(\frac{\omega}{\xi}\right)^{1/2}\sum_{r=1}\frac{(-1)^r}{r^{1/2}}\,r\,\frac{\cos\left(2\pi r\frac{\xi}{\omega}-\frac{\pi}{4}\right)}{\sinh 2\pi^2 r\,kT/\omega}+O\left(\frac{\omega}{\xi}\right)\right] \tag{54}$$

<u>(iii) Transverse magnetic field (B E).</u> In this case the velocity operators of interest are v_x and v_y. These have finite matrix elements only between the states $\alpha = (n, k_z, X)$ and $\alpha + 1 = (n+1, k_z, X)$ and

$$\langle\alpha|v_x+iv_y|\alpha'\rangle = i\sqrt{\frac{2\omega}{m}}\sqrt{n}\,\delta_{\alpha'\alpha-1}$$

$$\langle\alpha'|v_x-iv_y|\alpha\rangle = -i\sqrt{\frac{2\omega}{m}}\sqrt{n}\,\delta_{\alpha',\alpha-1} . \tag{55}$$

We write

$$J_x+iJ_y = (\sigma_1-i\sigma_2)(E_x+iE_y) \tag{56}$$

and calculate $\sigma_1 - i\sigma_2$.

$$\sigma_1 - i\sigma_2 = -\frac{e^2}{2}\int_0^\infty dt \sum_{\alpha\beta} \left(v_x(t) + i v_y(t)\right)_{\alpha\beta} \frac{f_\beta^a - f_\alpha^a}{\varepsilon_\beta - \varepsilon_\alpha} (v_x - i v_y)_{\beta\alpha}$$

$$= -\frac{e^2\omega}{m}\int_0^\infty dt \sum_\alpha e^{i\omega t - (\Gamma_\alpha - \Gamma_{\alpha-1})t} \frac{f_\alpha^a - f_{\alpha-1}^a}{\omega} \tag{57}$$

$$= -\frac{e^2\omega}{m}\sum_\alpha n \frac{\bar{\bar{\tau}}_\alpha (1 + i\omega\bar{\bar{\tau}}_\alpha)}{1 + (\omega\bar{\bar{\tau}}_\alpha)^2} \frac{f_\alpha^a - f_{\alpha-1}^a}{\omega} .$$

Here is now

$$\frac{1}{\bar{\bar{\tau}}_\alpha} = \frac{1}{\tau_\alpha} + \frac{1}{\tau_{\alpha-1}} = \Gamma_\alpha + \Gamma_{\alpha-1} . \tag{58}$$

It can again easily be shown that

$$\sum_\alpha n\omega \left[\frac{f_\alpha^a - f_{\alpha-1}^a}{\omega}\right] = n_e . \tag{59}$$

From equs. (57) and (59) we finally obtain - in the limit $\omega\bar{\bar{\tau}}_\alpha \gg 1$ - the Hall coefficient

$$R = -\frac{1}{\sigma_2 B} = \frac{1}{e n_e c} . \tag{60}$$

Furthermore for $T \to 0$, using again the Poisson summation formula

$$\sigma = \frac{\sigma_2^2}{\sigma_1}$$

$$= \frac{e^2 n_e \tau_0(\xi)}{m}\left[1 - \frac{9\sqrt{2}}{16\pi}\left(\frac{\omega}{\xi}\right)^{3/2} \sum_{r=1} \frac{(-1)^r}{r^{3/2}} \sin\left(2\pi r\frac{\xi}{\omega} - \frac{\pi}{4}\right)\left(1 - \frac{5}{4}\frac{\omega}{\xi}\right)\right] . \tag{61}$$

In the crudest approximation $\tau_\alpha = \tau$ = const., we obtain the well known elementary result

$$\sigma_1 - i\sigma_2 = \frac{e^2 n_e \tau}{m} \frac{1 + i\omega\tau}{1 + (\omega\tau)^2} . \tag{62}$$

In this approximation both the longitudinal and the transverse measurable conductivity are independent of the magnetic field. Thus, the magnetoresistance is in this case of free electrons, a purely oscillatory quantum effect and is due to the simple fact that the relaxation time is inversely proportional to the density of states, which has large equidistant peaks on the 1/B scale. Whenever an occupied electron energy level passes through the Fermi surface the conductivities take extremal values. As we shall see presently this simple picture has to be modified somewhat in the transverse case.

3.4. Importance of initial correlations

Our simple calculations performed in the preceding section predict - in contradiction to the experiment - qualitatively different behaviours for the longitudinal and transverse conductivities. The reason for this contradiction is an unpermitted simplification made in the derivation of the conductivity formula in the transverse case. Indeed, the replacement $f^a(H_0 + V) \rightarrow f^a(H_0) = f^0$ in equ. (45) is only justified for vanishing and for longitudinal magnetic fields, since in these cases the conductivities are proportional to V^{-2} (compare equ. (62)). In the transverse case, however, the conductivities for high magnetic fields ($\omega\tau \gg 1$) are proportional to V^2 and, therefore, the leading of the neglected terms, which describe correlations in the initial state f^a, is of the same order of magnitude as the one we have taken into account.

It can easily be shown that the replacement $f^a(H_0 + V) \rightarrow f^0(H_0)$ in the Kubo's formulae (38) is only permissible when H_0 and $\underline{j}_p$ commute. In the presence of a magnetic field the transversal components of the velocity do not commute with each other and, therefore, with the Hamiltonian either. One can also easily verify that this initial correlation effect is nothing more than the interference effect discussed in section 2.3. but described in the framework of linear response theory. Taking into account the initial correlation (or interference) effect amounts, roughly speaking, to the substraction of

$\frac{f_\alpha - f_{\alpha-1}}{\omega}$ (for $\omega \neq 0$) from $\frac{f_\alpha - f_{\alpha-1}}{\omega}$ (for all ω) in equ. (57), when calculating σ_1. The result is

$$\sigma_1 = -\frac{e^2\omega}{m}\sum_\alpha n \frac{\bar{\tau}_\alpha}{1+(\omega\bar{\tau}_\alpha)^2}\frac{df^a}{d\varepsilon_\alpha}. \tag{63}$$

Using again the Poisson summation formula, we obtain for $\omega\tau \gg 1$ and $T \rightarrow 0$

$$\sigma = \sigma_2^2/\sigma_1 = \frac{e^2 n_e \tau_0(\xi)}{m}\left[1 - \frac{5}{2}\frac{\sqrt{2}}{2}\left(\frac{\omega}{\xi}\right)^{1/2}\sum_{r=1}\frac{(-1)^r}{r^{1/2}}\cos\left(2\pi r\frac{\xi}{\omega} - \frac{\pi}{4}\right)\right]. \tag{64}$$

This result exactly holds for not too small quantum numbers (cf. the following section).

3.5. Calculation of the transport coefficients for point impurities

The quantity $\phi = j_p(t)\left[f^a, r_q\right]$ obeys the Liouville equation $i\dot{\phi} = [H, \phi]$. Using the same damping theoretical approximation as in section 2.2., this equation can be converted into a kind of quantum transport equation. For point scatterers this equation can be solved exactly. The final results for the transport coefficients are

$$L_{pq}^{xx} = -\frac{e^2\omega}{m}\sum_\alpha n\bar{\tau}_\alpha^2 \frac{\tau_{\alpha-1}^{-1} f_{\alpha-1}^{0\prime}\varepsilon_{\alpha-1}^{p+q-2} + \tau_\alpha^{-1} f_\alpha^{0\prime}\varepsilon_\alpha^{p+q-2}}{1+(\omega\bar{\tau})^2}, \quad (65)$$

$$L_{pq}^{xy} = -\frac{e^2}{m}\sum_\alpha n\frac{f_\alpha^0 - f_{\alpha-1}^0}{\omega}\left(\varepsilon_\alpha + \frac{\omega}{2}\right)^{p+q-2}, \quad (66)$$

$$L_{pq}^{zz} = -\frac{e^2}{m}\sum_\alpha \tau_\alpha k_z^2 f^{0\prime}, \quad (67)$$

where the prime on $f^{0\prime}$ denotes derivative in respect to ε_α. These expressions can further be developed using Poissons summation formula.

3.6. Titeica-typ formulae

For arbitrary impurities the equation mentioned above can not exactly be solved. However, it can easily be iterated in the transverse case for high magnetic fields ($\omega\tau \gg 1$). The results for the transport coefficients are then the Titeica-typ formulae

$$L_{pq}^{xx} = 2\pi\sum_{\alpha\beta}\left\langle|\langle\alpha|V|\beta\rangle|^2\right\rangle_s \delta(\varepsilon_\alpha-\varepsilon_\beta)(-f_\alpha^{0\prime})\varepsilon_\alpha^{p+q-2}(X_\alpha - X_\beta)^2. \quad (68)$$

Being linear in the transition probability

$$W_{\alpha\beta} = 2\pi\,|V_{\alpha\beta}|^2\,\delta(\varepsilon_\alpha - \varepsilon_\beta)$$

these expressions show clearly that the transversal current in the direction of the applied driving forces are - for large magnetic fields - due to the migration (or diffusion) of the center of the electronic cyclotron motion.

4. Local equilibrium currents (l.e.c.)

4.1. Importance of the l.e.c.

In the limit of vanishing electron-impurity interaction the simple calculations of section 3.3. yield

$$\sigma_{xy} = -\frac{ecn_e}{B} \tag{69}$$

in accordance with the well known classical result. For the corresponding thermal conductivity component, however, the same theory yields the entirely unphysical results that (apart from an oscillating term)

$$\varkappa_{xy} \propto \frac{B}{T}$$

In comparison with this, the classical result for is such that the Wiedemann-Franz law

$$\frac{\varkappa_{\mu\nu}}{\sigma_{\mu\nu}T} = \frac{\pi^2}{3}\left(\frac{k}{e}\right)^2 \tag{70}$$

is satisfied. Also for the antisymmetric components of the thermal e.m.f. tensor the theory leads to unphysical expressions. The reason for this unexpected failure of the theory is the fact that we forgot to take into account the local equilibrium currents

$$\underline{\mathcal{J}}_q^{\ell} = \mathrm{Tr}\{f^{\ell}\underline{J}_q\}, \tag{71}$$

or more accurately

$$\underline{\mathcal{J}}_q^{\ell} = \int d^3x\, \mathrm{Tr}\{f^{\ell}(\underline{x})\underline{J}_q(\underline{x})\}, \tag{72}$$

where

$$\underline{J}_1(\underline{x}) = e\underline{v}(\underline{x}),\quad \underline{J}_2(\underline{x}) = \frac{1}{2}(H\underline{v}(\underline{x}) + \underline{v}(\underline{x})H) \tag{73}$$

are the electric and energy current densities,

with

$$\underline{v}(\underline{x}) = \frac{1}{2}\left(\underline{v}\,\delta(\underline{r}-\underline{x}) + \delta(\underline{r}-\underline{x})\,\underline{v}\right). \qquad (74)$$

Expanding in $f^{\ell}(\underline{x})$ the local functions $\zeta(\underline{x})$ and $T(\underline{x})$ in Fourier series and assuming that all Fourier components with $\underline{k} = 0$ are sufficiently small, we obtain

$$f^{\ell}_{\alpha\beta} = f^{a}_{\alpha}\delta_{\alpha\beta} + \frac{f^{a}_{\alpha} - f^{a}_{\beta}}{\varepsilon_{\alpha} - \varepsilon_{\beta}} \sum_{\underline{k} \neq 0} \left(\zeta_{\underline{k}} - \frac{\zeta}{T} T_{\underline{k}} + \frac{\varepsilon_{\alpha} + \varepsilon_{\beta}}{2} \frac{T_{\underline{k}}}{T}\right)(n_{-\underline{k}})_{\alpha\beta} \qquad (75)$$

where $n_{\underline{k}} = \exp(i\underline{k}\underline{r})$.

4.2. Properties of the l.e.c.

The l.e.c. have two important properties.

(i) the l.e.c. are antisymmetric and hence non-dissipative. This follows from the transformation properties of $\underline{j}_q(\underline{x},\underline{B})$ and $f^{\ell}(\underline{x},\underline{B})$ under time reversal, $T\underline{j}(\underline{x},\underline{B}) = -\underline{j}^q(\underline{x},-\underline{B})$ $Tf^{\ell}(\underline{x},\underline{B}) = f^{\ell}(\underline{x},-\underline{B})$. Consequently

and in our linearezed theory

$$J^{\ell}_{p\mu} = K^{p1}_{\mu\nu\lambda}(B^2)B_{\nu}\mathring{X}_{1\lambda} + K^{p2}_{\mu\nu\lambda}(B^2)B_{\nu}\mathring{X}_{2\lambda}, \; K^{12} = K^{21}. \qquad (76)$$

Here $\mathring{\underline{X}}_p = \underline{X}_p$ for $\underline{B} = 0$. Since $\underline{J}_p$, $\nabla\zeta$ and ∇T are radial vectors and $\underline{B}$ is an axial vector, K^{pq} must be an antisymmetric tensor. In the isotropic case considered here,

$$K^{pq}_{\mu\nu\lambda} = K^{pq}\,\varepsilon_{\mu\nu\lambda} \qquad (77)$$

must hold. Thus, we end up with

$$\underline{J}^{\ell}_{p} = \sum_{q} K^{pq}\,\underline{B} \times \mathring{\underline{X}}_{q}\,. \qquad (78)$$

This formula shows that the local currents are antisymmetric and that therefore they do not contribute to the entropy production. Since $\underline{B}$ is constant, $\underline{J}^{\ell}_{p}$ can also be written in the form

$$\underline{J}^{\ell}_{p} = \mathrm{curl}\left(C^{p1}\zeta(\underline{x})\underline{B}\right) + \mathrm{curl}\left(C^{p2}T(\underline{x})\underline{B}\right) \qquad (79)$$

(ii) The l.e.c. are of quantum mechanical origin. To see this notice that in the classical limit

$$\underline{\mathcal{J}}_q^{\ell} = \int d^3r \int d^3v\, f^{\ell}(\underline{r},\underline{v})\, \underline{j}_q(\underline{r},\underline{v}) \tag{80}$$

holds, and the last integral vanishes since f^{ℓ} is even and $\underline{j}_q$ is odd in $\underline{v}$.

4.3. Total non-dissipative currents

The total non-dissipative current I_q^a is the sum of $\underline{J}_q$ and $\underline{J}_q^a$, the antisymmetric part of $\underline{J}_q$,

$$\underline{I}_q^a = \underline{\mathcal{J}}_q^{\ell} + \underline{\mathcal{J}}_q^{a}. \tag{81}$$

It is not difficult to show that in the limit the currents $\underline{I}_q^a$ for $\underline{E} = 0$ can be expressed in the form

$$\underline{I}_q^a = c\, \mathrm{curl} \langle \underline{M}_q \rangle_{\ell}, \tag{82}$$

where

$$\underline{M}_1 = \frac{e}{2c}\, \underline{r} \times \underline{v} \tag{83}$$

is the gauge invariant magnetic moment,

$$\underline{M}_2 = \frac{1}{2c}\frac{1}{2}\left(H\underline{r}\times\underline{v} + \underline{r}\times\underline{v}H\right) \tag{84}$$

and $\langle \ldots \rangle_{\ell} = \mathrm{Tr}(f^{\ell}(x)\ \ldots\ )$ denotes the average in respect to the local equilibrium distribution function. It is now very important to notice that I_1^a and I_2^a are not purely diamagnetic surface currents. Indeed, the macroscopic magnetization is not the average of the gauge invariant magnetic momentum but

$$\mathfrak{M} = -\left(\frac{\partial \psi}{\partial \underline{B}}\right)_{\xi, T, \Omega} = \chi \underline{B}. \tag{85}$$

For free electrons the thermodynamic potential is

$$\psi = -\frac{2}{\beta}\sum_{\alpha} \ln\left\{1 + e^{\beta(\xi - \varepsilon_{\alpha})}\right\} \tag{86}$$

and χ is the Landau susceptibility. It is easy to verify that $\underline{\mathfrak{M}}$ and $\langle \underline{M}_1 \rangle$ are related by

$$\underline{B}\,\underline{\mathfrak{M}} = \underline{B}\cdot\langle \underline{M}_1\rangle - \frac{\Psi}{\Omega} \tag{87}$$
$$= n_e\left(2\langle \varepsilon_{\parallel}\rangle - \langle \varepsilon_{\perp}\rangle\right).$$

Here is $\varepsilon_{\parallel} = \frac{k_z^2}{2m}$ the longitudinal and $\varepsilon_{\perp} = (\hbar)\omega(n+1/2)$ the transverse part of the electron energy, In the classical limit $\underline{\mathfrak{M}}$ vanishes since by the equipartition principle the transversal energy (corresponding to two degrees of freedom) is twice as large as the longitudinal one.

If we take into account the l.e.c. then we get instead of the unphysical results for the antisymmetric part of the thermal conductivity and thermal e.m.f. tensors mentioned in section 4.1. quite reasonable results. In particular, for $\omega\tau \to \infty$ the Wiedemann-Franz ratio turns out to be

$$\frac{\varkappa_{xy}}{\sigma_{xy}T} = \frac{\pi^2}{3}\left(\frac{k}{e}\right)^2\left(1 - \frac{3\sqrt{2}}{4}\left(\frac{\omega}{\xi}\right)^{1/2}\cos\left(2\pi\frac{\xi}{\omega} - \frac{\pi}{4}\right) + \dots\right) \tag{88}$$

This result is fairly close to the experimental one which, however, doesn't seem to show any oscillating terms present in formula (88). Furthermore, taking into account the l.e.c. the antisymmetric parts of the electric conductivity tensor and the diffusion tensor become different, i.e. the Einstein relation for these components is violated. This is rather disappointing from the theoretical point of view.

5. Measurable charge and heat transfer

Because the magnetization is discontinuous at the surface of the conductor, diamagnetic surface currents arise which, however, do not contribute to the charge and heat transfer through a cross section of the conductor. These diamagnetic currents are given by

$$\underline{I}_{sp} = c\,\mathrm{curl}\,\underline{\mathfrak{M}}_p\,. \tag{89}$$

The formula for p = 1 is very well known from elementary electrodynamics. These diamagnetic surface currents have to be substracted from the calculated total currents

$$\underline{\underline{I}}_1 = \underline{\underline{I}}_{m1} + \underline{\underline{I}}_{s1}$$

$$\underline{\underline{I}}_2 = \frac{\xi}{e}\underline{\underline{I}}_{m1} + \underline{\underline{I}}_{m2} + \underline{\underline{I}}_{s2} . \tag{90}$$

This leads in the limit $\omega\tau \to \infty$ to the following expressions for the measurable currents obtained first by P.S. Zyryanov:

$$\underline{\underline{I}}_{m1} = \frac{c e n_e}{B^2}\left(\underline{E} - \frac{1}{e}\nabla\xi\right)\times\underline{B} + \frac{cTs}{B^2}\left(-\frac{\nabla T}{T}\times\underline{B}\right) \tag{91}$$

$$\underline{\underline{I}}_{m2} = \frac{cTs}{B^2}\left(\underline{E} - \frac{1}{e}\nabla\xi\right)\times\underline{B} + \frac{cT}{eB^2}\int_{-\infty}^{\xi} T\left(\frac{\partial s}{\partial T}\right)_{\xi} d\xi \left(-\frac{\nabla T}{T}\times\underline{B}\right).$$

Here

$$s = -\frac{1}{\Omega}\left(\frac{\partial\psi}{\partial T}\right)_{\xi,\Omega,\underline{B}} \tag{92}$$

is the equilibrium entropy of the system. Short examination of equs. (91) shows that the Einstein relation is now satisfied. Furthermore, the thermal e.m.f. (being proportional to the entropy) vanishes for $T \to 0$. Finally it also turns out that the Wiedemann-Franz law is exactly satisfied for the measurable quantities. Thus, complete agreement is achieved between theory on the one side and physical expectation and experimental results on the other.

6. Divergence problem and its resolution

The Titeica-typ formulae (68) contain two summations over energy eigenstates. Therefore, two density of states factors (51) occure and hence the final energy integral diverges for $n_\alpha = n_\beta$. The physical reason for this divergence is that in the Born approximation an electron with $k_z = 0$ hits one and the same scattering center infinitely often. Furthermore, the longitudinal transport coefficients given by equ. (66) periodically vanishes in the corresponding approximation. Thus, in order to obtain convergent or physically realistic results we must go beyond the Born approximation. This program has been worked out. The results are in the (i) in the transverse case for high magnetic fields $(\omega\tau \gg 1)$ the Titeica-typ formulae (68) still hold, the only difference being that the Born approximation transition probability

$$w_{\alpha\beta}=2\pi\,|V_{\alpha\beta}|^2\,\delta(\varepsilon_\alpha-\varepsilon_\beta)$$

has to be replaced by the exact transition probability

$$w_{\alpha\beta}=2\pi T^{\pm}_{\alpha\beta}(\varepsilon_\alpha)\,\delta(\varepsilon_\alpha-H_0)\,T^{\mp}_{\beta\alpha}(\varepsilon_\alpha), \tag{93}$$

where $T^{\pm}$ are scattering operators defined by the integral equation

$$T^{\mp}(E)=V+V\frac{1}{E-H_0\mp i\eta}T^{\mp}(E) \tag{94}$$

and

(ii) in the longitudinal case (in an appropriate approximation) formula (66) still holds, only in $1/\tau$ the Born approximation transition probability has to be replaced by

$$w_{\alpha\beta}=2\pi n_s\int\frac{d^3R}{\Omega}\,t^{+}_{\alpha\beta}(\varepsilon_\alpha)\,\delta(\varepsilon_\alpha-H_0)\,t^{-}_{\beta\alpha}(\varepsilon_\alpha), \tag{95}$$

where $t^{\mp}$ are single scatterer operators defined by the equation

$$t^{\mp}(E)=v(\underline{r}-\underline{R})+v(\underline{r}-\underline{R})\frac{1}{E-H_0\mp i\eta}t^{\mp}(E) \tag{96}$$

7. Inclusion of the Coulomb interaction between electrons

Using the diagonality property of the many-electron Green's function in the Landau representation and a generalized Ward identity the Coulomb interaction between the conduction electrons can be taken into account. It turns out that if one goes beyond the Hartree-Fock approximation the transport coefficients can again be expressed by the same formulae as for independent electrons. The only difference is that the free electron quantities have to be replaced by corresponding quasi-particle quantities. In other words, Landau's quasi-particle-picture is correct in this case also for high (quantizing) magnetic fields.

Note. In the present survey the influence of the electron spin, the electron-phonon interaction, the phonon drag, the anisotropy of the Fermi surface and the influence of some other interesting effects on the transport coefficients have not been discussed. Also d.c. fields have been considered only.

References. In the following excellent and comprehensive review articles the interested reader can find further references and comparisons with experimental results:

1. Laura M. Roth and Petros N. Argyres, and

2. S.M. Puri and T.H. Geballe, Semiconductors and Semimetals (Ad. by Willardson and Beer), Vol. 1, Acad. Press, New York and London, 1966.

3. R. Kubo, N. Hashitsume, and S.J. Miyake, Solid State Physics, 17, 269 (1966).

4. C. Herring, Proc. Int. Conf. Phys. Semicond. Kyoto, 1966, J. Phys. Soc. Japan, Vol. 21, Supliment, 1966.

Compare furthermore

P.S. Zyryanov and V.I. Okulov, Sov. Phys. - Solid State 7, 1411 (1965),

V.G. Bar'yakhtar and S.V. Peletminskii, Sov. Phys. JETP 21, 126 (1965)

J. Hajdu, Semin. Phys. Plasmas, Inst. H. Poincaré, Paris, Preprint 1968, for sections 4 and 5

" Göttinger Nachrichten 95 (1964), for sec. 2

" Z. Phys. 181, 87 (1964) for secs. 2.2, 2.3, 3.2, 3.4

" and S. Fischer, Z. Phys. 181, 479 (1964), for secs. 1.1, 3.5, 3.6, 4

" and H. Keiter, Z. Phys. 201, 507 (1967) for secs. 2.2, 2.3, 3.4

H. Keiter, Z. Phys. 198, 215 (1967), for sec. 7

E. Bangert, Z. Phys. in press, for sec. 6

THE KONDO EFFECT

Konrad Fischer

Institut für Festkörper- und Neutronenphysik

der Kernforschungsanlage Jülich

1. INTRODUCTION

The resistivity minimum and other anomalies observed in some metals at low temperatures have been a puzzle for a long time. The explanation of the resistivity anomaly given by Kondo /1/ in his famous paper turned out to be extremely stimulating for both, theoretists and experimentalists. I would like to discuss in Sec. 2 some of the recent experimental results and derive in Sec. 3 Kondo's result. Sec. 4 deals with a theory based on temperature dependent Green's functions which turned out to be rather successful in explaining the resistivity, thermopower, and specific heat at all temperatures. Finally I will mention some of the open problems.

2. EXPERIMENTS

a) Resistivity

In many cases one measures at low temperatures T (about $10^{o}K$) a steady increase in resistivity with temperature. Part of the resistivity is temperature independent and is caused by the scattering of conduction electrons by lattice defect (impurities, interstitials, vacancies, dislocations, etc.). The temperature dependent part is caused by the electron-phonon interaction and varies at low temperatures about proportional T^5.

If one dissolves in a metal magnetic impurities (Cr, Mn, Fe, etc. in Cu, Ag, Au), one observes however a resistivity minimum /2/, /3/ at about 10°K (Fig. 1), in some cases /4/, /5/ a minimum followed by a maximum at lower temperatures, and sometimes a steady decrease /6/ in resistivity with decreasing temperature at temperatures where the resistivity due to the phonon scattering is neglible small. This indicates that the magnetic impurities produce an additional temperature dependent resistivity term. This term varies near the resistivity minimum as ln T. The minimum is produced by the sum of the resistivities due to phonons and magnetic impurities.

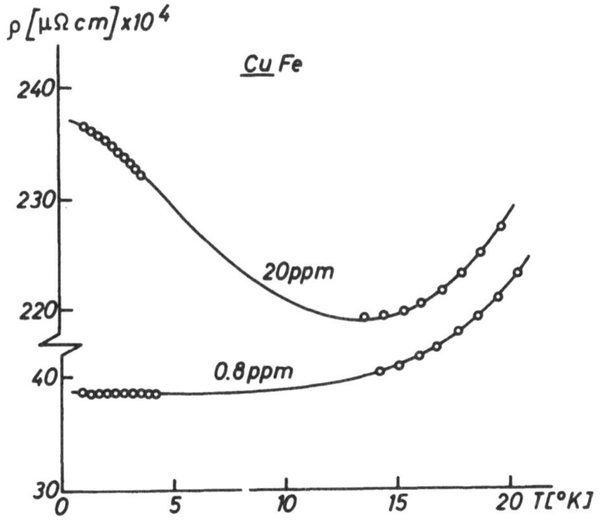

Fig. 1: The resistance minimum; the data are from Ref.2.

More recent experiments /7/ on Cu with a small concentration (20 to 100 ppm) of Fe or Cr impurities show the temperature dependence of the impurity resistivity given in Fig. 2, plotted on a logarithmic scale. The temperature dependent part amounts to between 10 percent and 50 percent of the total impurity resistivity and is explained by the interaction between the spins of the magnetic impurities (the d-electron spins in the case of transition metals) and the spins of the conduction electrons (see Sec. 3).

It has its maximum at T = 0 and varies strongest near a characteristic temperature T_K which will be called the Kondo temperature. In addition one has always an ordinary (spin independent) interaction which produces a temperature independent resistivity.

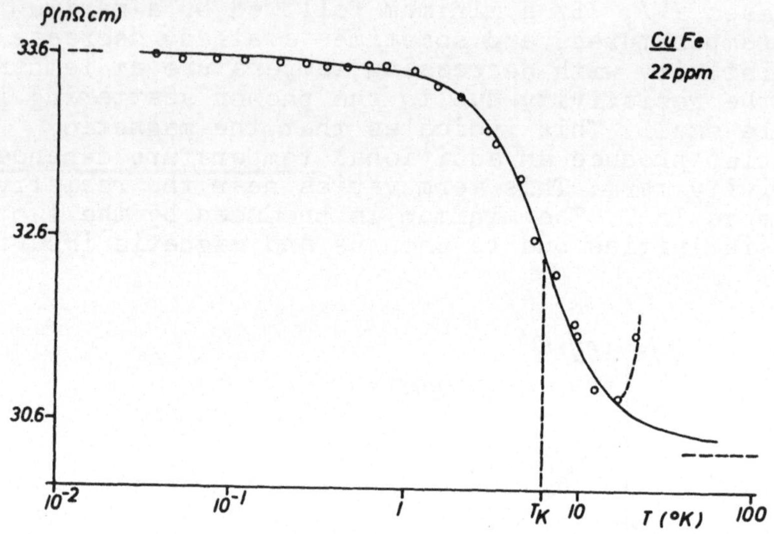

Fig. 2: Resistivity of CuFe at low temperatures, plotted on a logarithmic temperature scale. The data are from Ref. 7; the drawn line is from Eq.(57) with the phase δ , the Kondo temperature T_K, and the impurity spin S fitted.

A concentration dependent resistivity maximum is observed at higher impurity concentrations and may be explained by the magnetic interactions between the impurities.

b. Thermoelectric Power

The thermopower S is defined by ($\underset{\sim}{j}$ is the current density)

$$\underset{\sim}{E} = S\nabla T \quad (\underset{\sim}{j} = 0),$$

where the temperature gradient ∇T generates in an isolated sample the electrical field $\underset{\sim}{E}$. It consists of a "diffusion" term S_d and a "phonon drag" term S_{ph} which arises from the deviation of the phonon distribution from its thermal equilibrium. The term S_{ph} becomes neglible small at temperatures where the impurity resistivity is large compared to the phonon resistivity.

The diffusion term S_d is proportional T for non magnetic impurities. For magnetic impurities one measures a large broad peak /7/, /8/ near T_K which may be a factor 100 times larger than the value for non magnetic impurities at the corresponding temperature. Fig. 3 shows the thermopower of Fe in Cu as a function of ln T. The steep increase of S(T) at T 20°K is caused by the phonon scattering and by the phonon drag component. A large diffusion term S_d can be explained only by a relaxation time of the conduction electrons which varies strongly with the electron energy near the Fermi surface. It will be shown in Sec. 4 that for the anomaly in thermopower both, the spin dependent and the spin independent interactions are important.

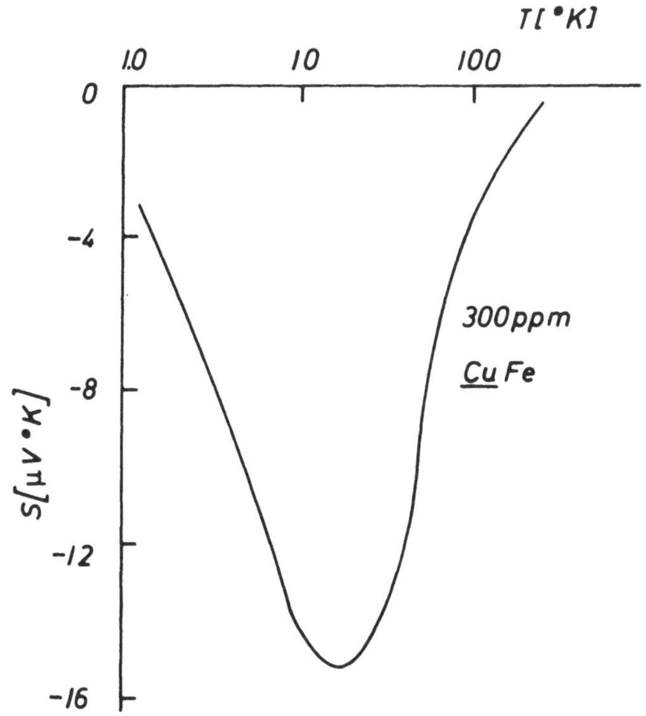

Fig. 3: Thermopower of CuFe (Ref. 7).

c) Magnetoresistance

We found the interaction between impurity spin and conduction electron spin to be responsible for the resistivity minimum, and expect therefore a strong dependence of the resistivity on an applied magnetic field. One observes indeed besides a "normal" positive term (which arises from the anisotropy of the scattering and of the Fermi surface) two effects:

(a) The impurity spins align in an external magnetic field. The spin ordering produces a negative term which saturates at high fields.

(b) A reduction in the resistivity anomaly. One measures a field dependent maximum /9/ below the minimum or simply a levelling off below the minimum /7/ (Fig. 4). This can be explained if spin flip processes are important. These become inelastic in a magnetic field and therefore freeze out at low temperatures. Kondo's theory shows that this interpretation is correct at temperatures $T \gg T_K$. However at present there seems to exist no rigorous theory which predicts the magnetoresistance at $T \leq T_K$.

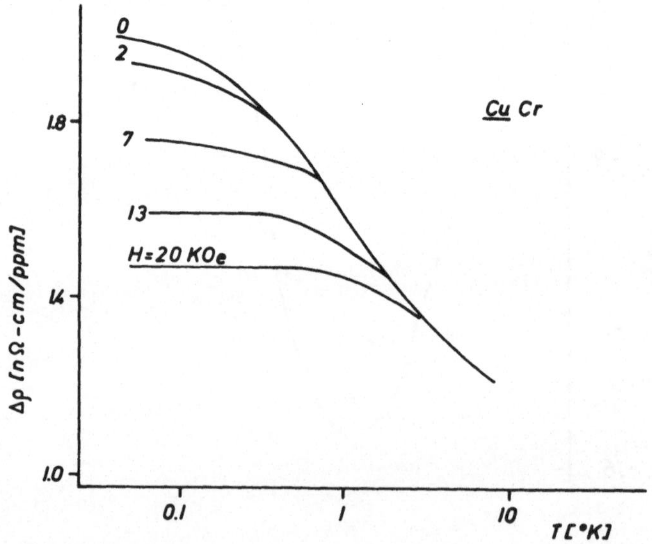

Fig. 4: Magnetoresistivity for 28 ppm Cr in Cu for various magnetic fields with the normal positive magnetoresistivity subtracted (Ref. 7).

d) Susceptibility

The impurity spin susceptibility $\Delta\chi$ is expected to follow at zero magnetic field, sufficiently small impurity concentration, and $T \gg T_K$ a Curie law

$$\Delta\chi = \frac{C}{T}$$

where the constant C is proportional to the squared of the localized magnetic moment. One observes /10/ however in CuFe and AuFe for $T \gg T_K$ a Curie-Weiss law $\Delta\chi \propto (T+\Theta)^{-1}$ with fairly concentration independent Curie temperatures $\Theta = -32^{o}K$ and $\Theta = -10^{o}K$, respectively. For AuFe, the Kondo temperature seems to be extremely low /11/ (below 100 $m^{o}K$). For CuFe one finds near $T_K \approx 6^{o}K$ a marked decrease in $\Delta\chi$ which may be interpreted as a decrease in the effective magnetic moment. This decrease is most likely caused by the magnetic screening of the impurity spin by the surrounding conduction electrons. The system has some similarity with a Cooper pair in the superconductivity where two electrons form a bound singlet state. However, the resistivity maximum and correspondingly strong scattering at T = 0 indicate more similarity with a scattering resonance just at the Fermi energy ε_F, as confirmed by current theories /12/ - /16/. For $T \leq T_K$ one finds /7/ in the CuFe system $\Delta\chi \propto T^{-0.5}$ for zero magnetic field (Fig. 5) and $\Delta\chi$ = const for large magnetic fields. This temperature dependence has not yet been explained.

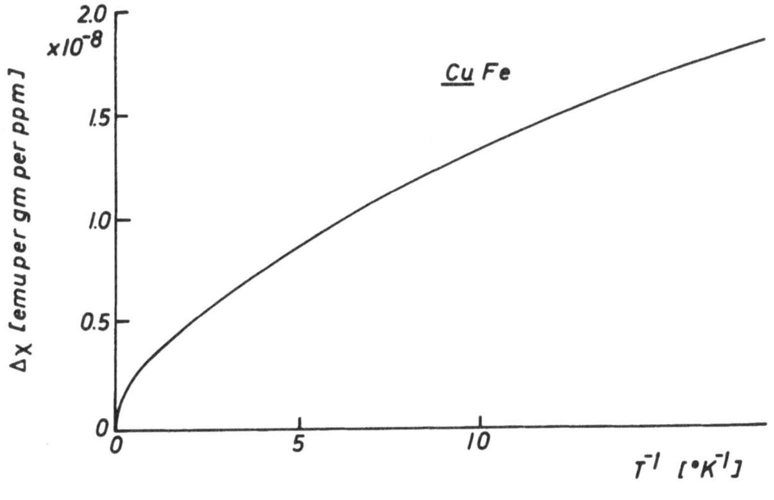

Fig. 5: Zero field impurity susceptibility for 110 ppm Fe in Cu versus reciprocal temperature (Ref. 7).

e) Specific Heat

One observes a peak in the specific heat at T_K in CuFe /17/, /18/ and CuCr /18/, /19/ which indicates a change in the internal structure of the "impurity" (including the screening cloud), usually referred to as "breaking up of the quasibound state". The entropy change associated with this is about R ln 2 and R ln 2.5, respectively. (R is the gas constant per mole).

3. PERTURBATION THEORY

a) Second Order Perturbation Theory

The discussion given above leaves open completely the question why spin dependent scattering may be strongly temperature dependent. It is this point which has been answered by Kondo /1/ using second order perturbation theory and taking into account the statistics of the conduction electrons (by using the second quantization formalism). He used the simplest possible model which describes this effect, well aware that for some experiments more complicated models are needed.

We consider a system consisting of N independent conduction electrons and a single magnetic impurity at the origin of the coordinate system, thus neglecting effects due to impurity interactions. The interaction between the electrons of the impurity with spin S and the conduction electrons is assumed to be of the exchange type with the exchange integral replaced by a constant J (J is negative for antiferromagnetic coupling). This corresponds to a contact interaction which gives rise to s-scattering only. The spin independent scattering is included by the interaction constant V. For most transition metal impurities in noble metals one has

$$\frac{J}{\varepsilon_F} \approx 10^{-1}, \qquad \frac{V}{\varepsilon_F} \approx 1. \tag{1}$$

The corresponding Hamiltonian is

$$H = \sum_k \varepsilon_k c^+_{k\mu} c_{k\mu} - \frac{J}{N} \sum_{kk'} c^+_{k\mu} c_{k'\nu} \underset{\sim}{S} \cdot \underset{\sim}{S}^e_{\mu\nu} + \frac{V}{N} \sum_{kk'} c^+_{k\mu} c_{k'\mu}. \tag{2}$$

The operator $c^+_{k\mu}$ creates an electron with momentum k, energy ε_k and z component of spin μ, $\underset{\sim}{S}^e = \frac{1}{2}\underset{\sim}{\sigma}$ is the spin operator for a conduction electron which may be expressed by the Pauli matrices $\underset{\sim}{\sigma}$, and repeated spin indices are summed over. We neglect first the ordinary interaction and put V = 0.

Following Kondo we calculate the transition probability for the scattering of a conduction electron in second Born approximation /20/. The T-matrix as a function of the energy E is given by /21/

$$T(E) = H' + H' \frac{1}{E - H^0 + i\eta} H' + \ldots, \quad \eta = 0^+, \tag{3}$$

with the unperturbed Hamiltonian H^0 (the first term in (2)) and the perturbation H'. We assume that the initial state is given by a single electron in the state k↑ outside the filled Fermi sphere $|F\rangle$. The impurity spin state is described by one of the (2S+1) eigenstates $|M\rangle$ of the z-component S^z. We consider a scattering event with the final state k'↑. The corresponding matrix element is given by

$$T_{k\uparrow,M;k'\uparrow M}(E) = \langle F,M| c_{k'\uparrow}\left(H' + H' \frac{1}{E - H^0 + i\eta} H' + \ldots\right) c^+_{k\uparrow} |F,M\rangle \tag{4}$$

We introduce the spin flip operators $S^{\pm}$

$$\underset{\sim}{S} \cdot \underset{\sim}{S}^e = S^z S^{ez} + \tfrac{1}{2}(S^+ S^{e-} + S^- S^{e+}) \tag{5}$$

and express H' by means of the Pauli matrices $\underset{\sim}{\sigma} = 2\underset{\sim}{S}^e$ in the form

$$H' = -\frac{J}{2N}\sum_{kk'}\left[(c^+_{k\uparrow}c_{k'\uparrow} - c^+_{k\downarrow}c_{k'\downarrow})S^z + c^+_{k\downarrow}c_{k'\uparrow}S^+ + c^+_{k\uparrow}c_{k'\downarrow}S^-\right] \tag{6}$$

Inserting (6) into (3) we find to second order

(a) processes without spinflip in the intermediate state ($\mu = \uparrow$); these terms are shown to be unimportant below;

(b) spin flip processes $\mu = \downarrow$.

The two possible spin flip processes are shown in Fig. 6

$$\frac{J^2}{N^2}\langle F,M|\, c_{k'\uparrow}\, c^+_{k'\uparrow}\, c_{q\downarrow}\, S^- \frac{1}{\varepsilon_k - \varepsilon_q + i\eta}\, c^+_{q\downarrow}\, c_{k\uparrow}\, S^+\, c^+_{k\uparrow}\,|F,M\rangle \quad (7a)$$

and

$$\frac{J^2}{N^2}\langle F,M|\, c_{k'\uparrow}\, c^+_{q'\downarrow}\, c_{k\uparrow}\, S^+ \frac{1}{\varepsilon_k - (\varepsilon_k + \varepsilon_{k'} - \varepsilon_{q'}) + i\eta}\, c^+_{k'\uparrow}\, c_{q'\downarrow}\, S^-\, c^+_{k\uparrow}\,|F,M\rangle \quad (7b)$$

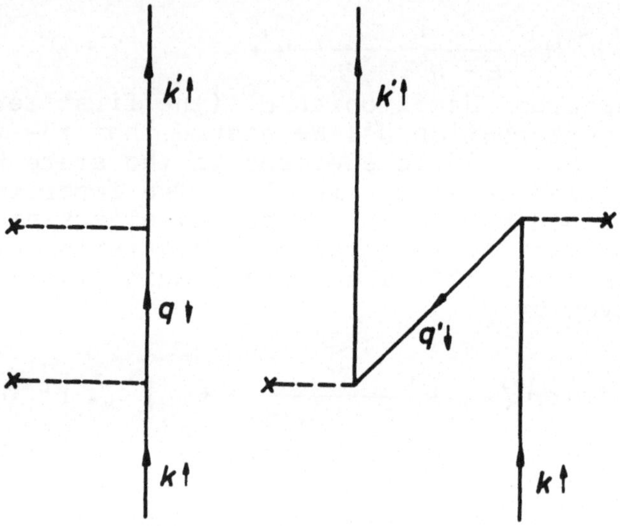

Fig. 6: Second order spin flip scattering processes

The second order T-matrix element becomes (f_q is the Fermi function for T = 0)

$$T^{(2)}_{k\uparrow,M;\,k'\uparrow M} = \frac{J^2}{N^2}\sum_q \left[\frac{\langle M|S^-S^+|M\rangle(1-f_q)}{\varepsilon_k - \varepsilon_q + i\eta} - \frac{\langle M|S^+S^-|M\rangle f_q}{\varepsilon_q - \varepsilon_{k'} + i\eta} \right] \quad (8)$$

We find with

$$\varepsilon_k = \varepsilon_{k'}\,, \quad [S^+, S^-]_- = 2S^z, \quad S^z|M\rangle = M_z|M\rangle$$

$$T^{(2)}_{k\uparrow,M;\,k'\uparrow M} = \frac{J^2}{N^2}\sum_q \left[\frac{2f_q - 1}{\varepsilon_k - \varepsilon_q} M_z + \frac{1}{2}\, \frac{\langle M|[S^+, S^-]_+|M\rangle}{\varepsilon_k - \varepsilon_q} \right. \quad (9)$$

where we neglected the imaginary part since it cancels if one calculates the transition probability

$$P_{k\uparrow,M;k'\uparrow,M} = \frac{2\pi}{\hbar} \left| T_{k\uparrow,M;k'\uparrow,M} \right|^2 \delta(\varepsilon_k - \varepsilon_{k'}) \qquad (10)$$

to order J^3.

The second term in (9) diverges for free electrons because of our unrealistic choice of the interaction. However, a natural cut off is given by the width of the conduction band for true metals. The first term has a very unusual property which is specific for the Kondo effect: It contains the Fermi function f_q in the intermediate state because the spin flip operators S^+, S^- do not commute. Since f_q varies strongly with temperature, we can expect anomalies in the temperature dependence of at least all transport properties of the system. The Fermi function cancels in the expression corresponding to (8) for non spin processes, where S^-S^+ and S^+S^- are replaced by M_z^2 (if one neglects again the term $i\eta$), and for scattering by an ordinary potential V. More generally a Kondo effect exists if (a) the scattering center possesses an internal degree of freedom (the impurity spin) which interacts with the conduction electrons, and (b) if the scattering is elastic. In the case of inelastic scattering (for instance by applying a magnetic field) the denominator in (9) no longer vanishes for $\varepsilon_q = \varepsilon_k$. A similar term containing f_q in the intermediate state is obtained by calculating the transition probability $P_{k\uparrow,M;k'\downarrow,M'}$ for true spin flip processes.

The sum

$$g(\varepsilon_k) \equiv -\frac{1}{N} \sum_q \frac{f_q}{\varepsilon_k - \varepsilon_q} \qquad (11)$$

can easily be calculated at T = 0 and for free electrons (P means "principle part")

$$g(\varepsilon_k) = \frac{\Omega_{At}\, m}{\pi^2 \hbar^2} P \int_0^{k_F} dq \, \frac{q^2}{q^2 - k^2} = N(\varepsilon_F)\left[1 + \frac{k}{2k_F} \ln\left|\frac{k - k_F}{k + k_F}\right|\right] \qquad (12)$$

$$\Omega_{At} = \frac{\Omega}{N} = \frac{3\pi^2}{k_F^3}, \quad N(\varepsilon_F) = \frac{3}{2}\,\varepsilon_F^{-1}, \qquad (13)$$

where $N(\varepsilon_F)$ is the density of states at the Fermi surface for one spin direction. Inserting (12) and (9) into (10) we find that the transition probability diverges for $\varepsilon_k = \varepsilon_F$ and T = 0, making the perturbational calculation useless in this limit.

The calculation for finite temperatures is straight forward, and one has only to replace the step function f_q by the Fermi function. In order to calculate the resistivity one has to solve the Boltzmann equation (which is trival for s-scattering), taking into account all possible scattering processes. The details can be found in Kondo's paper /1/. One finds for a concentration c of impurities the resistivity to order J^3.

$$\rho(T) = \rho_B(1+4J\int_0^\infty g(\varepsilon,T)(-\frac{df}{d\varepsilon})d\varepsilon) = \rho_B\left(1+2J\,N(\varepsilon_F)\ln\frac{KT}{\varepsilon_F}\right) \quad (14)$$

$$\rho_B = \frac{3\pi m}{2e^2\hbar\,\varepsilon_F} J^2 S(S+1)c \quad (15)$$

with the resistivity ρ_B in Born approximation. (K is the Boltzmann constant). We neglected all temperature independent terms to order J^3 since $J \ll \varepsilon_F$. The double integral (14) with (11) for T = 0 can be calculated in the approximation ($\omega = \varepsilon - \varepsilon_F$)

$$f(\omega) - \frac{1}{2} = \begin{cases} \frac{1}{2} & \omega < -2KT \\ -\frac{1}{2}\frac{\omega}{KT} & -2KT < \omega < 2KT \\ 0 & \omega > 2KT \end{cases} \quad (16)$$

This is Kondo's result and yields for $J < 0$ a resistivity minimum. The temperature dependence $\rho \propto \ln T$ has been verified in many cases.

If one takes into account the ordinary interaction V to arbitrary order, one finds /22/ instead of (14)

$$\rho(T) = \rho_0 + \rho_B\left[\cos^4\delta(1-4\sin^2\delta) + 2J\,N(\varepsilon_F)\ln\left(\frac{KT}{\varepsilon_F}\right)\cos^6\delta\cos 2\delta\right] \quad (17)$$

where ρ_0 is the ordinary scattering for J = 0 and δ is the corresponding phase shift. The ordinary interaction reduces the ln T term through a complicated interference mechanism between ordinary and exchange scattering and may even lead to the reversed sign for the slope of ρ(T).

b) Higher Order Terms and Kondo Temperature

The simple result (14) holds only for temperatures T at which higher order terms in the Born series may by neglected. An estimate of the temperature $\mathring{T}_K$ at which the perturbation series diverges may be obtained by setting

$$J\, N(\varepsilon_F)\, \ln \frac{K\mathring{T}_K}{\varepsilon_F} = 1 \;, \quad K\mathring{T}_K = \varepsilon_F\, e^{\frac{1}{J N(\varepsilon_F)}} \;, \quad J < 0 \qquad (18)$$

This corresponds to summing up a geometrical series for the T-matrix. (The factor 2 in (14) enters by taking $|T|^2$.). The temperature $\mathring{T}_K$ is the Kondo temperature for V = 0 besides small corrections. For V ≠ 0 one has to replace the exchange interaction J by an effective interaction /22/ $J_{eff}(\varepsilon) = J\cos^2\delta(\varepsilon)$ and finds for $\varepsilon = \varepsilon_F$

$$K T_K = \varepsilon_F\, e^{\frac{1}{J N(\varepsilon_F)\cos^2\delta(\varepsilon_F)}} \qquad (19)$$

The so defined Kondo temperature is characteristic for all physical properties of the system.

4. EQUATION-OF-MOTION METHOD

Since Kondo's discovery of the scattering anomaly /1/ a large number of theoretical papers appeared. Part of these deal with the scattering of conduction electrons by magnetic impurities at all temperatures (especially $T \leq T_K$), applying methods more powerful than finite order perturbation theory. Among them, dispersion theory /23/, infinite order perturbation theory /24/, variational methods, and the equation of motion method /12/, /13/, /25/ have been most successful. Besides this other physical properties as the susceptibility, specific heat, magnetoresistance, and thermopower were calculated to order J^3, following Kondo's treatment of the resistivity. It was realized very soon that finite order perturbation theory breaks down at low temperatures because it predicts in each order logarithmic singularities of the T-matrix. The problem turned out to be a true many body problem (which cannot be solved exactly), since the impurity spin correlates the originally independent conduction electrons.

a) Calculation of the Non-Spin-Flip T-Matrix

We restrict ourselves to the equation-of-motion method

and consider first the case V = 0. The temperature Green's function for the conduction electrons is defined /26/ as

$$G_{kk'}(\tau) = -\frac{1}{2}\langle T_\tau(c_{k\mu}(\tau)\, c^+_{k'\mu}(0))\rangle, \quad -\beta \le \tau \le \beta, \tag{20}$$

where $\beta = (KT)^{-1}$ and $\tau = it$ denotes a fictitious "imaginary time". The time ordering operator T_τ for fermions is defined by

$$T_\tau(c_k(\tau)\, c^+_{k'}(0)) = \begin{cases} c_k(\tau)\, c^+_{k'}(0) & \tau > 0 \\ -c^+_{k'}(0)\, c_k(\tau) & \tau < 0 \end{cases} \tag{21}$$

and $c_k(\tau)$ is an "Heisenberg" operator ($\hat{N}$ is the particle number operator and μ the chemical potential)

$$c_k(\tau) = e^{\mathcal{H}\tau}\, c_k(0)\, e^{-\mathcal{H}\tau}, \quad \mathcal{H} = H - \mu\hat{N} \tag{22}$$

The symbol $\langle\cdots\rangle$ denotes the thermal average over a grand-canonical ensemble

$$\langle\cdots\rangle = \frac{Tr(e^{-\beta\mathcal{H}}\cdots)}{Tr\, e^{-\beta\mathcal{H}}}. \tag{23}$$

The Green's function (20) can be expanded into a Fourier series /26/

$$G_{kk'}(\tau) = \frac{1}{\beta}\sum_n e^{-i\omega_n\tau}\, G_{kk'}(i\omega_n), \quad \omega_n = \beta^{-1}\pi(2n+1) \tag{24}$$

$$n = 0, \pm 1 \ldots,$$

$$G_{kk'}(i\omega_n) \equiv \frac{1}{2}\langle\langle c_{k\mu}|c^+_{k'\mu}\rangle\rangle = \int_0^\beta e^{i\omega_n\tau}\, G_{kk'}(\tau)\, d\tau. \tag{25}$$

We apply the equation of motion $\dot{c}_k(\tau) = [\mathcal{H}, c_k]_-$ which follows from (22) and ($\Theta(\tau)$ is the step function)

$$\frac{\partial}{\partial\tau}[T_\tau(c_k(\tau)\, c^+_{k'}(0))] = \frac{\partial}{\partial\tau}[(c_k(\tau)\, c^+_{k'}(0) + c^+_{k'}(0)\, c_k(\tau))\,\Theta(\tau) - c^+_{k'}(0)\, c_k(\tau)] = \delta(\tau)\,\delta_{kk'} + T_\tau\left(\frac{\partial}{\partial\tau}c_k(\tau)\, c^+_{k'}(0)\right). \tag{26}$$

One finds with the Hamiltonian (2) and (6) for V = 0 after Fourier transformation the equation of motion

$$(i\omega_n - \tilde{\varepsilon}_k)\, G_{kk'}(i\omega_n) = \delta_{kk'} - \frac{J}{N}\sum_q \Gamma_{qk'}(i\omega_n)\,, \quad \tilde{\varepsilon}_k = \varepsilon_k - \mu\,. \tag{27}$$

The Green's function

$$\Gamma_{kk'}(i\omega_n) = \langle \underset{\sim}{S} \cdot \underset{\sim}{S}^e_{\mu\nu}\, c_{k\nu} | c^+_{k'\mu} \rangle \tag{28}$$

contains correlations between the impurity spin S and the conduction electrons. The equation of motion for $\Gamma_{kk'}(i\omega_n)$ includes Green's functions with three c-operators and the spin operator S. The equations of motion of these generate again higher order Green's functions, etc. We retain the coupling in lowest order only and truncate higher order correlation terms in the equation for $\Gamma_{kk'}(i\omega_n)$; for instance

$$\langle T_\tau (c_{k\uparrow}(\tau)\, c^+_{q\uparrow}(\tau)\, c_{q'\downarrow}(\tau)\, S^- c^+_{k'\uparrow}(0))\rangle \tag{29}$$

$$= \langle c_{k\uparrow} c^+_{q\uparrow}\rangle \langle T_\tau (c_{k'\downarrow}(\tau)\, S^- c^+_{k'\uparrow}(0))\rangle + \langle c^+_{q\uparrow} c_{q'\downarrow} S^-\rangle \langle T_\tau (c_{k\uparrow}(\tau) c^+_{k'\uparrow}(0))\rangle$$

There is no a priori justification for this approximation. Surprisingly, the results are nearly equivalent with those obtained by summing up an infinite series of the most divergent diagrams (i.e. the terms of highest order in lnT for a given order of J), and to the results of dispersion theory in the approximation of single particle intermediate states.

In the approximation (29) one obtains the equation of motion

$$(i\omega - \tilde{\varepsilon}_k)\, \Gamma_{kk'}(i\omega_n)$$

$$= \frac{J}{2N}\Big[(m_k - S(S+1))\sum_q G_{qk'}(i\omega_n) + (1 - 2n_k)\sum_q \Gamma_{qk'}(i\omega_n)\Big] \tag{30}$$

with the thermal averages

$$n_k \equiv \frac{1}{2}\sum_q \langle c^+_{q\mu} c_{k\mu}\rangle = \sum_q G_{kq}(0^-) = \frac{1}{\beta}\sum_{q,n} e^{i\omega_n\eta} G_{kq}(i\omega_n)\,, \quad \eta = 0^+\,, \tag{31}$$

$$m_k = 2\sum_q \langle c^+_{q\mu} S \cdot S^e_{\mu\nu} c_{k\nu} \rangle = 2\sum_q \Gamma_{kq}(0^-) = \frac{2}{\beta}\sum_{qn} e^{i\omega_n \eta}\Gamma_{kq}(i\omega_n). \tag{32}$$

We applied

$$[S^z, S^\pm]_- = \pm S^\pm \;, \quad [S^+, S^-]_- = 2S^z, \tag{33}$$

and used the symmetries (in the absence of a magnetic field)

$$\langle S^z \rangle = \langle S^+ \rangle = \langle S^- \rangle = 0, \tag{34a}$$

$$\langle c^+_{k\uparrow} c_{k'\uparrow} \rangle = \langle c^+_{k\downarrow} c_{k'\downarrow} \rangle \tag{34b}$$

$$\langle c^+_{k\uparrow} c_{k'\downarrow} S^- \rangle = \langle c^+_{k\downarrow} c_{k'\uparrow} S^+ \rangle$$

$$= 2\langle c^+_{k\uparrow} c_{k'\uparrow} S^z \rangle = -2\langle c^+_{k\downarrow} c_{k'\downarrow} S^z \rangle \tag{34c}$$

since we average over all possible impurity spin directions.

The solution of the four self-consistent coupled equations (27) and (30) to (32) is rather tedious and shall not be considered in detail. It turns out to be expedient to introduce the T-matrix element t(z) for non spin flip s-scattering (note that the thermal average t(z) is different from T(z) Eq. (3))

$$G_{kk'}(z) = G^0_k(z)\,\delta_{kk'} + G^0_k(z)\,t(z)\,G^0_{k'}(z) \;, \quad z = i\omega_n \tag{35}$$

with the Green's function for free electrons

$$G^0_k(z) = (z - \tilde{\varepsilon}_k)^{-1}. \tag{36}$$

One finds a complicated non linear integral equation for t(z), first derived by Hamann /13/

$$t(z) = \phi^{-1}(z)\Big[\Big(\frac{J}{4}\Big)^2 S(S+1)F(z) + \frac{J}{\beta}\sum_n \frac{F(i\omega_n) - F(z)}{z - i\omega_n}\, e^{i\omega_n \eta}\, t(i\omega_n)\Big], \tag{37}$$

with

$$\phi(z) = 1 + J R(z) - \Big(\frac{J}{4}\Big)^2 S(S+1)F^2(z) + \frac{J}{\beta}\sum_n \frac{(F(i\omega_n) - F(z))^2}{z - i\omega_n}\, e^{i\omega_n \eta}\, t(i\omega_n); \tag{38}$$

$$F(z)=\frac{1}{N}\sum_k \frac{1}{z-\tilde{\varepsilon}_k}\;;\quad R(z)=-\frac{1}{\beta}\sum_n \frac{F(i\omega_n)-F(z)}{z-i\omega_n}e^{i\omega_n\eta}=\int_{-\infty}^{\infty} N(\omega)\frac{f(\omega)-\frac{1}{2}}{z-\omega}d\omega \quad (39)$$

The function $R(z) + \frac{1}{2} F(z)$ reduces for $z = \tilde{\varepsilon}$ to the Kondo integral $-g(\varepsilon)$ Eq. (11). The second part of (39) is calculated applying the Cauchy theorem: For an arbitrary function $h(z)$, except for poles at the imaginary axis and with $zh(z)\, f(z) \to 0$ as $|z| \to \infty$, holds

$$\frac{1}{\beta}\sum_n h(i\omega_n) = -\frac{1}{2\pi i}\int_c f(z')\,h(z')\,dz' \qquad (40)$$

since the Fermi function $f(z)$ has poles at $i\omega_n = \beta^{-1} i\pi(2n+1)$ with the residue $-\beta^{-1}$. The integration contour c is shown in Fig. 7

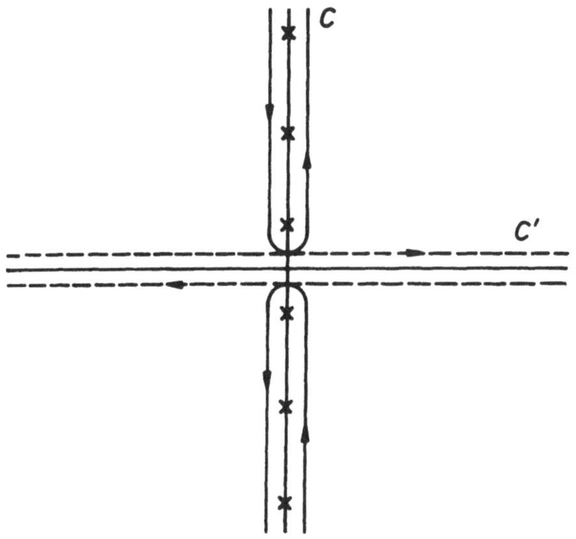

Fig. 7: Integration path for the integral Eq. (40).

The matrix element $t(z)$, and $F(z)$ and $R(z)$ are analytical functions except for a cut along the real axis with

$$t(z) = \begin{cases} t_R(z) & \mathrm{Im}\, z > 0 \\ t_A(z) & \mathrm{Im}\, z < 0 \end{cases} \qquad (41)$$

where $t_R(z)$ and $t_A(z)$ are "retarded" and "advanced" functions, respectively. Therefore we can replace the contour c by the contour c' along the real axis. (The contribution of the half circles with $|z| \to \infty$ vanish). If we assume the density of states $N(\varepsilon)$ to be analytical in a small strip around the real axis and if we continue $F_R(z)$ ($F_A(z)$) into the lower (upper) half plane, we have

$$F_R(z) - F_A(z) = -2\pi i\, N(z) \tag{42}$$

Applying (40) to (42) to R(z) one verifies the second part of (39).

The integral equation (37) has been solved exactly /26/. One finds

$$S_R(z) \equiv 1 - 2\pi i\, N(z)\, t_R(z) = X_R(z)\, e^{Q_R(z)} \quad (\mathrm{Im}\, z > 0) \tag{43}$$

with

$$Q_R(z) = \frac{1}{2\pi i} \int_{-\infty}^{\infty} \frac{d\omega}{z-\omega} \ln K(\omega) \tag{44}$$

$$K(z) = X_R(z)\, X_A(z) + S(S+1)\,(\pi J N(z))^2, \tag{45}$$

$$X_R(z) = 1 + J R_R(z) - \frac{J^2}{4} S(S+1)\, F_A(z) F_R(z) + J^3 \chi(z) \tag{46}$$

where $\chi(z)$ is proportional to N(z). We apply (43) with (44) to (46) to the calculation of the Kondo temperature, the resistivity, and the thermopower.

b) The Kondo Temperature

The S-matrix element (43) is not unitary since it includes the non spin flip part of the scattering only. One can show that $S_R(z) \to 1$ for $|z| \to \infty$ and all temperatures, $S_R(0) = 1 + O(J)$ for large T, and $S_{R_0}(0) = -1$ for $T = 0$. Therefore there exists a temperature T_K with $S_R(0, T_K) = 0$ which is specific for the system. Since $K(\omega)$ is real, symmetric in ω and positive definite, we can write

$$S_R(\omega) = \frac{X_R(\omega)}{\sqrt{K(\omega)}} \exp\left\{\frac{1}{2\pi i} P\int_{-\infty}^{\infty} \frac{d\omega'}{\omega-\omega'} \ln K(\omega')\right\}. \tag{47}$$

The Kondo temperature is then defined by

$$X_R(0,\mathring{T}_K) = 0 \quad , \quad -J\,R_R(0) + O(J^2) = 1 \tag{48}$$

and with (39)

$$1 = J\int_{-\infty}^{\infty} \frac{d\omega}{\omega - i\eta}\left(f(\omega)-\tfrac{1}{2}\right)N(\omega) + O(J^2). \tag{49}$$

One finds for the simplified density of states ($\ln\gamma = 0.577$ is Euler's constant, and D is the width of the conduction band)

$$N(\omega) = \frac{D^2}{\omega^2 + D^2} \tag{50}$$

$$\mathring{T}_K = 2\gamma D \pi^{-1} \exp\left\{\frac{1}{J\,N(\varepsilon_F)}\right\}, \tag{51}$$

in fair agreement with (18) where we used the density of states for free electrons. For $J < 0$, $|J|N(\varepsilon_F) = 10^{-1}$, and $D = 5$ eV one has $T_K = 3^{\circ}K$.

c) Resistivity and Thermopower

We assume that the density of states $N(\omega)$ is symmetrical with respect to the Fermi energy $\omega = 0$. It can then be shown that

$$F^*(z) = -F(-z^*) \quad , \quad t^*(z) = -t(-z^*) \tag{52}$$

or especially for $z = \omega$,

$$\mathrm{Re}\, t_R(\omega) = -\mathrm{Re}\, t_R(-\omega) \;;\; \mathrm{Im}\, t_R(\omega) = \mathrm{Im}\, t_R(-\omega). \tag{53}$$

The resistivity $\rho_i(T)$ and the thermopower $S_d(T)$ are essentially given by

$$\rho_i(T) \sim \int_{-\varepsilon_F}^{\infty} d\omega\, \tau(\omega)\frac{df(\omega)}{d\omega} \;;\quad S_d(T) \sim \int_{-\varepsilon_F}^{\infty} d\omega\, \omega\, \tau(\omega)\frac{df(\omega)}{d\omega}. \tag{54a,b}$$

From (35) follows for a small impurity concentration c a relation /22/ between t(z) and the self-energy $\Sigma(\omega)$

defined by $G^{-1}_{kk}(z) = \overset{o}{G}{}^{-1}_{k}(z) - N^{-1}\Sigma(z)$, the imaginary part of which can be expressed by the relaxation time $\tau(\omega)$, and finally

$$(2\tau(\omega))^{-1} = -c \, \mathrm{Im} \, t_R(\omega) \tag{55}$$

Inserting (55) with (53) into (54b) we find $S_d(T) = 0$. A more general (non symmetric) density of states $N(\omega)$ yields only a term of the order as it is observed for non magnetic impurities, but not the large peak measured for instance in CuFe. However, one can explain this peak by taking into account the ordinary interaction V. The calculation is similar to that for V = 0, and we give the results only. The T-matrix element t(z) is replaced /27/ by

$$t_{J+V}(z) = t_V(z) + e^{2i\delta(z)} \cos^2\delta(z) \, \tilde{t}(z)$$

where $t_{J+V}(z)$ includes ordinary and exchange interactions; $t_V(z)$ is the T-matrix element for ordinary scattering (with J = 0), and $\delta(z)$ is the corresponding phase shift; $\tilde{t}(z)$ is obtained from t(z) by replacing the unperturbed density of states $N(\omega)$ by $\tilde{N}(\omega) = N(\omega)\cos^2\delta(\omega)$ in the corresponding integrals. Assuming a Lorentzian density of states we find /14/ the impurity resistivity for s-scattering (n is the density of conduction electrons, m is the electron mass; $\rho_0 = mc\,(ne^2\pi\, N(\varepsilon_F))^{-1}$)

$$\rho_i(T) = \rho_0 \left[1 - \cos 2\,\delta(\varepsilon_F) \frac{\ln T/T_K}{\sqrt{\ln^2 T/T_K + \pi^2 S(S+1)}} + O(V])^2 \right] \tag{57}$$

The Kondo temperature T_K is (besides minor corrections) defined by (51) with $N(\omega)$ replaced by $\tilde{N}(\omega)$. It can be obtained from experimental data from $\rho_i(T_K) = 1/2\,[\rho_i(0) - \rho_i(\infty)]$. At T_K the resistivity ρ_i has its largest slope as a function of ln T.

The diffusion term of the thermopower is given by (ρ_{tot} is the total resistivity including the resistivity by electron-phonon interaction, etc.; $e < 0$)

$$S_d(T) = \frac{K}{2e} \frac{\rho_0}{\rho_{tot}} \frac{\pi^3 S(S+1) \sin 2\delta(\varepsilon_F)}{(\ln^2 T/T_K + \pi^2 S(S+1))^{3/2}} \tag{58}$$

and shows a peak at $T = T_K$ which depends strongly on the ordinary interaction.

In deriving (57) and (58) we used the Sommerfeld expansion. This seems to be correct for $|\ln T/T_K| \gg 1$. Near T_K correction terms become important and give rise to a more complicated temperature behavior.

5. CONCLUSIONS

The equation-of-motion method and similarly the dispersion theory and infinite order perturbation theory yield results in reasonable agreement with experimental data for the resistivity, thermopower, and for the specific heat. There is no rigorous theory available at present which takes into account an external magnetic field at all temperatures, or for instance at the most interesting temperature range around T_K. However, one has derived high temperature expressions for the susceptibility and magnetoresistance, applying perturbation theory to order J^3 (and perhaps J^4).

We neglected magnetic interactions between the impurities, restricting ourselves to extremely dilute alloys. Finally by applying the Hamiltonian (2) we tacitly assumed that the impurity spin is well localized, and that its magnitude does not change with temperature. It is still an open question whether or not these assumptions hold.

References

/1/ J. Kondo, Progr. Theoret. Phys. (Kyoto) 32, 37 (1984)
/2/ See for instance the review article by M.D. Daybell and W.A. Steyert, Rev. Mod. Phys. 40, 380 (1968)
/3/ D.K.C. Mac Donald, W.B. Pearson and I.M. Templeton, Proc. Roy. Soc. 266, 161 (1962); B. Knook, thesis, Leiden 1962
/4/ R.W. Schmitt and I.S. Jacobs, J. Phys. Chem. Solids 3, 324 (1957)
/5/ A.N. Gerritsen and J.O. Linde, Physica 18, 877 (1952)
/6/ R.B. Coles, Physics Letters 8, 243 (1964)
/7/ M.D. Daybell and W.A. Steyert, Phys. Rev. Letters 18, 398 (1967); 20, 195 (1968), Phys. Rev. 167, 536 (1968)
/8/ A. Kjekshus and W.B. Pearson, Can. J. Phys. 40, 98 (1962)
/9/ P. Monod, Phys. Rev. Letters 19, 1113 (1967)

/10/ C.M. Hurd, J. Phys. Chem. Solids 28, 1345 (1967); Phys. Rev. Letters 18, 1127 (1967)
/11/ A.J. Croft, E.A. Faulkner, J. Hatton and E.F.W. Seymour, Phil. Mag. 44, 289 (1953)
/12/ Y. Nagaoka, Phys. Rev. 138 A 1112 (1965); Progr. Theoret. Phys. (Kyoto) 37, 13 (1967)
/13/ D.R. Hamann, Phys. Rev. 158, 570 (1967); P.E. Bloomfield and D.R. Hamann, Phys. Rev. 164, 856 (1967)
/14/ K. Fischer, J. Phys. Chem. Solids, 29, 1227 (1968)
/15/ S.V. Maleev, Zh. Eksp. Teor. Fiz. 51, 1940 (1966) English translation: Soviet Physics JEPT 24, 1300 (1967)
/16/ H. Suhl and D. Wong, Physics 3, 17 (1967)
/17/ J.P. Franck, F.D. Manchester and D.L. Martin, Proc. Roy. Soc. London A 263, 494 (1961)
/18/ F.J. du Chatenier and J. de Nobel, Physica 32, 1097 (1966)
/19/ F.J. du Chatenier and A.R. Miedema, Physica 32, 403 (1966)
/20/ H. Suhl in "Rendiconti della Scuola Internazionale di Fisica "Enrico Fermi" 1966, Academic Press, London, 1967
/21/ A. Messiah, "Quantum Mechanics" John Wiley, New York 1962, p. 830
/22/ K. Fischer, Phys. Rev. 158, 613 (1967)
/23/ H. Suhl, Phys. Rev. 138, A 515 (1965), 141, 483 (1966), Physics 2, 39 (1966)
/24/ A.A. Abrikosov, Physics 2, 5 (1965)
/25/ J. Zittartz and E. Müller-Hartmann, Z. Physik 212, 380 (1968)
/26/ A.A. Abrikosov, L.P. Gorkov and I.E. Dzyaloshinski, "Methods of Quantum Field Theory in Statistical Physics", Prentice-Hall, Inc. London 1963, p. 97, 121
/27/ K.D. Schotte, Z. Physik 212, 467 (1968)

NONLINEAR OPTICAL SUSCEPTIBILITIES

N. Bloembergen

Harvard University

The first lecture gave a classical introduction to nonlinear optical properties in terms of an anharmonic oscillator model. A phenomenological description of various nonlinear phenomena was outlined. This material may be found in Chapter 1 of a lecture note volume on Nonlinear Optics.[1]

The second lecture described the theory of parametric up-conversion and parametric down-conversion. It was shown how the nonlinear polarization term, quadratic in the electric field amplitudes, leads to a coupling between three electromagnetic waves, for which the conditions of conservation of energy $\omega_1 + \omega_2 = \omega_3$, and momentum, $\vec{k}_1 + \vec{k}_2 = \vec{k}_3$, are satisfied. This material is treated in Chapter 4 of reference 1. More recent review papers which discuss this question are given in references 2 and 3. There the reader may find numerous citations to the original research literature. A very recent experimental realization of a continuous, tunable parametric light oscillator, utilizing a Nd^{3+} : YAG (yttrium aluminum garnet) laser and barium sodium niobate as the nonlinear optical crystal is described by Smith et. al.[4]

In the third lecture the quantummechanical calculation of nonlinear susceptibilities was briefly outlined. The general theory is described in Chapter 2 of reference 1. When certain simplifications are introduced, i.e. the low frequency approximation and the closure approximation, numerical results may be obtained for simple crystal structures such as III - V and group IV semiconductors.[5] The results of the calculation based on tetrahedral bonding orbital electronic wave functions agree well with the experimental values. A discussion of the physical nature of the magnitude of the optical nonlinearities of solids has also been given by Robinson.[6]

Finally, some examples were given of the laws of geometrical optics in the nonlinear domain. A review of the effects of second harmonic generation of light in reflection[7] discusses also the application to the measurement of the dispersion of the nonlinearity in absorbing media. More recently the generation of second harmonics by a laser beam entering from a dense linear fluid onto a less dense nonlinear crystal has been described. Even though the laser beam may be totally reflected from the nonlinear crystal, second harmonic generation can be observed, because the exponentially decaying laser fields near the surface of the crystal still create second harmonic polarization.[8]

There was no time in the lectures to discuss the stimulated raman effect, self-focusing of light beams, third harmonic generation and the numerous other nonlinear effects which may be described by a polarization which is a cubic function of the electric field amplitudes. The interested reader is referred to a recent review article on the stimulated raman effect.[9] In addition to references 1 and 3, the reader should also consult a large number of papers in the proceedings of various international conferences on quantum electronics.[10-16] Collectively these give an interesting picture of the rapid development of the field of nonlinear optics.

References

1. N. Bloembergen, Nonlinear Optics, W. A. Benjamin, Inc., New York, 1965.
2. S. A. Achmanov and R. V. Khochlov, Soviet Physics Uspekhi 9, 210 1966.
3. R. W. Minck, R. W. Terhune and C. C. Wang, Proc. IEEE 54, 1357 1966.
4. R. G. Smith et. al., App. Phys. Letters 12, 308 (1968).
5. S. S. Jha and N. Bloembergen, Proceedings of the 5th International Conference on Quantum Electronics, held in Miami, May 1968, to be published in the IEEE Journal of Quantum Electronics, October 1968. See also Phys. Rev. 171, July 15, 1968.
6. F. N. H. Robinson, Bell System Technical Journal 56, 913 (1967).
7. N. Bloembergen, Optica Acta 13, 311 (1966).
8. N. Bloembergen and C. H. Lee, Phys. Rev. Letters 19, 835 (1967).
9. N. Bloembergen, Am. Journal of Physics 35, 989 (1967).
10. Quantum Electronics, edited by C. H. Townes, Columbia University Press, 1960.
11. Advances in Quantum Electronics, edited by J. R. Singer, Columbia University Press, 1961.
12. Proceedings of the 3rd International Conference on Quantum Electronics, edited by P. Grivet and N. Bloembergen, Columbia University Press, 1964.

13. Proceedings of the International School of Physics, E. Fermi, Course XXXI, Academic Press, 1965.
14. Proceedings of the Puerto Rico Conference on Physics of Quantum Electronics, edited by P. L. Kelley, B. Lax and P. E. Tannenwald, McGraw-Hill, 1966.
15. Proceedings of the 4th International Conference on Quantum Electronics, IEEE Journal of Quantum Electronics 2, Aug. - Sept., 1966.
16. Proceedings of the 5th International Conference on Quantum Electronics, IEEE Journal of Quantum Electronics, to be published (1968).

LIGHT SCATTERING FROM PLASMONS IN SEMICONDUCTORS*

A. L. McWhorter

Lincoln Laboratory, Massachusetts Institute of Technology

Lexington, Massachusetts

INTRODUCTION

Light scattering with high-power CW lasers has proved to be an extremely powerful way to study some of the fundamental collective and single-particle excitations in solids. This lecture will discuss recent experimental and theoretical work[1-3] carried out at Lincoln Laboratory on light scattering from plasmons in semiconductors, particularly GaAs. Not only have these studies confirmed that plasmons exist and couple to phonons as predicted,[4] but they have allowed the various photon scattering mechanisms to be determined as well. Brief mention will also be made of some current work on light scattering from single-particle excitations.[5-7]

EXPERIMENTAL RESULTS

The experiments were performed with a 1.06 micron YAG:Nd^{3+} laser as the light source. The experimental techniques are described elsewhere.[1,2] Polished single-crystal GaAs samples with (100) faces and dimensions about 3 x 3 x 5 mm were used. The polarized laser beam was incident along a <100> axis of the crystal and the scattered light was collected at 90° along another <100> axis. The range of electron concentrations available for quantitative measurements of the plasmon modes in GaAs lies between about 4×10^{17} cm^{-3} and 5×10^{18} cm^{-3}. The high concentration limit is set by the solubility of donors in GaAs, the low concentration limit by Landau damping of the plasmons.

*This work was sponsored by the U. S. Air Force.

Figure 1 shows the Raman spectra for a series of n-type GaAs samples of different carrier concentrations. The lower frequency peak in the semi-insulating sample is due to excitation of transverse optic (TO) phonons and the higher frequency peak to longitudinal optic (LO) phonons. Note that the position of the TO mode is independent of carrier concentration, while in the doped samples there are two broad peaks rather than one sharp LO peak. What is happening is that there are now two longitudinal modes, plasmons as

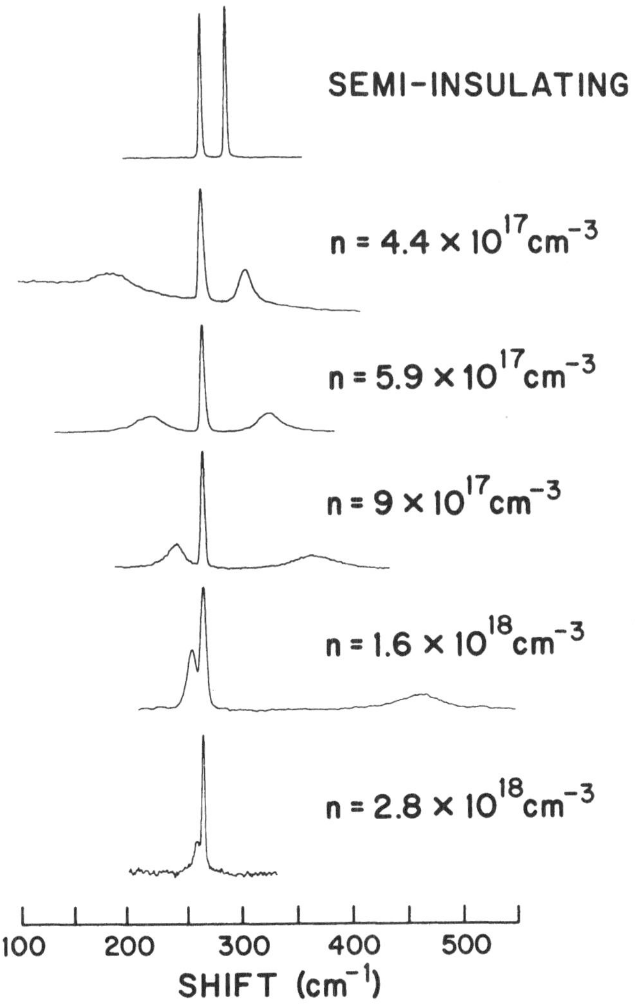

Fig. 1. Recorder traces of the Raman-scattered light from a series of GaAs samples of different electron concentrations.

well as LO phonons, and the two modes are coupling together. This is seen more clearly in Fig. 2, which is a plot of the Raman frequency shifts as a function of the square root of the electron concentration. In the absence of the phonons, the plasmon mode would have a frequency $\omega_p = (4\pi ne^2/\varepsilon_\infty m^*)^{1/2}$, where n is the electron concentration, m* the conduction band effective mass, and ε_∞ the optical dielectric constant. The eigenfrequencies of the coupled plasmon-phonon modes are given by the zeros of the total longitudinal dielectric function, which in the long wavelength limit has the form

$$\varepsilon(\omega) = \varepsilon_\infty \left[1 - \frac{\omega_p^{\,2}}{\omega(\omega + i/\tau)}\right] + \frac{(\varepsilon_o - \varepsilon_\infty)\omega_t^{\,2}}{\omega_t^{\,2} - \omega^2}\,, \qquad (1)$$

where ε_o is the static dielectric constant, ω_t is the transverse

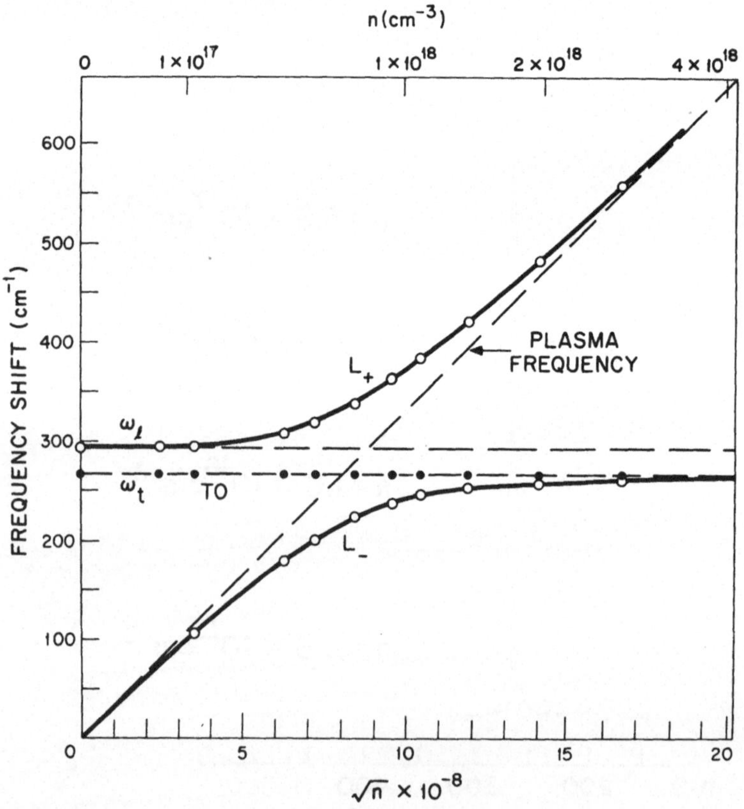

Fig. 2. Frequency shift of the Raman-scattered light in GaAs at room temperature as a function of the square root of the electron concentration. The solid curves labeled L_+ and L_- give the calculated frequencies of the mixed longitudinal plasmon-phonon modes.

optical phonon frequency, and τ is a phenomenological collision time. The solid lines in Fig. 2 are the zeros of $\varepsilon(\omega)$ with $\tau\to\infty$.

As previously noted the frequency of the TO mode is unaffected by the presence of free carriers; it also exhibits the polarization properties predicted by group theory for a zinc blende insulator.[8] However, the polarization properties of the upper and lower longitudinal modes, L_+ and L_-, differ from those found for a pure longitudinal optic mode in semi-insulating GaAs. The latter, in accordance with standard group-theory predictions, shows zero LO scattering at 90° when the incident and scattered beams are both polarized parallel to the plane of scattering (‖ , ‖) or both perpendicular to the plane of scattering (⊥ , ⊥). Figure 3 shows representative polarization traces for a sample with $n = 1.9 \times 10^{18} cm^{-3}$ taken at a temperature near that of liquid helium. The strong (⊥ , ⊥) scattering is apparent. Good optical alignment is assured by the nearly complete suppression of the TO mode in the (⊥ , ⊥) polarization configuration.

THEORY

As long as the Raman frequency shift ω is small compared with the incident frequency ω_1, the light scattering can quite generally be thought of as due to fluctuations in the dielectric susceptibility χ. The fluctuations in χ arise from three sources: thermal fluctuations in the electric field E, in the optical mode lattice displacement u, and in the electron concentration n. Hence the scattering cross section can be written as

$$\frac{d^2\sigma}{d\Omega d\omega} = \left(\frac{\omega_1}{c}\right)^4 \left|\hat{\varepsilon}_1{}^{\alpha}\delta\chi_{\alpha\beta}\hat{\varepsilon}_2{}^{\beta}\right|^2 \tag{2}$$

where

$$\delta\chi_{\alpha\beta} = \frac{\partial\chi_{\alpha\beta}(\omega_1)}{\partial E_\gamma}\delta E_\gamma(\omega) + \frac{\partial\chi_{\alpha\beta}(\omega_1)}{\partial u_\gamma}\delta u_\gamma(\omega) + \frac{\partial\chi_{\alpha\beta}(\omega_1)}{\partial n}\delta n(\omega) \tag{3}$$

and $\hat{\varepsilon}_1$ and $\hat{\varepsilon}_2$ are the incident and scattered polarization vectors. The first term in (3) arises from electro-optic coupling and the second from deformation potential coupling.

In the long wavelength limit and for frequencies near the coupled mode resonances

$$\delta u = \left(\frac{\varepsilon_0 - \varepsilon_\infty}{4\pi M}\right)^{1/2} \frac{\omega_t}{\omega_t^2 - \omega^2}\,\delta E \tag{4}$$

where M is the reduced mass density of the two sublattices. (For ω very near ω_t, one must use a more general expression[9] which includes the spontaneous fluctuation in δu.) From the fluctuation-dissipation theorem we have for the longitudinal fluctuations in electric field

$$\langle \delta E \delta E^+ \rangle_\omega = -4(n_\omega + 1)\,\mathrm{Im}\,[1/\varepsilon(\omega)] \tag{5}$$

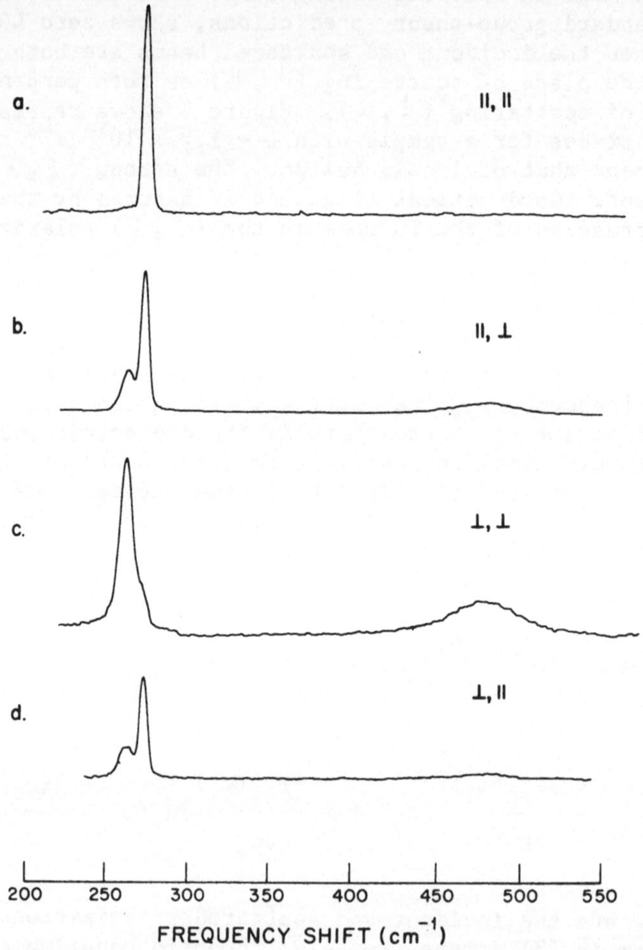

Fig. 3. Polarization recorder traces of the plasmon-phonon coupled modes for GaAs near liquid helium temperature with $n = 1.9 \times 10^{18}$ cm^{-3}. The scattering angle is 90°, with propagation along <100> directions. The polarization of the incident (scattered) light with respect to the plane of scattering is indicated by the first (second) symbol by each trace.

Hence for 90° scattering in zinc blende crystals with propagation along <100> axes, electro-optic and deformation potential coupling give for the longitudinal modes

$$\frac{d^2\sigma_\ell(\perp,\parallel)}{d\Omega d\omega} = \frac{d^2\sigma_\ell(\parallel,\perp)}{d\Omega d\omega} = -2(n_\omega+1)\left(\frac{\omega_1}{c}\right)^4$$

$$\times\left(1 + C_1\frac{\omega_t^2}{\omega_t^2-\omega^2}\right)^2 |b_{41}|^2 \operatorname{Im}\frac{1}{\varepsilon(\omega)}, \tag{6}$$

where

$$C_1 = \left(\frac{\varepsilon_0 - \varepsilon_\infty}{4\pi M}\right)^{1/2} \frac{1}{\omega_t}\frac{a_{41}}{b_{41}} \tag{7}$$

with $a_{41} = \partial\chi_{zy}/\partial u_x$ and $b_{41} = \partial\chi_{zy}/\partial E_x$. We shall treat C_1 as a constant to be determined from the data.

We next consider the scattering due the free carrier fluctuations. For $\omega_1 \ll E_G$, where E_G is the energy gap of the semiconductor, the contribution to $\chi(\omega_1)$ from the free carriers is $-(ne^2/m^*\omega_1^2)$. Hence $\partial\chi_{\alpha\beta}/\partial n = -(e^2/m^*\omega_1^2)\delta_{\alpha\beta}$. As ω_1 approaches E_G, this contribution is resonantly enhanced and we find

$$\frac{\partial\chi_{\alpha\beta}(\omega_1)}{\partial n} = -\frac{e^2}{m^*\omega_1^2}\frac{E_G^2}{E_G^2-\omega_1^2}\delta_{\alpha\beta}. \tag{8}$$

From the continuity equation $\nabla\cdot J + e\partial n/\partial t = 0$ and the equation of motion for the electrons, we also obtain for the fluctuation in n of wave vector q and frequency ω the expression

$$\delta n = \frac{q}{e\omega}\delta J = \frac{q}{\omega}\left(\frac{ne}{-i\omega m^*}\right)\delta E = \frac{iq}{4\pi e}\left(\frac{\varepsilon_\infty\omega_p^2}{\omega^2}\right)\delta E. \tag{9}$$

Combining (8) and (9) with (5) yields for 90° scattering

$$\frac{d^2\sigma_\ell(\perp,\perp)}{d\Omega d\omega} = \left(\frac{e^2}{m^*c^2}\right)^2(n_\omega+1)\left(\frac{E_G^2}{E_G^2-\omega_1^2}\right)^2\frac{q^2}{4\pi^2e^2}$$

$$\times\left(\frac{\varepsilon_\infty\omega_p^2}{\omega^2}\right)^2 \operatorname{Im}\frac{1}{\varepsilon(\omega)}, \tag{10}$$

with zero cross section for the free carrier scattering in all other polarizations.

Finally, for the TO modes the integrated (∥,∥) cross section is[8]

$$\frac{d\sigma_t(\parallel,\parallel)}{d\Omega} = \left(\frac{\omega_1}{c}\right)^4 \frac{n_t + 1}{2M\omega_t} \left|a_{41}\right|^2 \tag{11}$$

with 1/2 this amount for the (⊥, ∥) and (∥, ⊥) scattering, and zero for the (⊥, ⊥) scattering.

Since only the relative scattering cross sections are measured experimentally, it is convenient to work with the integrated cross sections normalized to the TO (∥, ∥) cross section. We then find for $\omega\tau \gg 1$

$$\frac{d\sigma_\ell^{\pm}(\perp,\parallel)/d\Omega}{d\sigma_t(\parallel,\parallel)/d\Omega} \approx \frac{n_{\omega_\pm} + 1}{n_t + 1} \frac{\varepsilon_0 - \varepsilon_\infty}{\omega_t(\partial\varepsilon/\partial\omega_\pm)} \left(\frac{1}{C_1} + \frac{\omega_t^2}{\omega_t^2 - \omega_\pm^2}\right)^2 \tag{12}$$

$$\frac{d\sigma_\ell^{\pm}(\perp,\perp)/d\Omega}{d\sigma_t(\parallel,\parallel)/d\Omega} \approx \frac{n_{\omega_\pm} + 1}{n_t + 1} \frac{\varepsilon_0 - \varepsilon_\infty}{\omega_t(\partial\varepsilon/\partial\omega_\pm)} \left(\frac{\omega_p}{\omega_\pm}\right)^4 \frac{1}{C_2^2} \tag{13}$$

where

$$C_2 = \left(\frac{\varepsilon_0 - \varepsilon_\infty}{4\pi M}\right)^{1/2} \frac{2^{3/2}\pi m^* \omega_1^2}{\varepsilon_\infty q e \omega_t} \left(\frac{E_G^2}{E_G^2 - \omega_1^2}\right)^{-1} a_{41} \tag{14}$$

is a second constant which will also be determined by fitting the data. Note that except for the Bose-Einstein factors, the relative scattering intensities really involve only two frequency ratios, ω_p/ω_t and ω_ℓ/ω_t, which can be evaluated directly from the Raman data, and the two dimensionless adjustable parameters C_1 and C_2.

DISCUSSION

In all samples the observed line shapes for the two longitudinal modes could be fitted by Lorentzians consistent with (1), (6) and (10) in the limit $\omega\tau \gg 1$. The value of τ was found to be of the order of 10^{-13} sec for the samples studied, which agrees with the collision time determined from dc mobility measurements.

The best fit between theory and experiment was obtained with $C_1 = -0.5$ and $|C_2| = 0.26$. Table I of Ref. 2 shows the measured and calculated results for two samples of relatively high electron concentration. In view of the experimental uncertainty (about $\pm$ 10% for the TO and L_- modes; $\pm$ 15 - 20% for the L_+ mode) the agreement is quite satisfactory, except for the ($\perp,\perp$) scattering of the L_+ mode for the $n = 1.9 \times 10^{18} cm^{-3}$ sample. The reason for this discrepancy is partly due to assuming $\omega \ll \omega_1$ in the theoretical expressions. From (7) and (14) we deduce $|a_{41}| = 8 \pm 3 \times 10^7 cm^{-1}$ and $|b_{41}| = 1.0 \pm 0.3 \times 10^{-6}$ esu at T = 4°K. Slightly lower values are obtained if the effect of resonant enhancement is treated more carefully.[10]

The existence of a strong ($\perp,\perp$) charge-density scattering for the L_- mode may seem puzzling until it is realized that in these samples we are dealing with a LO mode heavily screened by the conduction-band electrons. It is the screening charges that do the scattering. From the value obtained for a_{41} the deformation potential coupling should slightly dominate for the L_- mode when $n > 3 \times 10^{18}$ cm^{-3}. It is predicted from (12) and has been confirmed experimentally that around $n = 4 \times 10^{17}$ cm^{-3} the electro-optic coupling just cancels the deformation potential coupling for the L_- mode, giving zero scattering for the ($\parallel,\perp$) and ($\perp,\parallel$) configurations.

SINGLE-PARTICLE SCATTERING

Mooradian[5] has also observed light scattering from single-particle excitations in n-type GaAs. At T = 0°K this scattering arises from electrons that are excited from occupied states below the Fermi surface to unoccupied states above the Fermi surface, with the initial and final states being such that the energy and momentum of the electron-photon system are conserved. Figure 4 shows some of Mooradian's results[5] for a GaAs sample with $n = 1.4 \times 10^{18} cm^{-3}$. At very low temperatures and for infinite collision time, one would expect the Raman spectrum to be triangular with a sharp cutoff at $\omega = qv_F$, where q is the momentum transfer and v_F the Fermi velocity. At 5°K we see a tendency in this direction, but the rather short collision time (~10^{-13}sec) greatly smears out the cutoff.

The cross sections found by Mooradian[5] are between one and two orders of magnitude larger than those predicted on the basis of charge-density fluctuations, even taking into account the effects of nonparabolicity[11] and resonance enhancement. However, it has now been shown[6] that scattering from spin-density fluctuations can account for most of Mooradian's results. The coupling of the photons to the electron spin is through second-order p·A perturbation terms involving the spin-orbit splitting of the valence band. At high carrier concentrations, where the screening wave vector is much greater than the momentum transfer, the spin-density fluctuations are not screened out as the charge-density fluctuations are, and hence the scattering cross section can be much larger than that due

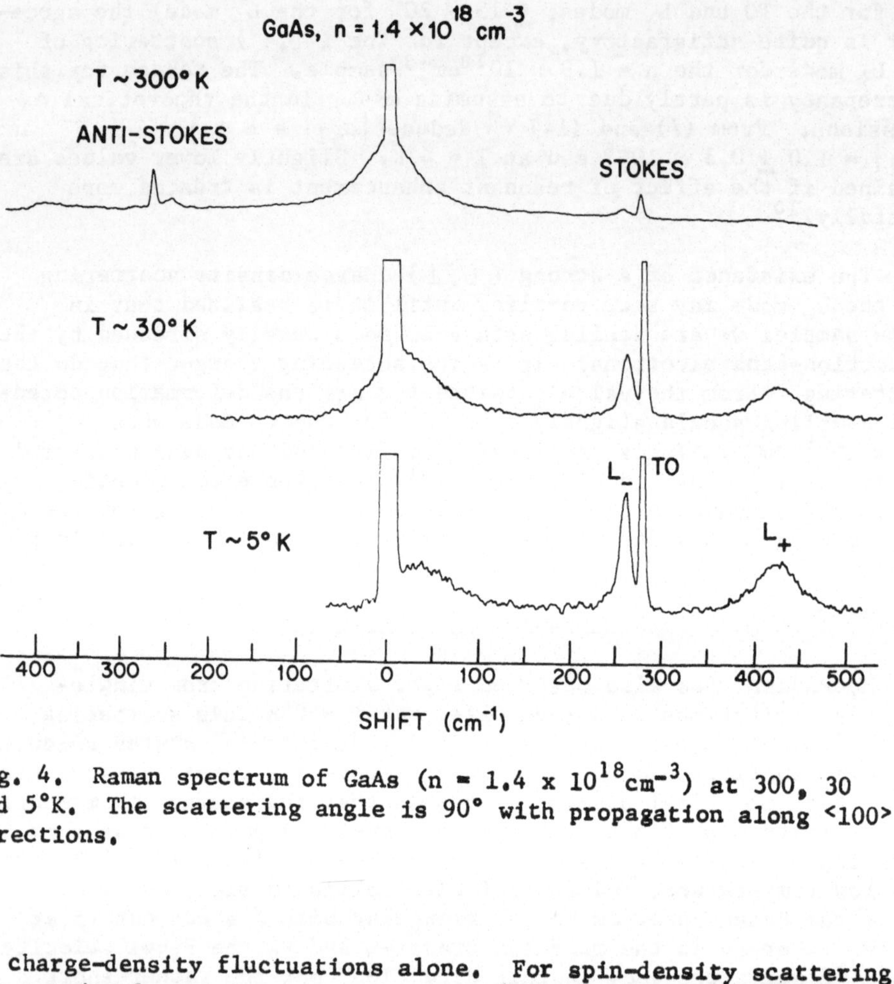

Fig. 4. Raman spectrum of GaAs ($n = 1.4 \times 10^{18} cm^{-3}$) at 300, 30 and 5°K. The scattering angle is 90° with propagation along <100> directions.

to charge-density fluctuations alone. For spin-density scattering the matrix element is proportional to $\vec{\sigma} \cdot (\hat{\epsilon}_1 \times \hat{\epsilon}_2)$. In addition Mooradian has found a large temperature-dependent cross section for $\hat{\epsilon}_1 \| \hat{\epsilon}_2$ in the high-concentration samples. This effect appears to be due to electron states of different energy being weighted by different resonant enhancement factors because of the k-dependence of the energy difference between conduction and valence band states.[7]

REFERENCES

1. A. Mooradian and G. B. Wright, Phys. Rev. Letters 16, 999 (1966).
2. A. Mooradian and A. L. McWhorter, Phys. Rev. Letters 19, 849 (1967).
3. A. L. McWhorter and P. N. Argyres, Bull. Am. Phys. Soc. 12, 102 (1967); also International Conference on Light-Scattering Spectra of Solids, New York, 1968 (to be published).

4. See, for example, B. B. Varga, Phys. Rev. 137, A1896 (1965).
5. A Mooradian, Phys. Rev. Letters 20, 1102 (1968).
6. D. C. Hamilton and A. L. McWhorter, International Conference on Light-Scattering Spectra of Solids, New York, 1968 (to be published.
7. A. L. McWhorter, A. Mooradian and D. C. Hamilton, to be published.
8. R. Loudon, Advan. Phys. 13, 423 (1964).
9. N. D. Strahm and A. L. McWhorter, International Conference on Light-Scattering Spectra of Solids, New York, 1968 (to be published).
10. A. Mooradian and A. L. McWhorter, International Conference on Light-Scattering Spectra of Solids, New York, 1968 (to be published).
11. P. A. Wolff, Phys. Rev. 171, 436 (1968).

TRANSFERRED ELECTRON PHENOMENA, INCLUDING THE GUNN EFFECT

C. Hilsum

Royal Radar Establishment, St. Andrews Road

Malvern, Worcestershire, England

This lecture will cover, rather shallowly, the concepts we need for an understanding of this topic. The first concept is that of a differential negative resistance in which dI/dV or dJ/dF is negative. There are two classifications of differential negative resistance - the S type in which current is a two or three valued function of voltage, and the N type with voltage a multi-valued function of current. We are here concerned with an N type d.n.r., the kind of characteristic we have seen published for devices like tunnel diodes. We should stress that we are here concerned not with interfacial effects, as in junctions, but with bulk effects. Let us then analyse the requirements for a bulk differential negative resistance of the N type.

The conductivity of a uniform sample is given by $ne\mu$. We need pronounced decreases in n or μ as the field increases. Cases where n decreases in this way are known - they are often the result of a field dependent trapping cross-section, such as we get in a charged centre - but such effects are slow and of less interest. Our attention is focussed on the case where μ changes markedly.

The Physics of Transfer

We can obtain such a marked change by postulating a material with a specific band structure. We need first low energy states of low effective mass, second high energy states of heavy effective mass and third a dominant scattering process which gives a rapid dependence of electron temperature on electric field. We can see what is likely to happen in this system. At low fields the carriers will all have low mass and therefore high mobility. At a high enough field the carriers will have sufficient energy to populate the higher energy states. Since these states have a high mass, their density will also be high, so at high energy there will be far more heavy states than light states available. Consequently most of the carriers will be heavy, and will contribute little to the conductivity. The transition region between the low and high field situations is the important region. If we have a rapid dependence of electron temperature or field, the transition is abrupt, and we get a d.n.r.

Let us give an example of this. GaAs has a conduction band structure with a central minimum in the conduction band having an effective mass of 0.07 m_o. In the (100) directions, 0.35 eV higher up, the mass is 0.35 m_o. The respective mobilities are 8000 and 150 cm^2/v.s., and there are 30 times as many heavy states as light. We can therefore draw a rough velocity-field characteristic. We also know that polar scattering dominates in pure GaAs, so that at a critical field near 3000 V/cm the electrons gain more energy from the field than they can lose to the lattice. We therefore get a rapid transfer of carriers, with a d.n.r.

Stability

An d.n.r. acquires its main interest from the effects which follow from it. In the region where the differential is negative, the dielectric relaxation time is negative. As a result departures from space charge neutrality will grow rather than decay. Consider a region in the sample with an incipient dipole layer, a slight accumulation nearer to the cathode, and a depletion on the anode side. The field in this layer will be slightly higher than outside it, because of Poisson's Law. The carriers are therefore travelling slower. Carriers therefore pile up against the accumulation layer, enhancing it, and draw away from the depletion layer, stretching it.

The field across the sample settles with a narrow high field region, and a field lower than threshold across the rest of the sample. It is as though the sample wants to move off the unstable bias region, and find two stable regions on the current-voltage characteristic.

This domain, the high field region, is simply a spatial variation in the electron density, and we must remember that the electrons are all moving with high speed from cathode to anode. Typically the speed will be near 10^7 cm sec. The domain also moves from cathode to anode at this speed, collapsing when it reaches the anode, and then reforming at the cathode. This is the Gunn effect.

Not all samples will show this behaviour. Clearly a domain can form only if the negative dielectric relaxation time is faster than the electron transit time. The transit time $\tau_T = \frac{\ell}{2.10^7}$ sec. The dielectric relaxation time is

$$\frac{\rho K}{4\pi} \text{ i.e. } \frac{K}{4\pi \cdot ne\mu}$$

$$\text{If } \tau_T > \tau_D,\ \frac{\ell}{2.10^7} > \frac{K}{4\pi ne\mu} \text{ or } n\ell > \frac{2.10^7}{1.6.10^{-19}\,3000.9.10^{11}} \text{ i.e } 4.10^{10}$$

(A more accurate analysis gives 2.10^{11}). Clearly though there will be a minimum $n\ell$ product for domain formation.

So we have established, by simple physical considerations, the consequence of the transferred electron effect - a series of dipole domains moving through the sample at high frequency. In a moment we shall examine in detail the domain shape, but before doing so I want briefly to refer to another form of spatial inhomogeneity which can occur. This is an isolated accumulation of electrons, with the sample divided into two regions, a high field region downstream, and a low field upstream. The accumulation layer moves from cathode to anode as a dipole domain does, but its velocity can be very high. Pure accumulation layers seem to form only in theoretical samples - in any practical case there are always fluctuations in impurity concentration, and these will convert accumulation layers into dipole domains.

Domain Shape

The simplest idea of a domain is that it is a rearrangement of electric field so that the heavy electrons within the domain travel at the same speed as the light electrons outside i.e.

$$F_R\, \mu_1 = F_D\, \mu_2 = v \tag{1}$$

and since in GaAs $\mu_1 \simeq 25\, \mu_2$ $F_D \simeq 25\, F_R$

$$\text{We also have } F_D - F_R = \frac{4\pi}{K}\, ned \tag{2}$$

for a fully depleted domain with depletion length d.

$$\text{Further } Fl = F_R l + (F_D - F_R)\,(x + \frac{d}{2}) \tag{3}$$

assuming heavy electrons occupy a length x.

This third equation may be difficult to satisfy. F_D must be near 50,000 volts/cm, so since n is known, d is fixed from (2). This gives us a minimum value for the R.H.S. of (3), yet we can reduce the L.H.S. by taking a small value for l. In fact there <u>is</u> a range of solutions with x = 0, for with x = 0, equation 1 has no validity. We no longer can determine F_D in this way, and so it can be smaller.

Whereas previously we consider the field configuration in the domain as flat-topped, now it will be triangular. Our flat topped domain had heavy electrons drifting stably in it, and so was represented by a point on the high field branch of our v-F characteristic. The triangular domain cannot be represented in this way. The relationship between v and F now lies in the region above the v-F curve we have been dealing with previously. We call these the dynamic and static characteristic respectively, and are interested in defining the dynamic characteristic.

The simplest way is as follows. Assume we have a domain in an infinite sample, moving with velocity v_D. Outside the domain the current is $I = n_0 e v_R$, and elsewhere

$$I = \underset{\substack{\uparrow \\ \text{conduction}}}{nev(F)} - e \underset{\substack{\uparrow \\ \text{diffusion}}}{\frac{\partial}{\partial x}} D(F).n + \frac{K}{4\pi}\frac{\partial F}{\partial t}$$

we also have $\frac{\partial F}{\partial x} = \frac{4\pi e}{K}(n - n_0)$

If we substitute $y = x - v_D t$ as a frame of reference

F and n are functions of y, $\frac{\partial F}{\partial x} = \frac{dF}{dy}$, $\frac{\partial F}{\partial t} = -v_D \frac{dF}{dy}$

Then $\frac{dF}{dy} = \frac{4\pi e}{K}(n - n_0)$, $\frac{d}{dy}[nD(F)] = n\,[v(F) - v_D] - n_0\,(v_R - v_D)$

or $$\frac{d}{dF} D(F).n = \frac{K}{4\pi e}\left[\frac{n[v(F) - v_D] - n_0\,[v_R - v_D]}{n - n_0}\right]$$

If $D(F) = D$

$$\frac{n}{n_0} - \ln\frac{n}{n_0} - 1 = \frac{K}{4\pi e D}\int_{F_R}^{F} dF\left\{\left[v(F) - v_D\right] - \frac{n_0}{n}\left[v_R - v_D\right]\right\}$$

The equation has two roots for n, one for the depletion layer and one for the accumulation layer. The two branches come together

$F = F_D$ and $F = F_R$. Since for $n = n_o$ the L.H.S. vanishes, the R.H.S. integral must vanish at $F = F_D$ when n has values on either branch. This is only possible if $v_R = v_D$. Moreover, the peak domain field is such that the $\int_{F_R}^{F_D} [v(F) - v_R]dF$ vanishes. This gives us our dynamic characteristic, a simple geometric construction which is now known as Butcher's equal area rule.

Working Characteristics

The last thing we need to know about the domain is its dependence on voltage across the sample. We can deduce the dependence of ϕ_D, the voltage drop across the domain, against F_R, since F_R determines F_D, and we can then calculate n and d. We plot ϕ_D against F_R, and on the same diagram plot the load line $\phi = \phi_D + F_R \ell$. This gives the working point.

With ℓ fixed we roll the line change ϕ.

If $\phi/\ell > F_T$ there is one working point.

If $\phi/\ell < F_T$ there are two working points, but no starting condition.

If we inject a domain it moves to the higher point - the lower one appears inaccessible. This is called triggered operation.

Circuit Interactions

The last subject to cover in the lecture is circuit interactions. The simplest case is a resistive circuit, where we see that the current appears as a series of spikes, separated in time by the electron transit time. This is the Gunn effect. What we see are transitions between the static and dynamic characteristics. If we have the sample in a reasonant circuit, we can have it operating at a frequency different from the transit frequency. Now we can consider the domain as exposed to the resultant field of the D.C. and the A.C. caused by the circuit ringing.

With large A.C. fields, and a circuit working at high frequency, the resultant field can drop below the sustaining field while the domain is in transit-the domain is quenched, and the resultant frequency generated is the cavity frequency. With a circuit at lower frequency, the domain researches the anode and is quenched, but nucleation is delayed until the A.C. field swings to threshold again. This is the delayed mode.

In this way the frequency can be varied slightly from the Gunn frequency. To obtain a frequency far removed from the Gunn frequency requires a strong cavity interaction. In this mode the A.C. field is large, and swings up and down so rapidly that the domains have no time to grow before the field is swung below threshold. For an appreciable time each cycle the resultant field is below threshold. The condition we must meet is that over a number of cycles there is no slow domain growth i.e. each cycle, decay exceeds growth. We can achieve this and still extract power from the sample. There is a second condition we must meet i.e. during the growth parts of the cycle the domain must not distort the field so appreciably that our conditions for domain decay are violated. These conditions may be expressed as limits on n/f, typically between 10^3 and 10^5. This mode of operation is called L.S.A. or Limited Space-Charge Accumulation. Its main interest is at high frequencies and high powers, because there is no link between the dimensions of the sample and the frequency - a long sample can be used at high frequencies. The importance for high power generation is that the sample impedance can be kept high while still using a large volume in order to inject high power.

At this stage we have covered the elementary theory of transferred electron phenomena, including the Gunn effect. I would stress that the concepts you have had presented are simplified, and are only approximate. But the complications are easy to introduce into this framework.

THEORY OF THE VELOCITY-FIELD CHARACTERISTIC OF SEMICONDUCTORS EXHIBITING THE GUNN EFFECT

P. N. Butcher

School of Physics, University of Warwick

Coventry, England

ABSTRACT

Recent theoretical work on the velocity-field characteristic of semiconductors exhibiting the Gunn effect is reviewed.

1. INTRODUCTION

It is now well established that the current oscillations first observed by J. B. Gunn[1] in n-type gallium arsenide and indium phosphide are due to the transfer of electrons in the conduction band from low-mass low-energy valleys to high-mass high-energy valleys at high electric fields. As a result of this transfer, the average electron drift velocity reaches a peak at a "threshold field" F_T and then decreases with increasing field. The negative differential mobility makes a uniform time-independent field $F > F_T$ unstable. Consequently, at high voltage bias levels, narrow domains of very high field build up and propagate periodically through the specimen to produce the current oscillations observed by Gunn.

The key to understanding the Gunn effect lies in the velocity-field characteristic. In this paper we describe the various calculations of this characteristic which have been made in the last few years. Further details of the early work and its comparison with experimental data can be found in the review articles by Butcher[2] and Bott and Fawcett[3] which also contain treatments of the nature of the space charge instability. The design and performance of microwave oscillators based on the Gunn effect has been discussed recently and compared with that of other semiconductor microwave generators by Hilsum[4].

2. FORMULATION OF THE PROBLEM

Most theoretical and experimental work on the Gunn effect has been done with gallium arsenide because of the advanced state of the technology of this material. The conduction band structure of gallium arsenide is shown schematically in Fig.1. The lowest energy valley is at the centre of the Brillouin zone (valley 1 - the "central" valley) and has a light effective mass $m_1 = 0.067$ m where m is the free electron mass. Surrounding the central valley are three "satellite" valleys (valley 2) in the (100) directions with minima 0.36 eV above the central minimum and an average effective mass in the order of 0.35 m. A multi-valley conduction band structure of this type (with high-energy, high-mass valleys to which hot electrons can transfer from low-energy, low-mass valleys) is essential for the Gunn effect.

To set up the equations for calculating the velocity-field characteristic let us introduce a positive integer i to label the various valleys and assume for simplicity that the energy in valley i may be approximated by

$$E_i(\underline{k}) = \frac{\hbar^2}{2m_i} (\underline{k} - \underline{K}_i)^2 + \Delta_i \qquad (1)$$

where $\underline{K}_i$ and Δ_i are respectively the wave vector and energy at the

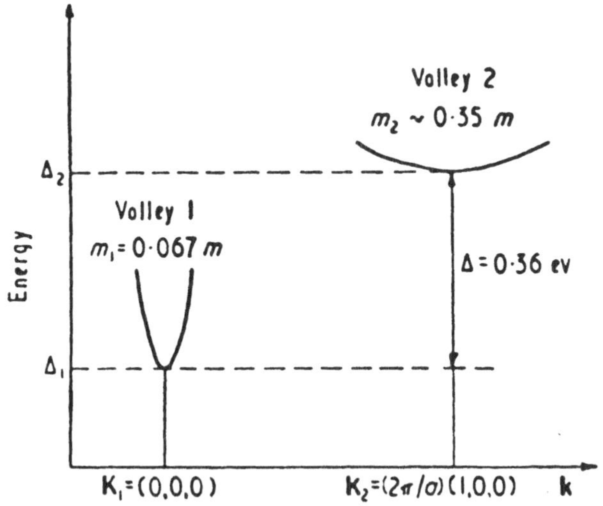

Fig. 1. Schematic conduction-band structure for gallium arsenide.

minimum of the valley and m_i is a scalar effective mass. Let $f_i(\underline{k})$ denote the distribution function giving the number of electrons in valley i per unit volume of $\underline{k}$-space per unit volume of the crystal. Then Boltzmann's equations for the distribution functions established in the presence of a constant electric field $\underline{F}$ may be written in the form

$$\frac{\partial f_i(\underline{k})}{\partial t} = \left(\frac{\partial f_i(\underline{k})}{\partial t}\right)_F + \sum_j \left(\frac{\partial f_i(\underline{k})}{\partial t}\right)_{ij} \tag{2}$$

The first term on the right-hand side is the rate of change of $f_i(k)$ due to the field:

$$\left(\frac{\partial f_i(\underline{k})}{t}\right) = \frac{e}{\hbar}\,\underline{F}\cdot\underline{\nabla}_k f_i(\underline{k}) \tag{3}$$

where e is the magnitude of the electronic charge. The second term is the rate of change of $f_i(\underline{k})$ produced by all the scattering processes. The summation is over all valleys; $(\partial f_i(\underline{k})/\partial t)_{ii}$ is the contribution from intravalley scattering in valley i and $(\partial f_i(\underline{k})/\partial t)_{ij}$ with $j \neq i$ denotes the contribution from the intervalley scattering between valleys i and j. In both cases we are primarily concerned with phonon scattering processes and the contribution to $\{\partial f_i(\underline{k})/\partial t\}_{ij}$ from phonons of type s may be written in the form

$$\begin{aligned}\left(\frac{\partial f_i(\underline{k})}{\partial t}\right)_{ij}^{(s)} = \int_j C_{ij}^{(s)}(\underline{k},\underline{k}')\Big(\{N_s f_j(\underline{k}') - \\ (N_s+1)f_i(\underline{k})\}\delta\{E_i(\underline{k}) - E_j(\underline{k}') - \hbar\omega_s\} \\ - \{N_s f_i(\underline{k}) - (N_s+1)f_j(\underline{k}')\}\delta\{E_j(\underline{k}') - E_i(\underline{k}) \\ - \hbar\omega_s\}\Big)d\underline{k}'\end{aligned} \tag{4}$$

for all i and j. The four terms in the integrand account for transitions from state $\underline{k}$ in valley i to state $\underline{k}'$ in valley j, and vice versa, due to the absorption and emission of phonons of type s with wave vector $\pm(\underline{k} - \underline{k}')$ and frequency $\omega_s(\underline{k} - \underline{k}')$. It has been assumed that the phonons remain in thermal equilibrium at the lattice temperature T_o and

$$N_s(\underline{k} - \underline{k}') = \left(\exp\left\{\frac{\hbar\omega_s(\underline{k} - \underline{k}')}{kT_o}\right\} - 1\right)^{-1} \tag{5}$$

denotes the thermal average number of phonons per mode. For the sake of brevity, the dependence of ω_s and N_s on $\underline{k} - \underline{k}'$ is not shown explicitly in equation (4).

The strength of the scattering enters equation (4) through the function $C_{ij}^{(s)}(\underline{k},\underline{k}')$ which depends on the type of phonon and whether we are concerned with intravalley or intervalley processes. We consider, first of all, intravalley scattering for which $\underline{k} - \underline{k}'$ is small so that long-wavelength phonons are involved. Nearly all the materials showing the Gunn effect have polar lattices and the dominant intravalley scattering mechanism at room temperature and low fields is polar-mode scattering. This mechanism is known to limit the low-field mobility of good quality n-type crystals of the III-V compounds at room temperature. It is due to the Coulomb interaction between electrons and the electric field associated with the longitudinal (polar) optic phonons. Non-polar intravalley scattering by the transverse optic phonons is negligible in comparison and will be ignored. We may therefore safely distinguish the long-wavelength longitudinal optic phonons by simply replacing s by o, for optical. The dependence of ω_o and N_o on the wave vector is sufficiently weak in the long-wavelength limit to be ignored. The dependence of $C_{ii}^{(o)}(\underline{k},\underline{k}')$ on k - k' has, however, the singular form expected for a Coulomb interaction:

$$C_{ii}^{(o)}(\underline{k},\underline{k}') = \frac{e^2\omega_o}{2\pi}\left(\frac{1}{\varepsilon_\infty} - \frac{1}{\varepsilon_o}\right)\frac{1}{|\underline{k} - \underline{k}'|^2} \tag{6}$$

where ε_o and ε_∞ are respectively the low frequency and high frequency dielectric constants of the lattice.

The wave vector dependence of $C_{ii}^{(o)}(\underline{k},\underline{k}')$ implies that polar-mode intravalley scattering becomes less effective at high fields because the electron distribution becomes hot and more spread out in $\underline{k}$ space. We shall see that this characteristic of polar-mode scattering plays an essential role in determining the threshold field for the Gunn effect in polar materials. It also implies that at high fields we must consider, in addition to polar-mode scattering, intravalley scattering due to collisions with acoustic phonons which become increasingly effective as the electrons become hotter (see equation (7)). The electron-phonon coupling in this case is through the lattice strain and may be characterised by a deformation potential tensor in the long-wavelength limit. For isotropic valleys the tensor reduces to a scalar and only the longitudinal acoustic phonons are involved. We may therefore safely label the long-wavelength longitudinal acoustic phonons by simply replacing s by a, for acoustic, and write Ξ_a for the scalar deformation potential. Then[5]

$$C_{ii}^{(a)}(\underline{k},\underline{k}') = \frac{\Xi_a^2}{8\pi^2 \rho s} |\underline{k} - \underline{k}'| \tag{7}$$

where ρ is the density of the material and s is the longitudinal sound velocity.

We consider now intervalley scattering between valleys i and j with $i \neq j$. In the cases to be discussed, for each pair of valleys, only one type of phonon is involved and it is convenient to label the phonon responsible for scattering between valleys i and j by replacing s by ij. The phonon wave vector $\pm (\underline{k} - \underline{k}')$ always lies in the immediate neighbourhood of the separation of the valley minima $\underline{K}_i - \underline{K}_j$, and the dependence of $C_{ij}^{(ij)}(\underline{k},\underline{k}')$, ω_{ij} and N_{ij} on $\underline{k}$ and $\underline{k}'$ may be ignored. The ij phonon may be either acoustic or optical, there being little distinction between the two at short wavelengths. In either case it is convenient to write $C_{ij}^{(ij)}(\underline{k},\underline{k}')$ in a form which is obtained from the right-hand side of (7) by replacing Ξ_a by aD_{ij}, s by $\omega_{ij}/|\underline{k} - \underline{k}'|$ and, finally, $|\underline{k} - \underline{k}'|$ by a^{-1}, where a is the lattice constant constant and ω_{ij} is the phonon frequency at $\underline{K}_i - \underline{K}_j$. Thus we have

$$C_{ij}^{(ij)}(\underline{k},\underline{k}') = \frac{D_{ij}^2}{8\pi^2 \rho\omega_{ij}} \tag{8}$$

where D_{ij} is the deformation potential field for the ij phonon which characterizes the strength of the scattering[6].

To calculate the velocity-field characteristic we have to find the time-independent solution of (2) subject to the normalization condition

$$\sum_i \int f_i(\underline{k})\, d\underline{k} = n \tag{9}$$

where n is the electron density. Then the average electron drift velocity $\underline{v}$ is given by:

$$\underline{v} = \frac{1}{n} \sum_i \int f_i(\underline{k})\, (\hbar/m_i)\, (\underline{k} - \underline{K}_i)\, d\underline{k} \tag{10}$$

The assumed spherical symmetry of $E_i(\underline{k})$ has the result that $\underline{v}$ is antiparallel to $\underline{F}$

3. THE DISPLACED MAXWELLIAN APPROXIMATION

The solution of Boltzmann's equations (2) at high-field strengths is difficult, even for a single valley, and it is usual to make some assumption about the form of the distribution function so as to simplify the calculation. The electron drift velocity is determined by the first moment of the distribution function and is not particularly sensitive to the detailed form of $f_i(\underline{k})$. One might hope, therefore, that a relatively crude approximation will yield good qualitative results and fair quantitative results. Fortunately, this is indeed the case.

The most successful approximation form for $f_i(\underline{k})$ is the displaced Maxwellian:

$$f_i(\underline{k}) = n_i \left(\frac{\hbar^2}{2\pi m_i kT_i}\right)^{3/2} \exp\left\{\frac{-\hbar^2(\underline{k} - \underline{K}_i - \underline{d}_i)^2}{2m_i kT_i}\right\} \tag{11}$$

where n_i, T_i and $\underline{d}_i$ are respectively the electron density, temperature and displacement in valley i. When these parameters have been determined in terms of $\underline{F}$, it follows immediately from (9), (10) and (11) that the velocity-field characteristic is given by

$$\underline{v} = \sum_i \frac{n_i}{n} \left(\frac{\hbar \underline{d}_i}{m_i}\right) \tag{12}$$

To determine the parameters in the steady state distribution function we use the steady state conservation equations

$$\int_i \frac{\partial f_i(\underline{k})}{\partial t} \phi_i(\underline{k})\, d\underline{k} = 0 \tag{13}$$

where $\partial f_i(\underline{k})/\partial t$ is given by (2) and we put $\phi_i(\underline{k}) = 1$ (number conservation), $\underline{k} - \underline{K}_i$ (momentum conservation) and $E_i(\underline{k})$ (energy conservation) in turn so as to obtain three equations for each valley involving n_i, T_i and $\underline{d}_i$. The equations are coupled by the intervalley scattering terms in (2), and must usually be solved numerically.

Considerable insight into the results to be expected in polar materials can be obtained from the early work of Frölich and Paranjape[7] and Stratton[8]. These authors were not of course concerned with the then undiscovered Gunn effect. Their interest was in the high-field behaviour of electrons in a single parabolic valley subject to scattering by longitudinal optic phonons alone. The total scattering rate of electrons out of state $\underline{k}$ in valley i due to polar mode scattering is, from (4) and (6),

$$\lambda_o(\underline{k}) = \int_i C_{ii}^{(o)}(\underline{k},\underline{k}') \Big((N_o + 1)\, \delta\{E_i(\underline{k}) - E_i(\underline{k}') - h\omega_o\} + N_o\, \delta\{E_i(\underline{k}') - E_i(\underline{k}) - \hbar\omega_o\}\Big) d\underline{k}'$$

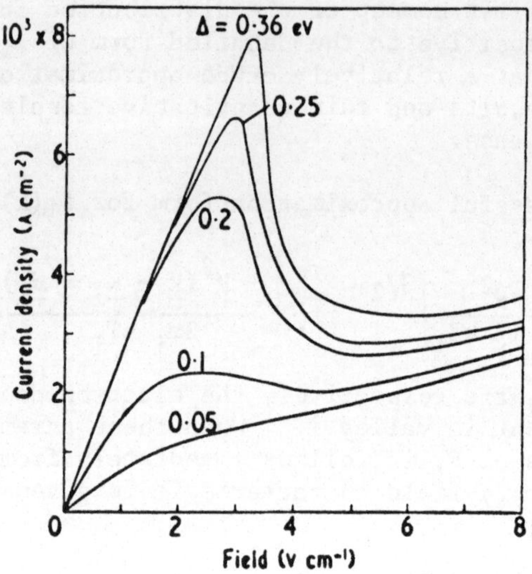

Fig. 2. Plots of current density against field in gallium arsenide for various values of the energy difference Δ between the central and satellite minima.[12]

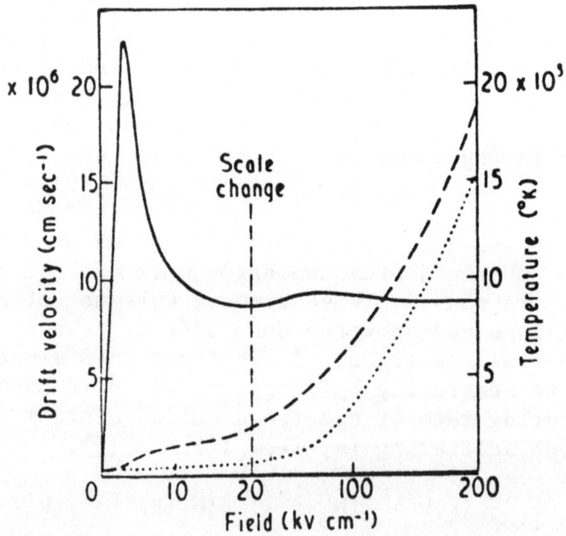

Fig. 3. Plots against field for gallium arsenide of average drift velocity (full curve), central valley temperature (broken curve) and satellite valley temperature (dotted curve).[2]

$$= \frac{2m_i e^2 \omega_o}{\hbar\left(2m_i E_i(\underline{k})\right)^{\frac{1}{2}}} \left(\frac{1}{\varepsilon_\infty} - \frac{1}{\varepsilon_o}\right) \left[(N_o + 1) \sinh^{-1}\left\{\frac{E_i(\underline{k}) - \hbar\omega_o}{\hbar\omega_o}\right\}^{\frac{1}{2}} + N_o \sinh^{-1}\left\{\frac{E_i(\underline{k})}{\hbar\omega_o}\right\}^{\frac{1}{2}} \right] \tag{14}$$

where the first term accounts for phonon emission processes and the second term accounts for phonon absorption processes. The rate of energy loss is equal to $\hbar\omega_o$ times the difference of the emissive and absorptive contributions to $\lambda_o(\underline{k})$.[9] When $E_i(\underline{k})$ is large the rate of energy loss falls of approximately as $\left(E_i(\underline{k})\right)^{-\frac{1}{2}}$. This ineffectiveness of polar mode scattering as an energy dissipation mechanism for high electron energies has the result that it is impossible to balance the energy input from the field by the energy dissipated to the polar phonons at high fields. Consequently, as Fröhlich and Paranjape and Stratton found, the conservation equations have no solution for F greater than a critical value (the polar mode breakdown field) which is about 3.5 kv cm^{-1} for the central valley of gallium arsenide, i.e. just above the threshold field for the Gunn effect. The very rapid transfer of electrons from the central valley to the satellite valleys which occurs in gallium arsenide as the field increases above the threshold (see Fig.3) is a consequence of this polar mode breakdown. The intervalley scattering mechanisms must come in to stabilise the distribution function for fields above threshold. A similar correlation between the threshold field for the Gunn effect and the polar mode breakdown field is found in other polar materials[10].

The first attempt at calculating a velocity field characteristic for gallium arsenide was made by Hilsum[11]. He used (12) and wrote $\underline{d}_i = -m_i\mu_i \underline{F}/\hbar$, taking the mobilities to be field-independent and equal to 10^4 $cm^2 v^{-1} sec^{-1}$ in the central valley and 500 cm^2 v^{-1} sec^{-1} in the satellite valleys. The electron concentrations in the valleys were obtained by assuming that the electrons had a Boltzmann distribution over the energy levels at the temperature calculated as a function of the field from the conservation equations for the central valley alone. A negative differential mobility was found between 3 kv cm^{-1} and the polar mode breakdown field of 3.5 kv cm^{-1}.

The scattering between the central and satellite valleys (which was ignored in Hilsum's calculations) of course plays a central role in the electron transfer process. Butcher and Fawcett[12] solved the conservation equations for the central and satellite valleys in gallium arsenide taking this scattering mechanism into account in addition to polar mode intravalley scattering. The resulting current density (n ev) - field characteristics at 300°K for an electron concentration $n = 2 \times 10^{15}\ cm^{-3}$ are shown in Fig.2 for various values of the energy separation Δ between the central and satellite minima. At zero pressure $\Delta = 0.36$ eV. The variation of the threshold field with pressure was determined from the other curves and found to be in good agreement with experimental data[2,3,12]. In subsequent calculations acoustic intravalley and inter-satellite valley scattering were included. These scattering mechanisms are of little significance at low fields but they give an enhanced scattering rate at high fields with the result that the velocity-field curve saturates as shown in Fig.3[13]. The rapid rise of electron temperature beyond threshold is also shown in Fig.3. The electrons transfer from the central valley to the satellite valleys extremely rapidly when the threshold field of 3.25 kv cm^{-1} is exceeded with the result that the velocity falls off with increasing field with a negative differential mobility of magnitude 3000 $cm^2\ v^{-1}\ sec^{-1}$.

4. THE APPROXIMATION OF NEARLY ELASTIC COLLISIONS

The displaced Maxwellian approximation is hard to justify à priori at the relatively low electron concentrations found in most samples of gallium arsenide used in experimental work on the Gunn effect. Conwell and Vassell[14] therefore used a completely different scheme of approximation in solving Boltzmann's equations. They assumed that the distribution function in valley i has the form

$$f_i(\underline{k}) = f_i^{\,o}(E) + |\underline{k} - \underline{K}_i|\ g_i(E) \cos\theta_i \tag{15}$$

where $f_i^{\,o}$ and g_i are functions of the electron energy E alone and θ_i is the angle between $\underline{k} - \underline{K}_i$ and $\underline{F}$.

Equations for $f_i^{\,o}$ and g_i may be derived by substituting equation (15) into equation (2) and keeping only terms which are either isotropic or proportional to $\cos\theta_i$. Conwell and Vassell suppose that, at high fields, the average electron energy is large enough for the electron-phonon collisions to be regarded as nearly elastic. In that case a relaxation time $\tau_i(E)$ exists; i.e. the anisotropic contribution to the scattering terms in equation (2) takes the simple form $|\underline{k} - \underline{K}_i|\ g_i(E) \cos\theta_i/\tau_i(E)$. Hence, picking out the term proportional to $\cos\theta_i$ in $(\partial f_i(\underline{k})/\partial t)_F$, we see that

$$g_i(E) = \frac{e\hbar F \tau_i(E)}{m_i} \frac{df_i^{\,o}}{dE} \tag{16}$$

This relation may be used to eliminate g_i from the isotropic terms in Boltzmann's equations so as to obtain coupled equations for f_i^0 in all valleys:

$$\frac{2e^2 F^2}{3m_i} (E - \Delta_i)^{-\frac{1}{2}} \frac{d}{dE} \left\{ (E - \Delta_i)^{3/2} \tau_i(E) \frac{df_i^0}{dE} \right\} + \sum_j \left(\frac{\partial f_i^0}{\partial t} \right)_{ij} = 0 \tag{17}$$

where $(\partial f_i^0/\partial t)_{ij}$ is the isotropic part of $(\partial f_i(\underline{k})/\partial t)_{ij}$. The contribution to $(\partial f_i^0/\partial t)_{ij}$ from the various scattering mechanisms and the associated relaxation times (which combine reciprocally)are easily derived from equations (4) to (8).

The contribution to $(\partial f_i^0/\partial t)_{ij}$ from phonons of type s (other than acoustic phonons) contains $f_i^0(E \pm \hbar\omega_s)$ in addition to $f_i^0(E)$ as a result of the energy-conserving δ-functions in equation (4). Equations (17) therefore have a difference-differential form and are difficult to solve. To circumvent this difficulty, Conwell and Vassell assume that, at high fields, the rate of variation of $f_i^0(E \pm \hbar\omega_s)$ is sufficiently slow for a Taylor expansion to order $(\hbar\omega_s)^2$ to be adequate. The resulting system of coupled differential equations was solved numericrlly for gallium arsenide using parameter values for the band structure and polar-mode, acoustic and satellite-to-satellite scattering which differ little from those used by Butcher and Fawcett[13].

Calculations were made for two values of the deformation potential field D_{12} for central-to-satellite scattering: 5×10^7 eV cm^{-1} and 5×10^8 eV cm^{-1}. The transfer of electrons to the satellite valleys was found to proceed extremely rapidly when $D_{12} = 5 \times 10^7$ eV cm^{-1}; 78% of the electrons are in the satellite valleys at a field of 2.4 kv cm^{-1}. This result is unacceptable in view of the nearly ohmic behaviour observed in gallium arsenide at fields of this order. For $D_{12} = 5 \times 10^8$ eV cm^{-1} only 30% of the electrons are in the satellite valleys at 2.4 kv cm^{-1} and Conwell and Vassell conclude that this value of D_{12} must be closer to the truth. It is very close to the value of $30/a = 5.3 \times 10^8$ eV cm^{-1} used by Butcher and Fawcett[13] in calculating the velocity-field characteristic shown in Fig.3.

In Fig.4, the calculated isotropic parts of the distribution functions in the central and satellite valleys are plotted against energy, measured in units of the polar optic phonon energy $\hbar\omega_0 = 0.036$ eV, for $D_{12} = 5 \times 10^8$ eV cm^{-1} and F = 2.4 kv cm^{-1} and 6 kv cm^{-1} (reference 15). The broken curve gives the Maxwellian function appropriate to any valley at zero field and the lattice temperature of 300°K. We see that the satellite valley distribution function remains Maxwellian with the lattice temperature at both the fields considered. The central valley distribution function has the

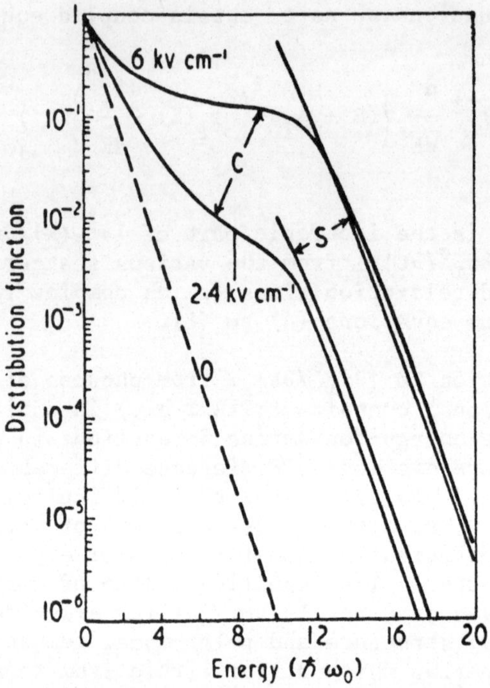

Fig. 4. Plots against energy for gallium arsenide of the isotropic part of the distribution function in the central valley (C) and the satellite valleys (S) for $D_{12} = 5 \times 10^8$ eV cm^{-1} and fields of 0, 2.4 and 6 kv cm^{-1}. (Reference 15).

same form for energies above 0.36 eV (i.e. 10 $\hbar\omega_0$) but changes its form below this energy. The Maxwellian form of the distribution functions above the energy of the satellite minima is due to the strong intervalley scattering.

The velocity-field characteristic for this case is shown in fig. 5[14]. The threshold field is close to 2.3 kv cm^{-1} and the threshold velocity is 1.1×10^7 cm sec^{-1}; the valley field is just under 10 kv cm^{-1}. By comparing Fig. 5 with Fig. 3, we see that the approximation of nearly elastic collisions yields a velocity peak with a similar shape to that obtained using the displaced Maxwellian approximation, but the threshold field and velocity are considerably smaller. The reduction of these quantities is due primarily to the large increase of electron transfer at low fields which occurs when the electron-phonon collisions are assumed to be nearly elastic. This assumption is in fact invalid at low fields and we shall see in the next section that the velocity-field characteristic obtained using the displaced Maxwellian approxima-

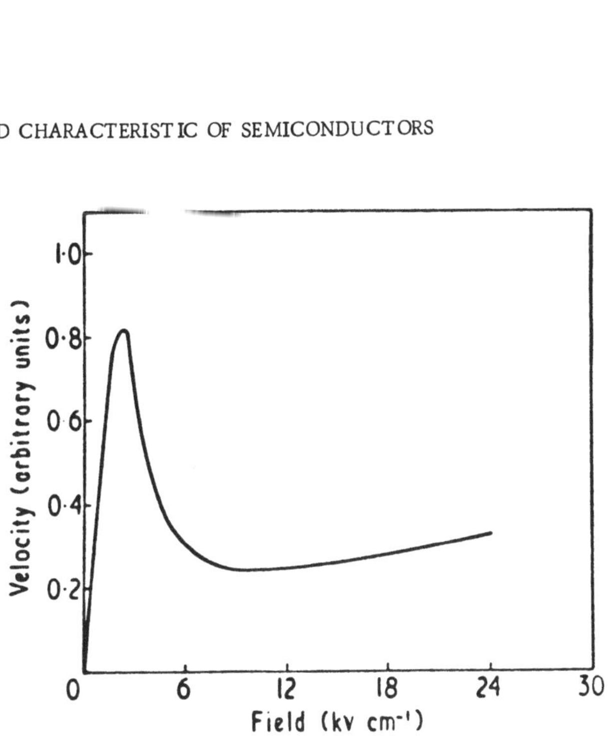

Fig. 5. Velocity-field characteristic of gallium arsenide.[14]

tion is in much better agreement with that obtained by a direct numerical solution of the problem.

5. MONTE CARLO CALCULATION

Recently Boardman, Fawcett and Reece[16] have calculated the velocity-field characteristic of gallium arsenide using the Monte Carlo technique. The principle of the method as applied to the determination of distribution functions is a computer simulation of the motion of a single electron in $\underline{k}$-space. After many scattering events and free flights in the electric field the time which the electron spends in an element of $\underline{k}$-space is proportional to the distribution function in that element.

The simulation is performed as follows. Suppose that the electron is at $\underline{k}$ at time zero. Then, it will be at $\underline{k}(t) = \underline{k} - e\underline{F}t/\hbar$ at time t provided that it is not scattered. The time t of scattering has, therefore, the probability density

$$p(t) = \lambda\left(\underline{k}(t)\right) \exp\left(- \int_0^t \lambda\left(\underline{k}(t')\right) dt'\right) \tag{18}$$

where $\lambda(\underline{k})$ is the total scattering rate out of the state $\underline{k}$ due to all scattering mechanisms. To determine the time of scattering we use the equation

$$P(t) \equiv \int_0^t p(t')dt' = r_1 \tag{19}$$

where r_1 is a random number taken from a collection of random numbers, generated in the machine, which are uniformly distributed between 0 and 1. The probability P(t) that the electron will have been scattered by the time t is obviously uniformly distributed between 0 and 1, and consequently the times generated by this procedure have the required probability density p(t). When the scattering time t has been determined, the precise scattering mechanism involved is determined with the correct probability by taking a second random number r_2 and using the inequality

$$\sum_{m<m'} \frac{\lambda_m\left(\underline{k}(t)\right)}{\lambda\left(\underline{k}(t)\right)} < r_2 < 1 - \sum_{m>m'} \frac{\lambda_m\left(\underline{k}(t)\right)}{\lambda\left(\underline{k}(t)\right)} \tag{20}$$

as the basis for deciding if the mechanism m' is operative. Here, it has been convenient to label the different scattering mechanisms by integers m and m' and $\lambda_m\left(\underline{k}(t)\right)$ is the contribution to $\lambda\left(\underline{k}(t)\right)$ from mechanism m. Lastly, having determined the scattering time and mechanism, the final wave vector reached after scattering from $\underline{k}(t)$ is determined in a similar way by generating further random numbers with distributions appropriate to the differential scattering rate of the scattering mechanisms involved. The final wave vector attained after the scattering event then becomes the initial wave vector for the next free flight.

The calculation of the contribution to $\lambda(\underline{k})$ from the various scattering mechanisms listed in section 2 is straightforward. The contribution from polar mode scattering, for example, is given by (14). The contribution which this mechanism makes to the exponent in (18) cannot, however, be evaluated analytically and the solution of (19) for t, given r_1, is therefore difficult. To circumvent this difficulty, Boardman, Fawcett and Rees introduce a non-physical "self-scattering" process from $\underline{k}$ into itself with the total scattering rate $\Gamma - \lambda(\underline{k})$ where Γ is an arbitrary positive number. When self-scattering is included in addition to the physical scattering mechanisms, $\lambda\left(\underline{k}(t)\right)$ is replaced by Γ in (18) and the solution of (19) for t is trivial. The price paid for this simplification is that there is now a finite probability of self-scattering which does not change $\underline{k}$ and does not contribute to the determination of the distribution function. In practice it was found that the distribution function stabilised to better than 1% after at most 2×10^5 collisions of which 75% are self-scattering events.

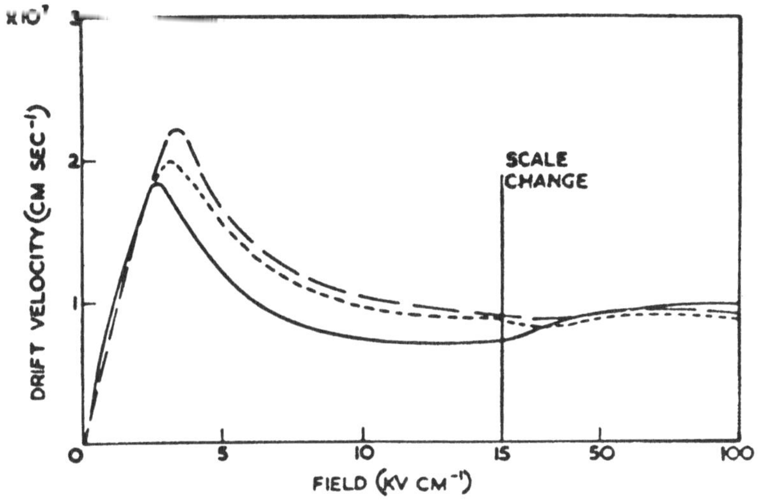

Fig. 6. Comparison of the calculated velocity-field relationship for gallium arsenide using the Monte Carlo method (full curve) and displaced Maxwellian distribution functions (broken curve). The dotted curve was obtained by the Monte Carlo method using different parameters.[16]

The full curve in Fig.6 shows the velocity-field curve for gallium arsenide at 300°K calculated by the Monte Carlo technique. The parameter values assumed were identical to those used in the calculation of the velocity-field curve shown in fig.3 which is reproduced as a dashed curve in Fig.6. The agreement between the Monte Carlo and displaced Maxwellian curves is remarkably good. Slight changes in the uncertain parameter values used in the Monte Carlo calculation yielded the dotted curve in Fig.6 which is in close agreement with the displaced Maxwellian curve.

The Monte Carlo distribution functions are not, however, displaced Maxwellians. Just below 0.36 eV there are relatively more electrons in the central valley than predicted by the displaced Maxwellian approximation. Above 0.36 eV, the strong intervalley scattering makes the distribution functions Maxwellian at the lattice temperature as was found by Conwell and Vassell (Fig.5). The distribution functions obtained by the Monte Carlo technique were, however, more strongly asymmetric at high fields than is suggested by the approximation of nearly elastic collisions.

6. COLLISION ITERATION CALCULATION

In a recent letter Rees[17] has outlined a technique for calculating the distribution function which exploits the stability of the steady state solution of Boltzmann's equations. Let $P_n(\underline{k},\underline{k}'t)dk'$ by the probability that an electron initially at $\underline{k}$ arrives in $(\underline{k}',d\underline{k}')$ at time t later after drifting in the field and suffering n collisions. For large n and t, $P_n(\underline{k},\underline{k}'t)$ will be independent of $\underline{k}$ and will depend on $\underline{k}'$ in the same way as the steady state distribution function $f(\underline{k}')$, where we have for simplicity dropped the label i which distinguishes the different valleys. Moreover, since the collisions do not occur with infinite frequency, $P_n(\underline{k},\underline{k}'t)$ is small at small times for large n. Consequently, for large n,

$$P_n(\underline{k},\underline{k}') \equiv \int_0^\infty P_n(\underline{k},\underline{k}',t)dt \tag{21}$$

will be dominated by the values of the integrand at large values of t. As $n \to \infty$, $P_n(\underline{k},\underline{k}')$ will be independent of $\underline{k}$ and will depend on $\underline{k}'$ in the same way as $f(\underline{k}')$. Hence, apart from a normalisation constant,we may write:

$$f(\underline{k}') = \lim_{n\to\infty} \int g(\underline{k})\, P_n(\underline{k},\underline{k}')d\underline{k} \tag{22}$$

where $g(\underline{k})$ is arbitrary.

Equation (22) is easily converted into an iterative procedure for the calculation of $f(\underline{k}')$. It is clear from the definition of $P_n(\underline{k},\underline{k}'t)$ that

$$P_{n+1}(\underline{k},\underline{k}'t) = \int_0^t dt' \int d\underline{k}'' \int d\underline{k}'''\, P_n(\underline{k},\underline{k}''t') \times S(\underline{k}'',\underline{k}''')\, P_o(\underline{k}''',\underline{k}',t-t') \tag{23}$$

In this equation: $P_o(\underline{k},\underline{k}',t)$ is just $P_n(\underline{k},\underline{k}'t)$ when $n = 0$; it is the probability per unit volume of $\underline{k}'$ that an electron initially at $\underline{k}$ arrives at $\underline{k}'$ a time t later having suffered no collisions and is equal to $[1 - P(t)]\delta[\underline{k}' - \underline{k}(t)]$ where P(t) is defined by (18) and (19). The quantity $S(\underline{k},\underline{k}')$ in (23) is the differential scattering rate from $\underline{k}$ to $(\underline{k}',d\underline{k}')$, i.e. when $\underline{k}$ is in valley i and $\underline{k}'$ is in valley j, $S(\underline{k},\underline{k}')$ is minus the coefficient of $F_i(\underline{k})$ in the integrand of (4) summed over the different scattering mechanisms s. When (23) is integrated over all t we obtain with the aid of (21):

$$P_{n+1}(\underline{k},\underline{k}') = \int d\underline{k}'' \int d\underline{k}'''\, P_n(\underline{k},\underline{k}'')\, S(\underline{k}'',\underline{k}''') \times P_o(\underline{k}'',\underline{k}') \tag{24}$$

Finally, by substituting (24) into (22), we see that the sequence of distribution functions which tend to $f(\underline{k}')$ in the limit may be generated by the iteration formula:

$$f_{n+1}(\underline{k}') = \int d\underline{k} \int d\underline{k}'' \int d\underline{k}''' g(\underline{k})\, P_n(\underline{k},\underline{k}'')S(\underline{k}'',\underline{k}''')\, P_o(\underline{k}''',\underline{k}')$$

$$= \int d\underline{k}'' \int d\underline{k}''' f_n(\underline{k}'')\, S(\underline{k}'',\underline{k}''')\, P_o(\underline{k}''',\underline{k}') \qquad (25)$$

where the zeroth order distribution function

$$f_o(\underline{k}') = \int g(\underline{k})\, P_o(\underline{k},\underline{k}')d\underline{k} \qquad (26)$$

is arbitrary since g(k) is arbitrary.

The iteration formula (25) is not difficult to programme for machine computation. The evaluation of $P_o(\underline{k},\underline{k}')$ which involves the exponential on the right-hand side of (18) can be rendered trivial by the introduction of a self-scattering mechanism as was discussed in section 5. Results obtained by this technique are in excellent agreement with those obtained using the Monte Carlo formalism[18].

7. SEMICONDUCTORS OTHER THAN GALLIUM ARSENIDE

The Gunn effect has been observed in gallium arsenide, indium phosphide, cadmium telluride, gallium arsenide-phosphide alloys and indium arsenide under pressure. Nearly all the theoretical and experimental work has been concentrated on gallium arsenide because of the advanced state of the technology of this material. Butcher and Fawcett[12] made displaced Maxwellian calculations for indium phosphide and cadmium telluride which yielded values for the threshold field close to both the polar mode breakdown field of the central valley and to the observed threshold fields for the Gunn effect. Similar calculations were made for gallium arsenide-phosphide alloys and the variation of threshold field with alloy composition was compared with experimental data[19].

The most interesting recent calculation carried out for a material other than gallium arsenide is the work of Fawcett and Paige on germanium[20]. The calculation was made using the displaced Maxwellian approximation and taking into account the four normally occupied (111) valleys and six (100) valleys at higher

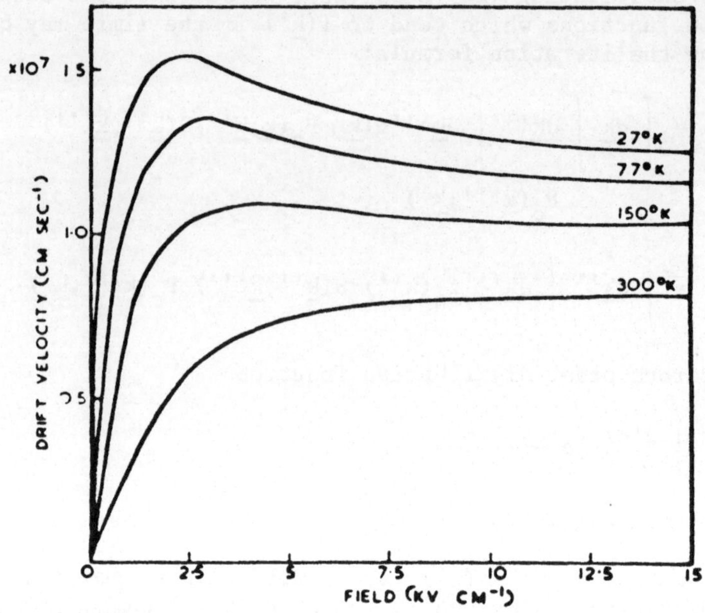

Fig. 7. Calculated velocity-field characteristics for germanium at various temperatures for a field in the (100) direction.[20]

energies with heavier effective masses. The anisotropy of the valleys was taken into account crudely by distinguishing between a density of states mass and a conductivity mass. The non-polar intravalley optical scattering which occurs in germanium was taken into account by replacing formula (6) for $C_{ii}^{(o)}(\underline{k},\underline{k}')$ by an obvious modification of (8). The calculated velocity-field curves for a field in the (100) direction are shown for several temperatures in Fig.7. We see that below 150°K a negative differential mobility appears as a result of electron transfer. The effect is not nearly as dramatic as that found in polar materials because there is no polar mode breakdown. Never-the-less, the results of these calculations suggest that germanium might exhibit the Gunn effect and they afford a possible explanation of the current oscillations[21] observed in this material. Fawcett and Paige have also made Monte Carlo calculations for germanium with qualitatively similar results[22].

8. CONCLUSION

The Monte Carlo and collision iteration techniques both have wide application to general problems in hot-carrier transport theory. They yield solutions of Boltzmann's equation with an error which becomes negligible when a large number of collisions are allowed to occur. The results obtained by applying these methods to gallium arsenide indicate that the displaced Maxwellian approximation yields a good approximation to the velocity-field characteristic but not to the distribution function. The approximation of nearly elastic collisions gives a less satisfactory velocity-field characteristic because it underestimates the power dissipated to the longitudinal optic phonons and consequently overestimates the average energy of the electron system and the fraction of electrons which transfer to the satellite valleys at any particular field.

The collision iteration technique may be adapted to study the time-development of distribution functions[18]. Both the collision iteration and Monte Carlo techniques are being employed to study the effects of anisotropy and non-parabolicity of the energy in the valleys, and of impurity scattering, on the velocity-field characteristic of a number of materials including gallium arsenide,[18,23]. The accuracy of these calculations is limited primarily by the lack of good experimental values for some of the energy band parameters and deformation potentials. Experimental data on the velocity-field characteristic itself is hard to obtain for materials exhibiting the Gunn effect and different workers have reported widely differing results for gallium arsenide[2,3]. Never-the-less, comparison of theoretical and experimental studies of electron transfer effects provides a useful tool for the investigation of energy-band parameters and scattering mechanisms in regions of the conduction band which are normally unoccupied.

REFERENCES

1. J. B. Gunn, 1964, I.B.M.Journ.Res.Dev., 8, 141-9.
2. P. N. Butcher, 1967, Rep.Prog.Phys. 30, Part I, 97-148.
3. I. B. Bott and W. Fawcett, 1968, Advances in Microwaves 3, 223-300.
4. C. Hilsum, 1968, Brit.J.Appl.Phys. (J.Phys.D) 1, 265-81.
5. E. G. S. Paige, 1964, Prog.in Semiconductors 8, 1-244 (London: Heywood, Eds. A. F. Gibson and R. E. Burgess).
6. H. G. Reik, 1964, Phonons and Phonon Interactions, pp.138-66 (New York: Benjamin, Ed. T. A. Bak).
7. H. Fröhlich and B. V. Paranjape, 1956, Proc.Phys.Soc. 69, 21-32.
8. R. Stratton, 1958, Proc.Roy.Soc. A 246, 406-22.
9. E. M. Conwell, 1966, Phys.Rev. 143, 657-8.
10. A. G. Foyt and A. L. McWhorter, 1966, Trans.Inst.Elect. Electron Engrs. ED 13, 79-87.
11. C. Hilsum, 1962, Proc.Inst.Radio Engrs. N.Y. 50, 185-9.
12. P. N. Butcher and W. Fawcett, 1965, Proc.Phys.Soc., 86, 1205-19.

13. P. N. Butcher and W. Fawcett, 1966, Phys.Letts., 1966, 21, 489-90.
14. E. M. Conwell and M. O. Vassell, 1966, Proc.Int.Conf. on the Physics of Semiconductors, Kyoto, Journ.Phys.Soc.Japan, 21 (Supplement) 527-31.
15. E. M. Conwell and M. O. Vassell, 1966, Phys.Letts., 21, 612-4.
16. A. D. Boardman W. Fawcett and H. D. Rees, 1968, Sol.State Comm. 6, 305-7.
17. H. D. Rees, 1968, Phys.Letts. 26A, 416-7.
18. H. D. Rees, private communication.
19. P. N. Butcher, W. Fawcett and C. Hilsum, 1966, Trans.Inst.Elec. Electronic Engrs. ED 13, 192-3.
20. W. Fawcett and E. G. S. Paige, 1967, Elec.Letts. 3, 505-7.
21. J. C. McGroddy and M. I. Nathan, 1967, I.B.M. J.Res.Dev.11, 337.
22. W. Fawcett and E. G. S. Paige, private communication.
23. W. Fawcett, private communication.

ACKNOWLEDGEMENT

The author would like to thank W. Fawcett and H. D. Rees for communicating the results of their recent work prior to publication.

THE ACOUSTOELECTRIC EFFECT

A. Many and I. Balberg

The Hebrew University, Jerusalem, Israel

1. INTRODUCTION

The acoustoelectric effect in piezoelectric semiconductors encompasses the various phenomena arising from the interaction between the rather strong electric field associated in these materials with acoustic waves and the electrons or holes present in the crystal. In 1960 Nine[1] observed a large conductivity-sensitive ultrasonic attenuation in photoconducting CdS crystals. Soon after, Hutson et al.[2] measured the ultrasonic attenuation in CdS under an externally applied electric field. They found that when the electron drift velocity exceeds the sound wave velocity, amplification rather than attenuation of the input acoustic signal takes place. Simultaneously with these experiments Hutson and White[3] developed a classical small-signal theory of the acoustoelectric effect, which accounted quite well for the dependence of the observed gain (or attenuation) constant on applied field, sample conductivity and acoustic-wave frequency.

Acoustic amplification occurs when the electrons (or holes) transfer energy to the waves in excess of their ohmic losses. Such transfer in energy should be reflected also in a change in the transport characteristics of the electrons. And indeed, in 1962, Smith[4] has observed non-ohmic behaviour related to acoustic amplification in CdS. For applied fields exceeding a critical value the current-voltage characteristics were found to depart from a linear relation, the current tending to a saturation level with increasing voltage.

These early investigations have stimulated a considerable research activity, and to date there is an extensive literature[5-9] covering the two major aspects of the acoustoelectric effect — the acoustic phenomena (ultrasonic amplification) and the electrical phenomena (non-ohmic behaviour). The acoustoelectric effect has been studied in other group II-VI compounds (CdSe, CdTe, ZnO) as well as in group III-V compounds (such as GaAs, GaSb, InSb) and in Te and Se. The methods used include transducer techniques,[10,11] current-voltage characteristics, electrical and optical probing,[8] microwave absorption[12] and emission[13] measurements and Brillouin scattering techniques.[9] A review of all this work is unfeasible within the limited frame of these talks. Rather, we shall attempt to present some of the main ideas and fundamental processes underlying the acoustoelectric effect. The discussion will be confined mostly to the non-ohmic behaviour. We begin by a discussion of piezoelectricity and show by simple physical arguments how the electrons in a piezoelectric crystal are coupled to the acoustic waves. Next we review briefly small-signal theories of the acoustoelectric effect. This is followed by a more detailed account of the spatial distribution of the amplified acoustic flux. The discussion here will be illustrated by work carried out in this laboratory on semiconducting CdS, a medium in which the acoustoelectric effect is most clearly displayed. Next we consider possible extensions of the theory to large-signal conditions. Finally, we present data obtained on non-linear effects encountered when high concentrations of amplified acoustic flux are present in the crystal.

2. PIEZOELECTRICITY AND THE COUPLING BETWEEN ELECTRONS AND ACOUSTIC WAVES

A crystal is termed piezoelectric[14] when a deformation causes a proportional induced electric dipole and, conversely, the application of an electric field causes a proportional deformation of the crystal. A necessary condition for piezoelectricity is the absence of a center of symmetry, a condition that is valid for 20 of the 32 point groups. For simplicity of presentation we shall write the tensorial piezoelectric equations[15] in the one dimensional case in the scalar form:

$$T = cS - eF \quad , \qquad (1)$$

$$D = eS + \varepsilon F \quad , \qquad (2)$$

where T is the stress, S the strain, F the electric field and D the electric displacement; c is the elastic stiffness for constant electric field, ε the electric permitivity for constant strain and e is the piezoelectric constant. It is apparent from these equations that the stress is determined by the applied field as well as by the mechanical strain and that the induced dipole moment is determined by the strain as well as by the electric field.

It can readily be shown that the ratio between the electrical energy that can be transferred to mechanical energy (or vice versa) and the geometric mean of the mechanical and electrical energies stored in the crystal is $K=e(\varepsilon c)^{-\frac{1}{2}}$. This electro-mechanical coupling constant is a measure for the "strength" of the piezoelectricity.[15]

In the dynamic case of an acoustic wave propagating in a piezoelectric crystal, the alternating strain (or stress) gives rise to an alternating electric field and hence to a potential, the variation of which in time and space follows that of the waves. The interaction between free electrons (if present in the crystal) and the piezoelectric potential gives rise to the acoustoelectric effect. This interaction can be described[16] in simple classical terms when the acoustic wave length λ is large compared to the electron mean free path ℓ for scattering. In this case the energy distribution of the electrons can adjust — by the normal scattering process — to the variations in the piezoelectric potential. Thus there would be a tendency for the electrons to bunch at the potential troughs, the more so the higher the wave amplitude (the deeper the potential troughs). As the acoustic wave propagates, the bunched electrons will lag behind the wave if their drift velocity v_d is less than the sound velocity v_s and they will tend to run ahead of the wave if $v_d>v_s$. In the first case the bunched electrons are forced into regions of the field which tend to increase their velocity and the sound wave is attenuated. In the absence of an external field, for example, such attenuation is accompanied by a dc current, the so-called acoustoelectric current.[17-19] In the other case, $v_d>v_s$, the bunched electrons are forced into regions of the field which tend to decrease their velocity and the sound waves are amplified. As amplification proceeds the potential troughs become deeper than kT/q, where kT is the thermal energy and q is the electronic charge, and most of the conduction electrons become bunched in the troughs. At this stage the electrons are forced to drift at a velocity which is approximately equal to the sound velocity and the current saturation mentioned above sets in.

The other limiting regime, $\lambda<<\ell$, can also be discussed classically, but the quantum-mechanical picture is more illustrative. Here one describes the coupling between the electrons and the waves in terms of electron-phonon interactions. One can then look upon the collective process of acoustic amplification as a phonon maser.[20] The phonon emission is analogous[21] also to the Cherenkov radiation of light which occurs when the electrons drift faster than the light velocity in the medium.

3. REVIEW OF SMALL-SIGNAL THEORIES

We begin by discussing the Hutson-White theory,[3] which is a classical theory valid in the regime $\lambda>>\ell$. Only the main steps will be reviewed, leading to the expression for the attenuation constant. One considers a plane acoustic wave propagating in a homogeneous n-type semiconductor. In order to obtain the necessary coupling between the electrons and the waves, the particle displacement should be such as to give rise to a piezoelectric field along the direction of wave propagation. In CdS, for example, this condition is satisfied[6] for shear waves if the direction of wave propagation is normal to the c-axis, and for longitudinal waves if the wave propagation is parallel to the c-axis. For simplicity the latter case will be considered, so that the particle displacement u and the piezoelectric field E_1 are all functions of x only, the direction of the applied dc field E_o. Recalling that $S=\partial u/\partial x$ and using eq.(1), one can write the wave equation in the form

$$\rho \frac{\partial^2 u}{\partial t^2} - c \frac{\partial^2 u}{\partial x^2} = -e \frac{\partial F}{\partial x} \quad , \qquad (3)$$

where ρ is the lattice density. In a non-piezoelectric crystal, e=0 and eq.(3) reduces to the ordinary wave equation. The current density J is given by

$$J = q\mu_n(n_0+n_1)(E_0+E_1)+qD_n(\partial n_1/\partial x) \quad , \qquad (4)$$

where μ_n is the electron mobility and D_n the diffusion constant; n_0 is the equilibrium electron density and n_1 the variation in density (positive or negative) due to the alternating piezoelectric field E_1. The other equations governing the electron transport are: the Poisson equation

$$\partial D/\partial x = -qn_1 \quad , \qquad (5)$$

and the continuity equation

$$\partial J/\partial x = q\partial n_1/\partial t \quad . \qquad (6)$$

If the particle displacement in the acoustic wave is written as

$$u = u_o \exp[i(kx-\omega t)] \quad , \qquad (7)$$

then the field $F=E_0+E_1$ acting on the electrons is given by

$$F = E_0+E_{10} \exp[i(kx-\omega t)] \quad , \qquad (8)$$

where u_o and E_{10} are the respective amplitudes, k the acoustic wave vector and ω the angular frequency of the acoustic wave. Both E_{10} and k will in general be complex, the first because of possible phase shifts between u and E_1 and the second because of the wave attenuation (or gain).

A linear, small-signal theory is obtained when the term n_1E_1 is neglected in eq.(4) and when, further, the attenuation constant (per unit time) is small compared to ω and $K^2 \ll 1$. Equations (4)-

-(6) can be combined to relate D and F, and by the use of eq.(2) one then obtains a linear relation between E_1 and S. The electric field can be eliminated from eq.(3) and the wave equation assumes the form

$$\rho \frac{\partial^2 u}{\partial t^2} = c' \frac{\partial^2 u}{\partial x^2} \quad , \qquad (9)$$

where c' is an effective elastic stiffness. Inserting eq.(7) into eq.(9) and writing k in the form $k=\omega/v_s - i\alpha_0/v_s$ (no dispersion) where v_s is the sound velocity, yields for the attenuation constant α_0 at frequency ω

$$\alpha_0(\omega) = \tfrac{1}{2}K^2\omega_c \frac{\gamma}{\gamma^2+(\omega_c/\omega+\omega/\omega_D)^2} \quad , \qquad (10)$$

where

$$\gamma = \mu_n E_0/v_s - 1 \quad , \quad \omega_c = \sigma/\varepsilon = q\mu_n n_0/\varepsilon \ , \ \omega_D = v_s^2/D_n \ . \qquad (11)$$

In this treatment we did not consider the effect of electron trapping in the semiconductor. If such trapping does occur, and if the thermal emission from the traps is sufficiently rapid to allow the trapped charge to follow the variation n_1 in free-electron density, eqs.(10) and (11) still remain valid provided that μ_n is replaced by an effective mobility $f\mu_n$, where f is the ratio of free to the sum of the free and trapped electron densities.[3]

It is apparent from eqs.(10) and (11) that for any acoustic wave frequency ω, amplification ($\alpha_0>0$) occurs for $\mu_n E_0>v_s$ and attenuation ($\alpha_0<0$) for $\mu_n E_o<v_s$. In either case the sound waves exert a net dc force on the electrons (see below). When the drift velocity far exceeds the sound velocity ($\mu_n E_o>>v_s$), the amplification decreases because then the electrons experience essentially only an ac field so that there is little energy transfer between the electrons and the waves.

The frequency of maximum gain (or attenuation) ω_m is given by $(\omega_c\omega_D)^{\frac{1}{2}}$, for which

$$\alpha_0(\omega_m) = \tfrac{1}{2}K^2\omega_c \frac{\gamma}{\gamma^2+4(\omega_c/\omega_D)} \quad , \quad \omega_m = (\omega_c\omega_D)^{\frac{1}{2}} \quad . \qquad (12)$$

That the frequency of maximum gain should lie in-between ω_c and ω_D is apparent from simple physical agruments. The relaxation frequency ω_c expresses the speed with which electrons in a medium of conductivity σ can be bunched at the troughs of the piezoelectric potential, while the diffusion frequency ω_D expresses the speed of de-bunching by diffusion processes. Thus the efficiency of bunching is small for $\omega>\omega_c$ and for $\omega<\omega_D$. In the first case the electron distribution cannot follow the piezoelectric-potential variation, whereas in the other case electron diffusion out of the potential troughs hampers the bunching process.

As to be expected, the higher the carrier density (higher ω_c) the greater is the gain. In the limiting case of $\omega_c \gg \omega_D$, corresponding usually to resistivities in the range of several ohm-cm, eq.(10) reduces to

$$\alpha_o(\omega) = G(\omega)(\mu_n E_o/v_s - 1); \quad \omega_c \gg \omega_D \quad , \tag{13}$$

where

$$G(\omega) \equiv \tfrac{1}{2}K^2\omega_c(\omega_c/\omega + \omega/\omega_D)^{-2} \quad , \tag{13a}$$

while eq.(12) reduces to

$$\alpha_o(\omega_m) = G_m(\mu_n E_o/v_s - 1) \quad ; \quad \omega_c \gg \omega_D \quad , \tag{14}$$

where

$$G_m = \tfrac{1}{8}K^2\omega_D \quad . \tag{14a}$$

The regime where $\lambda \ll \ell$ can be treated[22] classically using the Boltzmann transport equation together with a shifted Maxwellian distribution. Quantum mechanical treatments[6,23,24] make use of time-dependent perturbation theory and are based on the energy and wave-vector conservation laws in the phonon-electron interaction. The results obtained are similar to those derived by classical theory, indicating that the effect is essentially a classical one.[25]

So far we have discussed the effect of the wave/electron interaction on the wave propagation, considering its amplitude change. We now proceed to evaluate the effective force exerted by the waves on the electrons, the so-called acoustoelectric field.[18] The most convenient way of doing that is by considering the energy and momentum exchange in a phonon-electron interaction.[26] The rate of energy transferred from the waves to the electrons (or vice versa) can be expressed in the form

$$dW/dt = \hbar\omega(dN/dt) \quad , \tag{15}$$

where $\hbar$ is Plank's constant and N the number of phonons in the wave, all of which are assumed for simplicity to have the same angular frequency ω. The rate of momentum transfer is, correspondingly,

$$dP/dt = \hbar k(dN/dt) \quad . \tag{16}$$

The average force acting on each electron is $(1/n_o)dP/dt$. This force can be looked upon as due to a dc electric field E_{ae}, the acoustoelectric field. On the assumption that there is no dispersion

($\omega/k=v_s$), one obtains by eliminating dN/dt from eqs.(15) and (16)

$$E_{ae} = (1/qn_0 v_s)(dW/dt) \quad . \qquad (17)$$

As expected, the sign of E_{ae} depends on whether the waves are amplified (dW/dt>0) or attenuated. In the case of acoustic amplification the direction of the force acting on the electrons is opposite to that of the wave propagation, while in the case of attenuation the force is along the direction of propagation.

Since α_0 in eq.(10) refers to the amplitude attenuation (or amplification), the rate of energy change is $dW/dt = 2\alpha_0 W$. Hence

$$E_{ae} = (2\alpha_0/qn_0 v_s)W \quad . \qquad (18)$$

This, the so-called Weinreich relation, can also be obtained from the Hutson-White theory by considering the term n_1E_1 and using the expressions derived there for n_1 and E_1.

Equation (18) holds irrespective of whether an external field is or is not present.[27] In the former case, the measured current is a result of the ohmic current and the acoustoelectric current J_{ae} defined by $J_{ae}=\sigma E_{ae}$. In the absence of an external field, J_{ae} represents the current accompanying the attenuation of the sound wave and flowing in the direction of the wave propagation.[19,28]

4. ACOUSTIC FLUX DOMAINS

While it was observed quite early[10,19] that current saturation and other non-ohmic effects accompany acoustic amplification in piezoelectric semiconductors, some of the underlying processes were not entirely clear at first.[7] In particular, the early results reported on current oscillations,[7,29,30] the damping out of the oscillations with time and the manner in which steady-state conditions are established[31-33] were not always in accord and were sometimes even contradictory. Subsequent probe measurements of the internal field distribution[34-38] resolved most of these difficulties. Such measurements revealed that throughout its buildup, the amplified acoustic flux is concentrated: in propagating domains in the transient state and in a stationary domain bordering with the anode when the steady state is reached. Flux domains have been observed in most of the piezoelectric semiconductors studied. The results obtained for CdS, however, provide more quantitative information on the non-ohmic behaviour and the discussions that follow will be confined to this material.

The measurements to be described[38] were carried out on rectangular filaments oriented normal to the c-axis and ranging in resistivity between 2 and 10 ohm-cm. The field distribution along the filament was obtained by a pair of close potential probes mounted on a movable frame. The pulsed drift field was applied either under constant-voltage conditions (series resistance R_s small compared to the filament resistance R_o) or under nearly constant-current conditions ($R_s >> R_o$), the results being much the same in the two cases. Typical oscillograms obtained under intermediate conditions ($R_s = R_o$) are shown in Fig.1 for a drift field exceeding the threshold for acoustic amplification. The oscillations in the voltage V_f across the entire filament (Fig.1a) reflect the variations in filament resistance between the ohmic value (R_o) and some higher value ($R_o + \Delta R_{max}$) attained when the acoustic flux in the sample builds up to a sufficient level. R_o and $R_o + \Delta R_{max}$ correspond, respectively, to the crests and troughs of the current oscillations commonly reported in the literature (see below). Fig.1b displays the potential drop δV across the probes for three distances x of their center from the cathode. In each trace δV is initially small but after a certain time interval, a narrow high-resistance domain of acoustic flux arrives between the probes and δV exhibits a sharp rise to its peak value δV_p. As the domain moves out, δV drops and assumes again its low ohmic value, until a second domain arrives and the process is repeated, either indefinitely or with decaying amplitude (see below). The mode of propagation of the domains is apparent from the displacements along the time axis of the corresponding peaks in the three traces. Plots of the distance x between the pair of probes and the cathode against the arrival times of the first and second

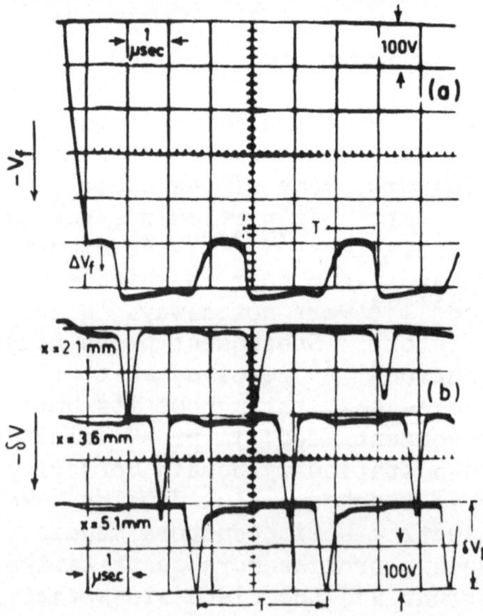

Fig.1. Oscillograms of potential drop (a) across entire filament and (b) across pair of probes (separation 0.25 mm) for three distances x of their center from cathode. Filament length 5.5 mm, resistivity 7.2 ohm-cm; $R_s = R_o = 560$ ohm.

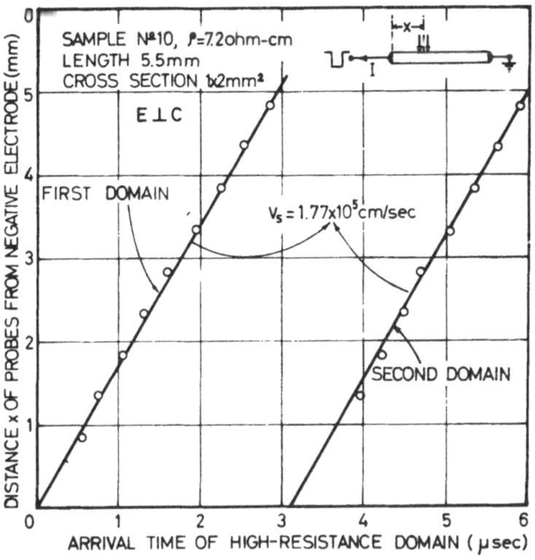

Fig.2. Plots of distance between pair of probes and cathode against arrival time of first and second high-resistance domains. Filament length 5.5 mm, resistivity 7.2 ohm-cm.

domains (measured from the onset time of the pulse) are shown in Fig.2. (The data were taken with a fast-risetime pulse — 0.1μsec — in order to enable an unambiguous determination of the onset time.) Such parallel straight lines are obtained for all subsequent domains. The common slope of these lines yields for the domain velocity a value of 1.77×10^5 cm/sec, which matches well the sound velocity for shear waves in CdS (1.75×10^5 cm/sec). The measured velocity is essentially independent of the pulse amplitude and risetime, the uniformity of the sample or its resistivity.

The first straight line in Fig.2 extrapolates to x=0 precisely at t=0, while the time-separation T between successive domains is equal to the sonic transit time (L/v_s) through the length L of the filament. It follows then that all the domains originate at the cathode, the first at the onset of the pulse (t=0) and each subsequent domain just when the preceeding one has moved out of the sample through the anode. These results are accounted for[34] by the fact that the application of the pulse as well as the exit of the high-resistance domains give rise to time-varying gradients in the electric field at the contact boundary which, as a result, acts as a site of ultrasonic generation.[39]

Since at any instant there is only one domain in the filament, it follows that if its width is smaller than the probe separation, the peak probe drop $\delta V_p(t)$ associated with it should represent the entire excess filament resistance ΔR at that instant of time. Hence

$$\delta V_p(t) = (1+R_0/R_s)\Delta V_f(t) \quad , \qquad (19)$$

where $\Delta V_f(t)$ is the voltage across the filament in excess of the ohmic drop. This relation is well satisfied in Fig.1 ($R_s=R_o$ and $\delta V_p \approx 2\Delta V_f$), as it does in fact for any other position of the probes and at any time before steady-state conditions are approached. Such one-to-one correspondence between $\Delta V_f(t)$ and $\delta V_p(t)$ was found for all samples studied. The domain width, however, varies somewhat from sample to sample (in the range 0.1 - 0.4 mm). In any one sample it is practically independent of the applied field, the sequential number of the domain or its position along the filament.

The "square-wave" shape of V_f in Fig.1 is typical to CdS samples of uniform resistivity and illustrates well the character of the flux buildup process. Each domain starts out at a very low flux level and initially ($x \lesssim 2$ mm, $t \lesssim 1$ μsec) it can neither be detected by the probes nor does it raise V_f above its ohmic value. The flux is strongly amplified by interaction with the drifting electrons and after a certain incubation time which decreases as expected with increasing pulse amplitude, ΔR rises rapidly to a fairly constant value ΔR_{max} which is maintained until the domain leaves the sample. Irrespective of the pulse amplitude, the value ΔR_{max} is always such that the domain absorbs the entire excess voltage, leaving the rest of the crystal under a field for which the electron drift velocity is very nearly equal to the sound velocity (see below). The field inside the domain can typically be 5×10^4 V/cm, whereas that outside the domain is only about 6×10^2 V/cm.

Relatively small inhomogeneities in resistivity (less than 10%) may give rise to large undulations in V_f within one sonic transit time. These often appear as "higher harmonies" of the fundamental frequency[7] but in fact they merely reflect the modulation of the amplified flux in a single domain due to the variation in field along the filament. The situation becomes more complicated when gross inhomogeneities are present since these can act as additional sources of acoustic flux.

In sufficiently uniform samples one usually observes that the pulsed nature of the acoustic flux gradually damps out and the steady-state condition (current saturation) is eventually reached. This is illustrated by the oscillograms in Fig.3 obtained for a CdS filament under a drift field exceeding the threshold for acoustic gain. The two traces represent the voltage drop across a pair of probes for two neighbouring positions along the filament. The decay of δV back to its initial ohmic level in Fig.3a is typical for probe-to-anode distances larger than about 0.5 mm and indicates clearly the absence of acoustic flux over most of the filament when the steady state is reached. The entire flux is then concentrated in a narrow stationary domain close to the anode, as can be seen from Fig.3b (probes-to-anode distance 0.30 mm), where δV now tends to a considerably higher, non-ohmic value. This is shown on a more quantitative basis in Fig.4, where the steady-state potential

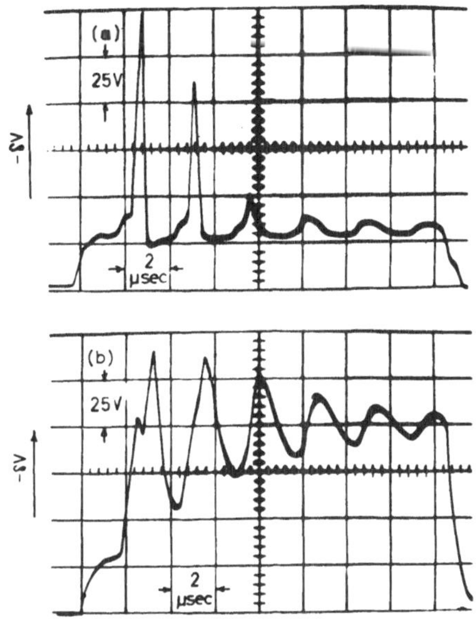

Fig.3. Potential drop across probes (separation 0.25 mm) for distance 0.55 mm (a) and 0.30 mm (b) of their center from anode. Filament length 4.4 mm, resistivity 4.5 ohm-cm.

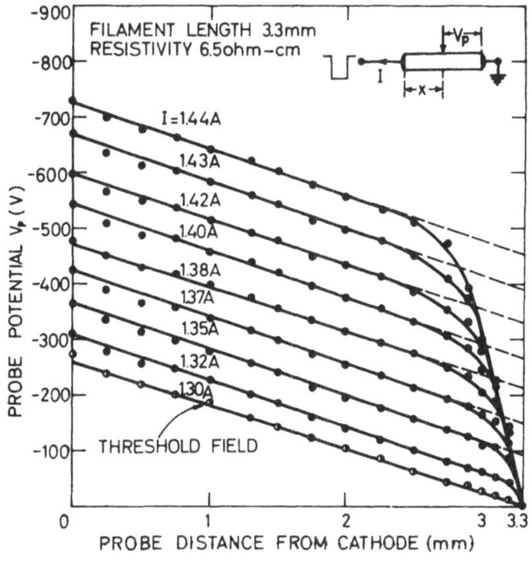

Fig.4. Steady-state potential distribution along filament for various drift currents.

distribution along the filament (measured with a single probe) is plotted for several values of the drift current I. The lower curve corresponds to conditions just below threshold, the others are for larger currents. The uniformity of the filament resistivity is apparent from the collinearity of the points under ohmic conditions. In the amplifying state (upper curves), the ohmic field distribution is maintained (the points lie on parallel straight lines) — except near the anode, where the entire excess voltage drop occurs. The width of the stationary domain is seen to increase with increasing current. This spreading out of the flux indicates that a non-linear loss mechanism becomes operative at high flux densities.

As the stationary domain builds up, the field in the rest of the filament gradually decreases until it falls below threshold and the generation and amplification of additional domains stops. At this stage, the steady-state flux must be derived from continuous amplification of background noise close to the anode (the only region where the field is above threshold). The constant flux level is maintained by a balance between such build up and non-linear loss.

The damping time characterizing the transition from conditions of propagating domains to those of a steady-state stationary domain (current saturation) is controlled by slight inhomogeneities in the resistivity of the anode region, and can assume practically any value between a few sonic transit times and infinity ("continuous oscillations"[40,41]). Thus there is not much point in attempting[33] to correlate the damping time with fundamental material properties. In fact, as is illustrated in Fig.5, even in one and the same filament dampled oscillations are often observed for one polarity of the pulse and continuous oscillations for the other, although the difference in resistivity between the two end regions does not exceed several percent. This behaviour can be reversed by changing the cross section of one end of the filament with respect to the other by a comparable percentage. In either configuration, continuous

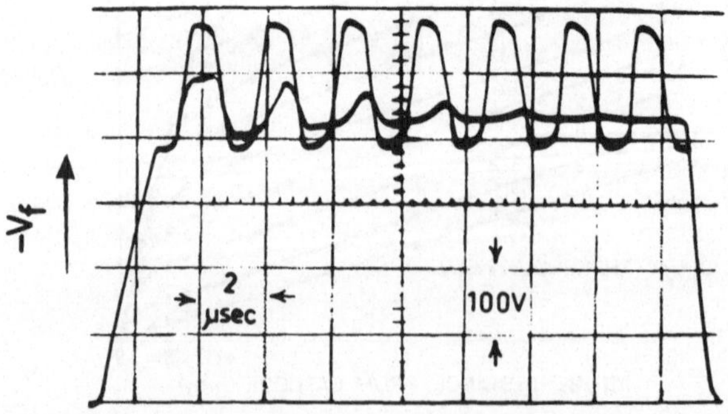

Fig.5. Superimposed oscillograms of potential drop across filament for opposite polarities of applied voltage pulse. Filament length 4.2 mm, resistivity 5 ohm-cm.

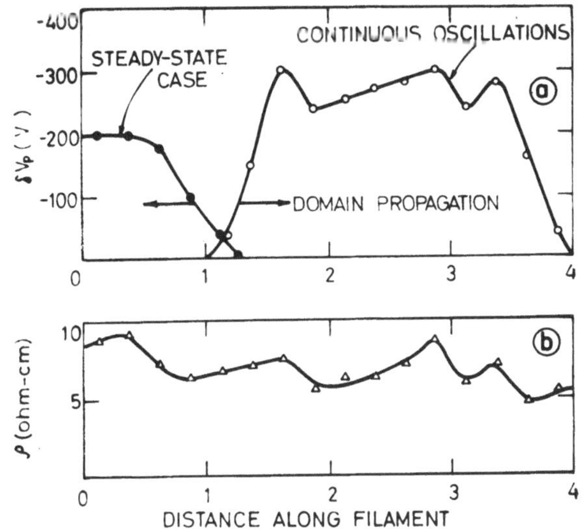

Fig.6. (a) Peak value δV_p of potential drop across pair of probes (separation 0.25 mm) against position along filament for opposite polarities of applied voltage pulse.
(b) Resistivity profile of filament.

oscillations obtain when the anode region has the lower resistance. Probe measurements reveal in this case that each flux domain is attenuated shortly before reaching the anode, as to be expected in view of the lower drift field in that region. This is shown in Fig.6, where the amplitude δV_p corresponding to the first flux domain is plotted against distance for the two polarities of the drift field. For the case of continuous oscillations the waves move from left to right. After a certain incubation time the domain becomes detectible and δV_p builds up rapidly to a fairly constant level. Close to the anode, however, the domain is attenuated and almost dies out completely by the time it reaches the anode boundary. Such attenuation does not occur in the reverse polarity (waves moving from right to left). The bottom curve shows the resistivity profile of the filament. Attenuation of the flux domain occurs when the anode is at the low-resistivity end. One is therefore led to conclude that the steady-state condition can be established only when the waves impinging on the anode boundary are sufficiently intense (see below).

The attenuation of the flux domains observed in the case of continuous oscillations is apparent also in the oscillograms of Fig.5 shown above. Here the period of the continuous oscillations is seen to be slightly shorter than that of the damped oscillations. Careful measurements of the type depicted in Fig.2 show in fact that this shorter period is slightly less than the sonic transit time (L/v_s). The reason for this is that each domain is formed — by the feedback mechanism discussed above – when its predecessor is attenuated, which occurs slightly before the anode is reached.

The observation that steady-state conditions can be attained only when the amplified flux impinges on the anode with sufficient intensity has been verified by measurements on many CdS samples. The most likely explanation for this phenomenon is that acoustic flux must be reflected from the anode boundary back into the filament. The reflected flux moves against the drift field and in the semiconducting samples under study it would be rapidly attenuated.[10] Small as the attenuation length might be,however,such a process can be accumulative. Consider,for example, the first flux domain hitting the anode. Suppose that no reflection takes place. In this case the electron drift velocity would increase from the saturated value to its higher ohmic value as soon as the domain moves out of the filament. Reflected flux, on the other hand, would prevent the drift velocity from attaining fully its flux-free ohmic value. As a result, the field near the anode boundary will remain higher – even if slightly so – than that in the rest of the filament, and the cathode generation efficiency of the second domain, as well as the domain amplification as it propagates from cathode to anode,will be reduced. The flux accumulated near the anode boundary gradually builds up until it absorbs the entire excess voltage and steady state conditions are established.

We have discussed the formation of the steady state in terms of propagating domains merely to show the contrast with the case of continuous oscillations where the absence of reflected flux prevents the attainment of the steady state. It has been shown,[38] however, that amplification of thermal noise plays an important, if not a dominant role even in the initial stages of the steady-state formation. The same kind of argument involving reflected flux can obviously be applied to this case as well. The two processes, cathode--domain and noise amplification, are initially competative and hamper each other, but noise amplification predominates as steady state conditions are approached.

The cathode domains can be almost entirely suppressed if the pulse risetime is increased sufficiently. This is seen by the superimposed oscillograms of Fig.7 which display the current response to fast-risetime (~0.04 μsec) and slow-risetime (~10 μsec) pulses of equal amplitude. In the former case shock excitation at the cathode boundary is strong so that amplification of propagating domains predominates and steady state conditions cannot be attained (continuous oscillations). For the slow risetime pulse, on the other hand, noise amplification predominates, the current rising essentially monotonically to its steady-state level as the stationary domain builds up.

Another problem that has received a fair amount of attention in the literature is whether the propagating domains originate from shock excitation[4,34,38] at the cathode boundary, as presumed above, or from thermal noise.[42,43] The following experiments provide

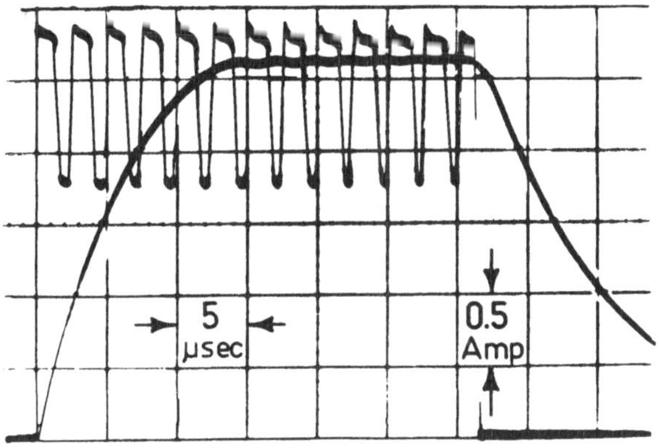

Fig.7. Oscillograms of the current through a filament for fast-risetime (0.04 μsec) and slow-risetime (10 μsec) pulses. Filament length 5.3 mm, resistivity 2.7 ohm-cm.

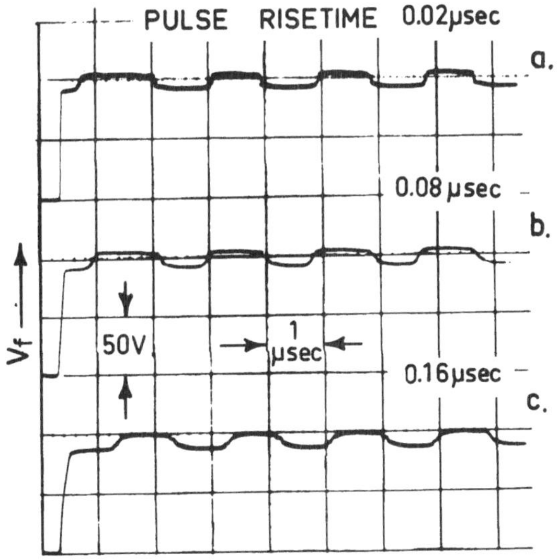

Fig. 8. Oscillograms of voltage drop across filament for three pulses of same amplitude but different risetime. Filament length 3.4 mm, resistivity 2 ohm-cm, $T=77^{o}K$.

strong evidence that in CdS the former mechanism is operative under normal conditions. The three traces of Fig. 8 represent (similarly to Fig. 1a) the filament voltage V_f for three applied pulses of identical amplitude but different risetime. The incubation time for the rapid build up of flux in the first domain (first rapid rise in V_f) is seen to be longer the longer the risetime. Since the incub-

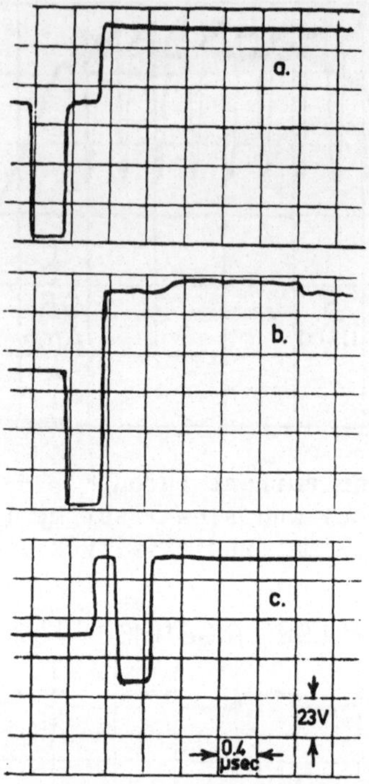

Fig.9. Oscillograms of potential drop across filament obtained by application of two successive voltage pulses of opposite polarity. Flux build up is detectible only when fall time of shorter pulse coincides with risetime of longer one (middle trace). Filament length 3.4 mm, resistivity 2 ohm-cm, T=77°K.

ation time is expected to vary inversely with the initial flux intensity at the cathode boundary (see also below), it follows that the initial flux intensity decreases with increasing rise time. This indicates that the initial flux is indeed generated by shock excitation arising from the time-varying field gradient set up at the onset of the voltage pulse. For the faster rise time (upper trace), the incubation time is seen to be shorter for the first domain than for the subsequent ones, indicating that in this case the shock produced by the pulse rise is larger than that produced by the rise in V_f attending the exit of an amplified domain through the anode. As was pointed out above, amplification of noise can become important for very slow risetimes. This can also be the case under high applied drift fields.[38]

Figure 9 shows the combined effect of two pulses of opposite polarity on the formation of amplified flux. The three traces are representative of a series of traces obtained by displacing (along the time axis) one pulse with respect to the other without changing either amplitude. By a suitable pre-adjustment of the amplitudes of the pulses one observes that only when the fall time of the shorter pulse coincides with the risetime of the longer one is sufficient initial flux generated at the cathode to enable detectable amplification by the longer pulse (see middle trace). The amplification disappears abruptly when the pulses are shifted even slightly out of coincidences. Since the shorter pulse is of opposite polarity, its only effect on the amplification (by the longer pulse) is by enhancement at coincidence of the time-varying field gradient and hence of the shock excitation.

5. PHENOMENOLOGICAL THEORIES FOR LARGE-SIGNAL CONDITIONS

As we have seen above the amplified acoustic flux in the transient case is concentrated in relatively narrow propagating domains in which the field may attain considerably higher values than the average drift field. So far there is no satisfactory theory applicable to such large flux densities and one must resort to the Hutson--White theory, which has proved so successful under small-signal conditions. If this theory is assumed to be valid for large signals as well, the question arises whether the gain should be assumed to depend on[44] $\mu_n E/v_s$ (Model A) or on[24,45,46] $\mu_n \bar{E}/v_s$ (Model B), where E and $\bar{E}$ are the field inside and outside the amplified domain, respectively. We shall now analyze these two possible models for semiconducting materials and compare the results with experiment.[47]

Let us consider the build-up process of a narrow flux domain (width ΔL) originating at the cathode at t=0 (with initial energy density W_o) and propagating towards the anode of the filament (length L). W_o may be the result of shock excitation at the onset of the drift pulse or it may consist of thermal noise. Consistently with experiment we assume that all amplified flux is contained within the domain and that $\Delta L \ll L$ and remains constant as the domain propagates. To simplify matters we further assume that the amplified flux is limited to a narrow frequency range around the frequency ω_m of maximum gain. For conducting samples ($\omega_c \gg \omega_D$), the amplification factor (per unit time) can then be written as (see eq.(14))

$$\alpha = G_m(f\mu_n F/v_s - 1) \qquad , \qquad (20)$$

where G_m and f are as defined in Section 3. In Model A, F represents the field E inside the domain, while in Model B, F represents the field $\bar{E}$ outside the domain. Since in the latter region ohmic conditions prevail, $\bar{E} = v_d/\mu_n$.

The rate of growth of the acoustic energy density, W, the Weinreich relation for the acoustoelectric field E_{ae} (see eq.(18)), the law of current continuity and the condition of constant voltage across the filament can be expressed as

$$dW/dt = 2(\alpha-\alpha_\ell)W \quad , \quad E_{ae} = 2\alpha W f\mu_n/\sigma v_s \quad ,$$
$$\sigma\bar{E} = \sigma(E-E_{ae}) \quad , \quad E_oL = E\Delta L+\bar{E}(L-\Delta L) \quad , \qquad (21)$$

where α_ℓ is the lattice (non-electronic) attenuation factor at frequency ω_m, σ the conductivity, and E_o the initial (uniform) field (prior to the build up of significant flux). We assume that α_ℓ is independent of flux density (see below) and that space-charge effects are negligible. Use of eqs.(20) and (21) leads to the following differential equations:

$$(1/2\beta_\ell)(d\beta/dt) = \pm|\beta-1|\,(\beta-\beta_\ell)(\beta/\beta_\ell) \quad , \qquad (22)$$

where $\beta \equiv \alpha/\alpha_o$, $\beta_\ell \equiv \alpha_\ell/\alpha_o$,and $\alpha_o = \alpha_o(\omega_m)$ (see eq.(14)). The upper sign applies to Model A, the lower to Model B. The solutions of these two equations are:

$$(\alpha_o-\alpha_\ell)t = (1/2\beta_\ell)\,\log\left[C_\pm\beta^{1-\beta_\ell}\,(\pm\beta\mp1)^{\beta_\ell}/(\beta-\beta_\ell)\right] \quad . \qquad (23)$$

The constants $C_\pm$ are determined by the values of β at t=0 which, for sufficiently small W_o, are given by (see eq.(21)) $\beta(0)\approx1+\eta_A$ and $\beta(0)\approx1-\eta_B$ for Model A and B respectively, where

$\eta_A = 2G_mW_o(f\mu_n)^2/\sigma v_s^2$, $\eta_B=\eta_A\delta$ and $\delta=\Delta L/(L-\Delta L)<<1$. ($\eta_A$ and η_B are assumed small compared to unity — see below.)

Inspection of eq.(23) shows that for Model A, β (which is a measure of the field E inside the domain) increases from its initial value of slightly above unity, first exponentially and then much faster than exponentially, tending to infinity at a finite value of t. The maximum, steady-state value of β is determined by the condition of constant voltage (see eq.(21)). The steep rise in β as this maximum value is approached is due to the regenerative process inherent in the gain mechanism of Model A. In Model B, β (which is a measure of the field $\bar{E}$ outside the domain) starts at a value slightly below unity and decreases towards β_ℓ as the steady state is approached.

The build up of a flux domain is accompanied by a decay of the current density $J = \sigma\bar{E}$ from its initial ohmic value of $J_o = \sigma E_o$ towards its minimum, saturated value of $J_{min} = \sigma\bar{E}_{min}$. In Model A, $\bar{E}_{min}\approx v_s/f\mu_n$ (see below) while in Model B, $\bar{E}_{min} = \bar{E}_\ell$, where $\bar{E}_\ell$ is defined by $\alpha_\ell \equiv G_m(f\mu_n\bar{E}_\ell/v_s-1)$. The "reduced" current $j \equiv (J-J_{min})/(J_o-J_{min})$ is given by $1-(\beta-1)\delta$ and $(\beta-\beta_\ell)/(1-\beta_\ell)$ for Model A and B, respectively. These two functions (as calculated from eq.(23) are plotted in Fig.10(a) against $t-T_{1/2}$, where $T_{1/2}$, the incubation time, is defined as the value of t at which $j = 1/2$. As expected, the current decay is much faster for Model A than for B.

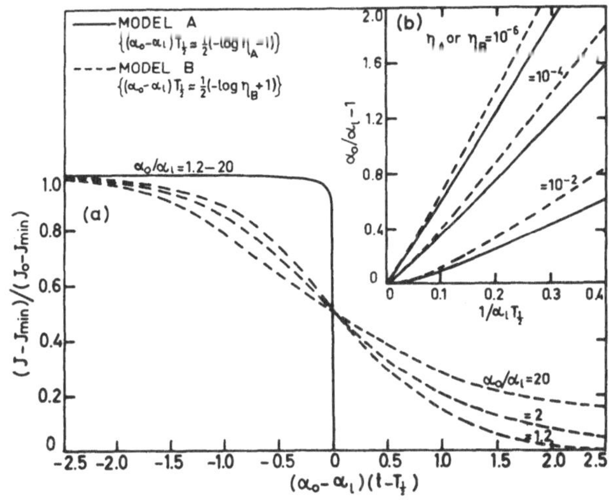

Fig.10. Current decay characteristics (a), and field dependence of incubation time $T_{\frac{1}{2}}$ (b), according to Model A (solid curves) and Model B (dashed). The decay curve for Model A corresponds to $\delta=0.05$ (those for Model B are independent of δ).

Except at very low drift fields, the product $(\alpha_o-\alpha_\ell)T_{\frac{1}{2}}$ is essentially independent of α_o. This is shown in Fig.10 (b) where $\alpha_o/\alpha_\ell-1$ is plotted against $1/\alpha_\ell T_{\frac{1}{2}}$. Departures from linearity are seen to occur only close to the origin $(\alpha_o=\alpha_\ell)$.

Some of the assumptions underlying the theory, particularly that concerning the constancy of α_ℓ , break down when J approaches J_{min} since in that region the flux density may attain very high levels. It can be shown, however, that this does not alter appreciably the conclusions of eq.(23), the main effect being to reduce somewhat the current decay rate for both models. In the case of Model A, $\gamma(=f\mu_n E/v_s-1)$ becomes very large as J approaches J_{min}, so that the neglect of γ^2 in the denominator of eq.(12) may not be justified in this range even for the case $\omega_c/\omega_D>>1$ being considered. It can be shown, however, that incorporation of the γ^2 term modifies the results of the analysis only slightly by rounding off the tail end of the current decay curve.

Comparison of the predictions of the analysis just presented with experiment will be made for semiconducting CdS. Typical oscillograms of current oscillations obtained for a filament of uniform resistivity are shown in Fig.11(a) for two applied voltage pulses. These oscillograms are similar to those shown in Fig.1(a), except that here the current rather than the voltage is displayed. As discussed there the period of the oscillations corresponds to a single flux domain originating at the cathode and propagating to the anode. The domain starts out at a very low flux level and initially

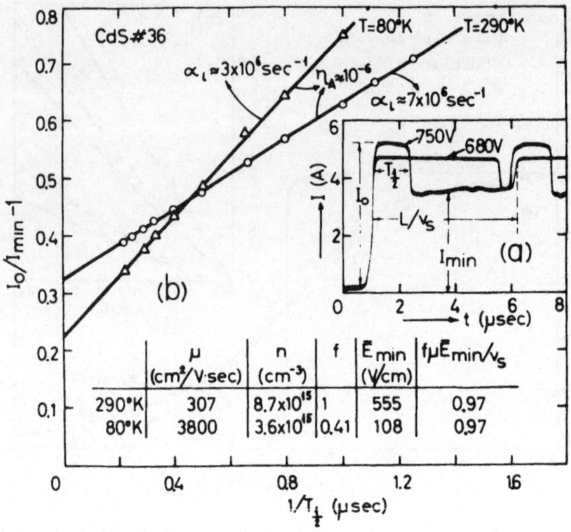

Fig.11. Oscillogram of current oscillations (at $T=290^{\circ}K$) for two applied voltages (a), and initial-current dependence of incubation time at two temperatures (b). Filament length 9 mm, room-temperature conductivity 0.43 $ohm^{-1}\ cm^{-1}$.

the current I assumes its ohmic value I_o. The flux is amplified by interaction with the drifting electrons and after a certain incubation time $T_{\frac{1}{2}}$, it builds up rapidly and I decays to its minimum, saturated value I_{min}. This latter value (which is independent of applied voltage — compare two traces) is maintained until the domain leaves the filament. Probe measurements show that the width ΔL of the domain is about 0.2 mm so that $\delta \approx 0.02$.

In Fig.11(b), $I_o/I_{min}-1$ is plotted against $1/T_{\frac{1}{2}}$ for two temperatures. According to the analysis above this function should have the form

$$I_o/I_{min}-1 = (\bar{E}_\ell/\bar{E}_{min}-1)+\left[v_s P/f\mu_n \bar{E}_{min} G_m\right](1/T_{\frac{1}{2}}) \quad , \qquad (24)$$

where $P \approx \frac{1}{2}(-\log\eta_A-1)$ and $P \approx \frac{1}{2}(-\log\eta_B+1)$ for Model A and B, respectively. The linear relationship exhibited by the experimental points at each temperature is thus compatible with both Models. However, the fact that the straight lines do not extrapolate to the origin immediately rules out Model B since in that Model $\bar{E}_{min} = \bar{E}_\ell$ and the free term of eq.(24) should vanish. In Model A, on the other hand, the precise value of $\bar{E}_{min}$ cannot be derived from the analysis. One would expect, however, that $\bar{E}_{min} = v_s/f\mu_n$ since at high flux densities all electrons should be bunched and thus moving with the sound velocity. This relationship is indeed borne out by conductivity and Hall effect measurements (see data in Fig.11). Hence, eq.(24) reduces to

$$I_o/I_{min}-1 = \alpha_\ell/G_m+\left[\tfrac{1}{2}(-\log\eta_A-1)/G_m\right](1/T_{\frac{1}{2}}) \quad . \qquad (25)$$

A fit of eq.(25) to the data in Fig.11(b) using the appropriate values of G_m for the sample under study, yields the values indicated in the figure for α_ℓ and η_A. The difference in the slopes of the two straight lines reflects mainly the difference in the values of G_m at the two temperatures, possible changes in W_o (and thus in η_A) entering only logarithmically into the expression for the slope.

Further support for the conclusion that the slope of the $I_o/I_{min}-1$ vs. $1/T_{\frac{1}{2}}$ line is proportional to the logarithm of the initial flux is provided by the following experiment. Two superimposed pulses are applied to the filament, resulting in the current trace shown by the oscillogram in Fig.12. The amplitude of the first voltage pulse is chosen so as to be lower than the threshold necessary for observing non-ohmic behaviour but sufficiently large to ensure that acoustic amplification does take place ($\alpha_o>\alpha_\ell$). The second voltage pulse is superimposed on top of the first after a time interval t_1. As a result, the current rises initially from I to I_o and after the incubation time $T_{\frac{1}{2}}$ decays, as usual, to its minimum value I_{min}. That the amplfied flux originates from the shock excitation produced at the onset of the first pulse is easily verified by noting that the time elapsed between the onset of the first pulse and the

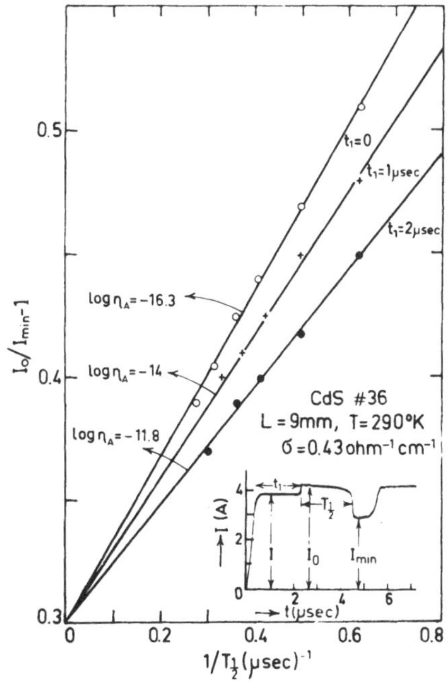

Fig.12. Initial-current dependence of incubation time for different values of t_1. Also shown is current-response trace in which variables are defined.

exit of the flux domain (~5.1 μsec) is just the sonic transit time through the filament (L/v_s). The three straight lines in the Figure represent measurements of $T_{\frac{1}{2}}$ as a function of I_0 for different values of the time interval t_1 but for the same amplitude of the first pulse (t_1=0 means, in effect, that only one pulse is used). The values of $\log \eta_A$ derived from the slopes are indicated near the respective straight lines and are seen to increase linearly with t_1. This is just what one would have expected, as can be seen from the following considerations. In the time interval 0 to t_1 no appreciable flux has yet been built up and the amplification factor is a constant α_0 given by the small-signal expression (eq.(14)). Thus, at $t = t_1$ the initial (dimensionless) flux density $\eta_A(0)$ produced by shock excitation at the onset of the first pulse is amplified to the level $\eta_A(t_1) = \eta_A(0)\exp[2(\alpha_0-\alpha_\ell)t_1]$. This new level constitutes the initial flux that is amplified by the second pulse. Thus the slope of the $(I_0/I_{min}-1)$ vs. $1/T_{\frac{1}{2}}$ line should be proportional to $\log \eta_A(t_1)$, and hence to t_1, as is indeed observed. It is further observed that all three straight lines in Fig.12 intersect the ordinate at the same point indicating that, at least under small-signal conditions, the non-electronic attenuation factor α_ℓ is a constant independent of flux density.

It should be noted that because of the rapid decay in current around $t = T_{\frac{1}{2}}$ (about 0.2 μsec), $T_{\frac{1}{2}}$ represents also, to a good approximation, the beginning of the flux growth process, before appreciable flux has been built up and non-linear effects have become important. Hence, the values of α_ℓ determined from the data of Figs. 11 and 12 correspond essentially to small-signal conditions.

Further evidence for the validity of Model A as opposed to Model B is provided by the current decay characteristics associated with the flux build up process. In Fig. 13, the normalized current $(I-I_{min})/(I_0-I_{min})$ obtained from the 680-V trace of Fig.11(a) is

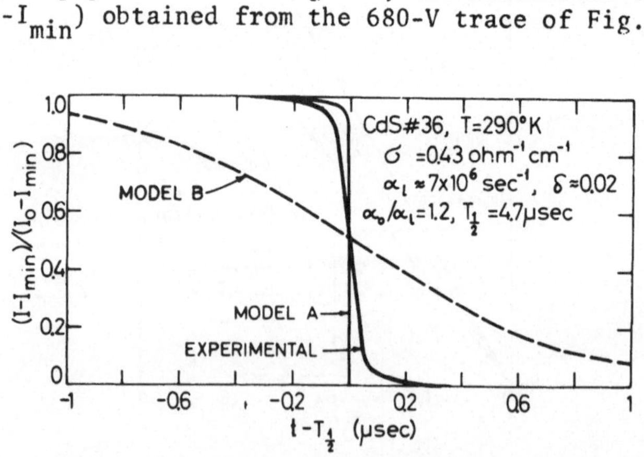

Fig.13. Current decay curve obtained by expansion of the 680-V trace of Fig.11(a), together with the corresponding theoretical curves predicted by the two Models.

plotted against $t-T_1$. Also shown are the corresponding theoretical plots based on the value of α_ℓ determined from the results of Fig. 11(b). It is seen that Model A fits the data much better than Model B. As was pointed out above, non-linear effects would be expected to slow down the current decay for both models. While such effects would enhance still further the discrepancy between experiment and theory in the case of Model B, they would act in the direction of improving the agreement in the case of Model A.

The data obtained for CdS lend strong support for the validity of Model A. It should be pointed out, however, that the situation in other piezoelectric semiconductors is not as clear cut. Thus, in GaAs[48] and ZnO[49] the results appear to fit Model B better than Model A (see below).

6. FREQUENCY SHIFTS DURING GROWTH OF ACOUSTIC FLUX IN CdS

In the preceeding Section we have discussed the build up of acoustic flux up to the time when the current decays to its minimum saturated value I_{min}. At this point all carriers become bunched at the piezoelectric potential troughs and drift with the sound velocity v_s. One may expect that the validity of Model A should break down in this regime. And indeed, as we shall see in a moment, one can derive[25] by quite general considerations (independently of the specific model assumed) an expression for the rate of energy gain which differs from that given by the predictions of Model A in that regime. The bunched electrons within a propagating domain absorb energy from the external source at a rate qn_oEv_s where E denotes, as before, the field in the domain. Short of the ohmic losses by the normal scattering process, the rate of which being $qn_o(v_s/\mu_n)v_s$, this energy is transferred to the acoustic waves. Hence, the rate of gain of acoustic energy W is given by

$$dW/dt = qn_ov_s(E-v_s/\mu_n) \quad . \qquad (26)$$

This means that the gain at high flux densities is independent of the energy of the amplified flux. On similar grounds one may also expect that the gain be relatively insensitive to the frequency of the flux.

These considerations indicate, then, that the acoustic flux can continue to grow after current saturation has been reached. Such growth has indeed been observed[50] by Brillouin scattering techniques. It was further observed that the frequency of maximum intensity (FMI) of the amplified flux decreases as the flux propagates through the filament. A detailed study of the shifts in frequency should be of considerable interest because of the light it might shed on the various non-linear processes that give rise to such shifts. In the

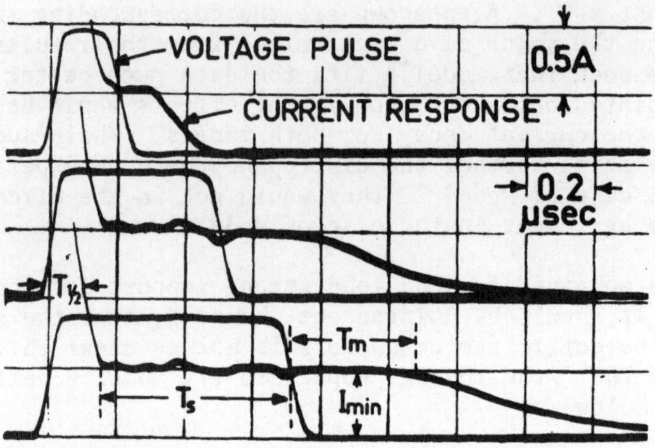

Fig.14. Current response (at 77°K) to three square voltage pulses of same amplitude (720 V) and different duration. Filament length 9 mm, resistivity (at 77°K) 1.2 ohm-cm.

present Section we describe[51] a fairly simple procedure for this purpose and discuss some of the more recent data obtained in this laboratory on the non-linear processes involved.

The measurements are similar to those leading to the oscillograms of current oscillation shown in Fig.11 above, except that here the voltage pulse is terminated well before the flux domain reaches the anode. Typical oscillograms of the current response to three square voltage pulses of the same amplitude and different duration are displayed in Fig.14. In all three traces the current decays from its ohmic value I_o (not shown) to I_{min} after the same incubation time $T_{½}$. After the termination of the pulse, the current maintains its minimum value for a certain time interval T_m following which it decays to zero. Both T_m and the decay time characterizing the "after-current"[28] are seen to be longer the larger the time interval T_s between the onset of current saturation ($t \approx T_{½}$) and the pulse termination. A similar behaviour[52] is observed for a constant T_s and increasing voltage: the higher the applied voltage the longer are T_m and the decay time.

Figure 15 is a semilog plot of the after-current decay for different values of T_s. It is apparent that the tails of the decay curves, which correspond to small-signal conditions, follow an exponential relation. The decay constant 2α derived from the tails are listed in the Figure.

As was shown above (eq.(26)) the acoustic gain after current saturation is expected to be proportional to the field inside the domain. Since throughout the saturation range this field remains essentially the same (see above) the gain should be constant with time. If the gain exceeds the loss, increasing T_s (and/or the

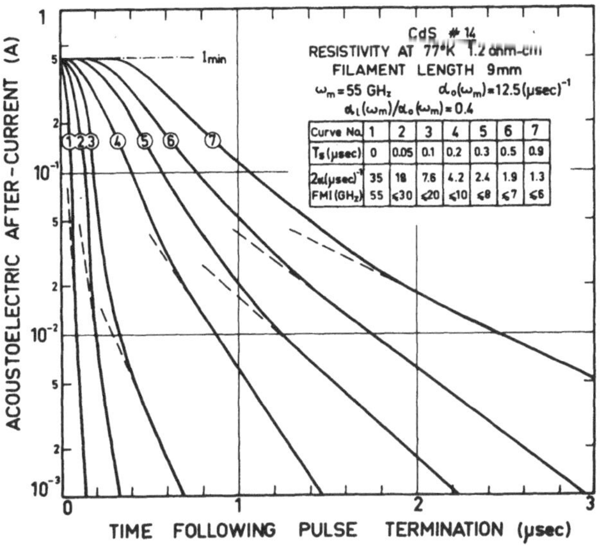

Fig.15. Semilog plot of acoustoelectric after-current vs. time as obtained from oscillograms of type shown in Fig.14 (with Y-axis sensitivity increased 10 - 100 times). The seven curves correspond to different time intervals T_s between onset of current saturation and pulse termination. The Table lists the respective values of T_s, the small-signal decay constant 2α derived from the tails of the decay curves, and the upper-limit values for the angular frequency of maximum intensity (FMI).

voltage) should result in an enhancement of the acoustic energy at the termination of the pulse. Our results that T_m increases with increasing T_s indicate that net amplification does indeed occur. That T_m is a measure of the acoustic energy contained in the domain just before the pulse is terminated follows from simple considerations.[4,25] At high flux densities, the potential troughs associated with the acoustic waves are deep compared to the thermal energy kT so that practically all the electrons in the domain are bunched in the troughs. After the pulse terminates, the flux is attenuated but the current maintains its saturated value I_{min} until the trough depth becomes comparable to kT. At this stage de-bunching begins to take place and the current decays.

Under small-signal conditions, the acoustoelectric after-current is proportional to the acoustic flux density (see eq.(18)). In this regime, acoustic flux of frequency ω decays with a time constant of $2[G(\omega)+\alpha_\ell(\omega)]$ where $G(\omega)$ is the field-free electronic attenuation factor given by the Hutson-White linear theory (eq.(13)) and $\alpha_\ell(\omega)$ is the non-electronic attenuation factor at that frequency. For the specimen under discussion the calculated values of the frequency

ω_m of maximum gain and of G_m (see eq.(14(a)) are 55 GHz and $12.5(\mu sec)^{-1}$, respectively. A plot of $(I_o/I_{min}-1)$ versus $1/T_{\frac{1}{2}}$ yields the value of 0.4 for the ratio $\alpha_\ell(\omega)/G(\omega)$. If this ratio is assumed to correspond to the frequency of maximum gain, then $2[G_m+\alpha_\ell(\omega_m)] = 35(\mu sec)^{-1}$, which is in good agreement with the small-signal decay constant 2α measured for $T_s \approx 0$ (curve 1 in Fig.15). Thus immediately after the rapid flux build-up, the frequency of maximum intensity is indeed very close to ω_m. As T_s increases, 2α becomes successively smaller, indicating that the FMI continually decreases as the flux propagates through the sample. An estimate for the FMI for each T_s can be obtained by calculating the frequency ω at which $2[G(\omega)+\alpha_\ell(\omega)]$ becomes equal to the measured small-signal decay constant 2α for that value of T_s. Since $\alpha_\ell(\omega)$ for $\omega \neq \omega_m$ is not known, only an upper limit for the FMI can be derived, that obtained by setting $\alpha_\ell(\omega) = 0$. This upper limit is not far-off since $\alpha_\ell(\omega)/G(\omega)$ is expected[53] to decrease with decreasing frequency, as is indeed found to be the case experimentally (see below). The upper-limit values derived from the data are listed in Fig.15 and show that the FMI decreases from $\omega_m = 55$ GHz for $T_s \approx 0$ to less than 6 GHz for $T_s = 0.9$ µsec.

The observed shift in the FMI must be due to a non-linear loss mechanism. In the presence of a large flux-density, a non-linear loss involving a three-phonon process may become operative in transforming the acoustic energy to higher, sum frequencies where the non-electronic loss exceeds the electronic gain.[53] Such losses are expected[54,55] to increase with flux density and frequency. In our case, the acoustic density just after its rapid build-up ($T_s \approx 0$) is still small so that electronic gain which favours ω_m appears to be dominant. After current saturation the losses increase faster with flux density and frequency than does the amplification so that the net gain favours lower frequencies and the FMI is shifted downwards.

An estimate for the frequency dependence of the small-signal non-electronic attenuation factor $\alpha_\ell(\omega)$ can be obtained by a double-pulse experiment similar to that described in connection with Fig. 12. Here, however, the two pulses are not superimposed on each other but separated by a fixed time delay t_d. The current response to such a double pulse is shown in Fig.16 for three different amplitudes of the second voltage pulse. Examination of the time relationships shows that in each case only one flux domain is involved throughout. The domain originates as usual at the cathode and is amplified by the first voltage pulse. At the termination of the pulse the domain decays and after a time interval t_d it is picked up by the second pulse and is re-amplified. From the tail of the current decay following the termination of the first voltage pulse one can determine the small-signal attenuation constant $2[G(\omega)+\alpha_\ell(\omega)]$ as described above. At the same time one measures the incubation time $T_{\frac{1}{2}}$ as a function of the initial current I_{o2} induced by the second voltage pulse. Plots of $I_{o2}/I_{min}-1$ vs. $1/T_{\frac{1}{2}}$ obtained in this manner for

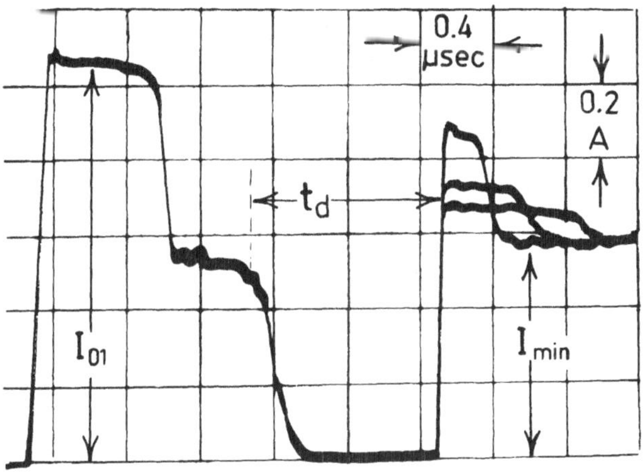

Fig.16. Oscillograms of current response to double voltage pulse for three amplitudes of second pulse. Filament length 8 mm, conductivity 1.25 ohm^{-1}-cm^{-1}, T = 77°K. (Sample No. 82).

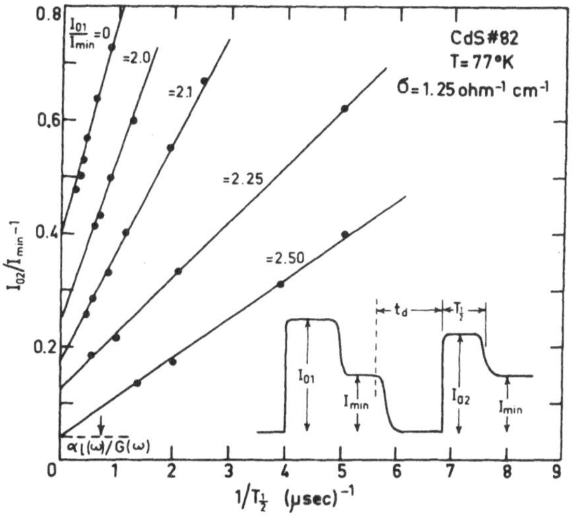

Fig.17. Initial-current dependence of incubation time for second pulse at different amplitudes of first pulse. Also shown is schematic diagram of current response to double pulse (Fig. 16.). (I_{01}/I_{min} = 0 means that only one pulse is applied.)

several values of the initial current I_{01} induced by the first pulse are shown in Fig. 17. As was shown above the higher I_{01} is, the higher is the flux density and the lower its FMI at the termination of the pulse. Thus, the signals picked up after the fixed time interval t_d by the second pulse are of different intensity and FMI for different values of I_{01}. As a result, both the slopes of the straight lines in Fig. 17 and their intercepts at the ordinate differ from one another. For each line the intercept yields $\alpha_\ell(\omega)/G(\omega)$. Combining this result with the value of $G(\omega)+\alpha_\ell(\omega)$ obtained from the current decay following the termination of the first pulse, one can determine both $G(\omega)$ and $\alpha_\ell(\omega)$ for each value of I_{01}. As pointed out above such measurements yield the small-signal value of $\alpha_\ell(\omega)$. Using eq.(13(a)) one can then derive the frequency dependence of the non-electronic attenuation factor α_ℓ, as shown in Fig.18. The experimental points are seen to follow fairly well a ω^2 dependence, as to be expected from the Akhiezer linear-loss mechanism.[53] Our findings thus indicate that this is the dominant small-signal loss mechanism in the frequency range studied (10 - 70 GHz).

In the phenomenological large-signal theories discussed in Section 5 we have tacitly assumed that the frequency of maximum intensity FMI coincides with the frequency of maximum gain $\omega_m = (\omega_c\omega_D)^{\frac{1}{2}}$. As was shown earlier in the present Section, this

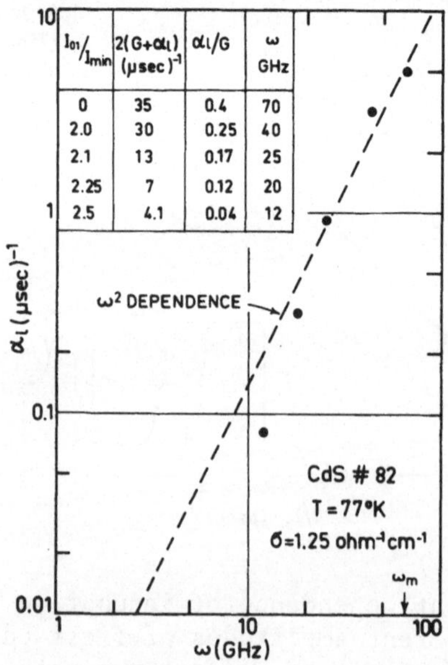

I_{01}/I_{min}	$2(G+\alpha_\ell)$ $(\mu sec)^{-1}$	α_ℓ/G	ω GHz
0	35	0.4	70
2.0	30	0.25	40
2.1	13	0.17	25
2.25	7	0.12	20
2.5	4.1	0.04	12

Fig.18. Angular-frequency dependence of small-signal non-electronic attenuation factor α_ℓ. Table lists values of $2[G(\omega)+\alpha_\ell(\omega)]$ and of $\alpha_\ell(\omega)/G(\omega)$ derived from data as explained in text and used in the determination of $\alpha_\ell(\omega)$.

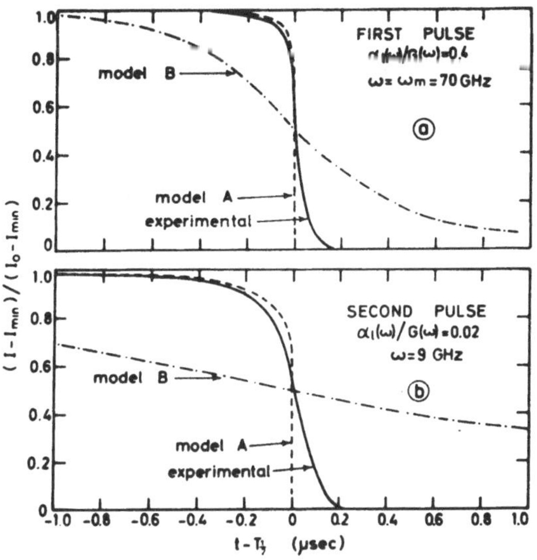

Fig.19. Current decay curves at two frequencies obtained on sample No.82 together with the corresponding theoretical curves predicted by the two Models. The theoretical curves for Model A correspond to $\delta = 0.05$ (those of Model B are independent of δ).

assumption is indeed valid for semiconducting CdS prior to current saturation. Thus the current decay characteristics from I_{o1} to I_{min} in Fig.16 can be compared with the predictions of Model A and B on this basis. Such a comparison is shown in Fig.19(a) which is similar to Fig.13 above. In the case of the current decay from I_{o2} to I_{min} (second pulse), however, the amplified acoustic flux is centered around a frequency that is lower than ω_m. One must therefore use in the theoretical calculations the values $G(\omega)$ and $\alpha_\ell(\omega)$ corresponding to this lower frequency and not G_m and $\alpha_\ell(\omega_m)$. This is done in Fig.19(b) where the characteristics of a current decay curve corresponding to the second pulse derived from a trace such as those shown in Fig.16 are compared with the predictions of Model A and Model B based on the values of $G(\omega)$ and $\alpha_\ell(\omega)$ determined independently above. (It should be noted that the theoretical curves are insensitive to the value of $\alpha_\ell(\omega)$ used because $\alpha_\ell(\omega) << \alpha_o$.) The fit between theory and experiment in both cases (Fig.19(a) and (b)) again points to the validity of Model A. It should be noted that, as expected, the flux build up is faster for the higher frequency (higher gain) than for the lower frequency (lower gain). It is quite likely that in ZnO and GaAs where the current decay is slower than in CdS and Model B appears[48,49] to give a better agreement with experiment, Model A is nevertheless valid but the frequency of the amplified flux is lower than ω_m even at the early stages of the flux build up process.

7. SUMMARY AND CONCLUSION

We have attempted in these talks to review some of the fundamental processes underlying the acoustoelectric effect, with special emphasis placed on the non-ohmic behaviour associated with the build up of acoustic flux. After a brief resume of small-signal theories of acoustic amplification and attenuation, we have discussed in some detail some of the experimental data obtained on semiconducting CdS. Probe measurements of the electric field distribution show that throughout its build up the acoustic flux is highly concentrated: in narrow propagating domains in the transient state (current oscillations) and in a stationary domain bordering with the anode at the steady state (current saturation). Whereas the propagating domains are generated at the cathode by shock excitation, the stationary domain consists of amplified thermal noise originating near the anode and re-inforced by reflections from the anode boundary. The current oscillations can be quantitatively accounted for in terms of the time-varying non-ohmic resistance associated with the propagating domains.

The occurance of flux domains emphasizes the fact that large--signal rather than small-signal conditions are the characteristic conditions pertaining to real situations. The two possible models suggested by the extension of the linear, Hutson-White theory to such conditions have been analyzed and the conclusions of each model were expressed in a form amenable to direct experimental test. Comparison of the predictions of the analysis with results obtained on semiconducting CdS indicates that of the two possibilities, Model A (the regenerative one) is applicable.

Non-linear effects associated with large concentrations of flux can be advantageously studied by measurements of the acoustoelectric after-current. Such measurements indicate that after current saturation the frequency of maximum intensity decreases continually as the domain propagates down the filament, typically reaching a value as low as one tenth of the frequency of maximum (linear) gain. This behaviour enables a rather simple determination of the frequency dependence of the non-electronic attenuation factor α_ℓ under small--signal conditions. In the frequency range 10 - 70 GHz, α_ℓ is found to increase quadratically with frequency, as to be expected from a linear Akhiezer loss mechanism.

In conclusion, the general features of the acoustoelectric effect in semiconducting CdS are fairly well understood. Many of the observed phenomena can be theoretically accounted for, at least phenomenologically, leading to a more or less self consistant picture of acoustic amplification and non-ohmic behaviour. The situation in the other piezoelectric semiconductors appears to be more complex and further work is required before as good a picture can be obtained. Detailed investigation of non-linear effects in both II - VI and III - V compounds should be particularly valuable in gaining a deeper insight into phonon-electron interactions in piezo-electric materials.

REFERENCES

1. H.D.Nine, Phys.Rev.Letters 4, 359 (1960).
2. A.R.Hutson, J.H.McFee and D.L.White, Phys.Rev.Letters 7, 237 (1961).
3. A.R.Hutson and D.L.White, J.Appl.Phys. 33, 40 (1962) and D.L.White, J.Appl.Phys. 33, 2547 (1962).
4. R.W.Smith, Phys.Rev.Letters 9, 87 (1962).
5. N.G.Einspruch, Solid State Physics Vol. 17 (Academic Press, 1965), pp.217-268. (References).
6. J.H.McFee, Physical Acoustics Vol. IVA, Chap.I (Academic Press, 1966). (References)
7. H.R.Carleton, H.Kroger and E.W.Prohofsky, Proc. IEEE 53, 1452 (1965). (References)
8. Proc. Int. Conf. on Semiconductor Physics, Kyoto (1966), pp.455-497. (References)
9. Proc. Int. Conf. on II-VI Semiconducting Compounds, Providence (1967), pp.910-935. (References)
10. J.H.McFee, J.Appl.Phys. 34, 1548 (1963).
11. H.Kroger, Appl.Phys.Letters 4, 190 (1964).
12. C.Hamaguchi, A.Ishida and Y.Inuishi, Proc. IEEE 53, 1259 (1965).
13. W.H. Haydl and C.F.Quate, Appl.Phys.Letters 7, 45 (1965).
14. W.G.Cady, Piezoelectricity (Dover Publications Inc. 1964).
15. D.A.Berlincourt, D.R.Curran and H.Jaffe, Physical Acoustics, Vol. 1A, Chap.III (Academic Press, 1964).
16. A.Rose, RCA Review 27, 98 (1966).
17. R.H.Parmenter, Phys.Rev. 89, 990 (1953).
18. G.Weinreich, Phys.Rev. 107, 317 (1957).
19. W.C.Wang, Phys.Rev.Letters 9, 443 (1962).
20. A.B.Pippard, Phil.Mag. 8, 161 (1963).
21. J.C.Ashley, J.Appl.Phys. 36, 528 (1965).
22. H.N.Spector, Phys.Rev. 127, 1084 (1963).
23. E.Conwell, Phys.Letters 13, 285 (1964).
24. J.Yamashita and K.Nakomura, Proc.Theor.Phys. 33, 1022 (1965).
25. A.Rose, RCA Review 28, 634 (1967).
26. N.H.Spector, Solid State Physics Vol. XIX (Academic Press, 1967) pp.291-361.
27. S.G.Eckstein, J.Appl.Phys. 35, 2702 (1964).
28. J.D.Maines and E.G.S.Paige, Phys.Letters 17, 14 (1965).
29. M.Kikuchi, Japan J.Appl.Phys. 2, 807 (1963).
30. J.Okada and H.Matino, Japan J.Appl.Phys. 3, 698 (1964).
31. W.C.Wang, Appl.Phys.Letters 6, 81 (1965).
32. A.Ishida, C.Hamaguchi and Y.Inuishi, J.Phys.Soc.Japan 20, 1946 (1965).

33. C.Hamaguchi, A.Ishida and Y.Inuishi, J.Phys.Soc.Japan 20, 1279 (1965).
34. P.O.Sliva and R.Bray, Phys.Rev.Letters 14, 372 (1965).
35. G.Quentin and J.M.Thuillier, Phys.Letters 19, 631 (1966).
36. J.D.Maines and E.G.S.Paige, Solid State Comm. 4, 381 (1966).
37. W.H.Haydl and C.F.Quate, Phys.Letters 20, 463 (1966).
38. A.Many and I.Balberg, Phys.Letters 21, 486 (1966); Proc. Int. Conf. on Semiconductor Physics, Kyoto (1966) p.474.
39. E.H.Jacobsen, J.Am.Acoust.Soc. 32, 949 (1960).
40. J.Okada and H.Matino, Japan J.Appl.Phys. 2, 736 (1963).
41. H.Murakami, M.Miya, S.Nagano, T.Ishiguro and I.Uchida, Japan J.Appl.Phys. 3, 299 (1964).
42. W.H.Haydl, Appl.Phys.Letters 10, 36 (1967).
43. C.Hervouet, J.Lebailly, P.Leroux-Hugon and R.Veillex, Solid State Comm. 3, 413 (1965).
44. G.Quentin and J.M.Thuillier, Phys.Letters 23, 42 (1966).
45. R.Bray, C.S.Kumar, J.B.Ross and P.O.Sliva, Proc. Int. Conf. on Semiconductor Physics, Kyoto (1966) p.483.
46. H.Ozaki and N.Mikoshiba, J.Phys.Soc.Japan 21, 2486 (1966).
47. A.Many and I.Balberg, Phys.Letters 24A, 705; 707 (1967).
48. P.O.Sliva and R.Bray, Private Communication.
49. N.I.Meyer, E.Mosekilde and M.H.Jørgensen, Proc. Int. Conf. on II-VI Semiconducting Compounds, Providence (1967) p.950.
50. J.Zucker, S.Zemon and J.H.Wasko, Int. Conf. on II-VI Semiconducting Compounds, Providence (1967) p.919.
51. I.Balberg and A.Many, Appl.Phys.Letters 13, 100 (1968).
52. W.H.Haydl, K.Harker and C.F.Quate, J.Appl.Phys. 38, 4295 (1967).
53. A.R.Hutson, Phys.Rev.Letters 9, 296 (1962).
54. B.Tell, Phys.Rev. 136, A772 (1964).
55. N.Mikoshiba, J.Phys.Soc. Japan 20, 2160 (1965).

SHIFT OF THE CUT-OFF MAGNETIC FIELD IN MAGNETRONS

J.M. ROCARD, R. CLERC and A. CAPDEVILA

Laboratoire de Physique des Milieux

Ionisés, Faculté des Sciences, TOULOUSE, FRANCE

INTRODUCTION

The motion of electrons in crossed electric and magnetic fields has been studied by many authors. A.W. HULL [1], in particular has investigated the special case of cylindrical magnetrons. He has shown that, in the case of a small cathode, (filament), in the axis of an anode of radius R, at potential V, the cut-off magnetic field is given by the relation

$$(1) \qquad B_c = \sqrt{\frac{8m}{e}} \; \frac{V^{1/2}}{R}$$

where m and e are the mass and the charge of an electron.

This critical field which does not depend on potential or charge distribution is obtained through the energy and momentum conservation laws applied to an electron leaving the cathode with $v_0 \simeq 0$ and reaching the anode with $v_R = 0$, $v_t = \sqrt{\frac{2eV}{m}} = R\frac{d\theta}{dt}$

The object of this report is to show both theoretically and experimentally that the cut-off field in magnetrons should be a function of the potential distribution in the tube and therefore of the space charge distribution. The classical expression (1) is valid when space-charge effects can be neglected (low emission level). The critical field is shifted towards lower values when space-charge effects are not negligible (space-charge limited emission) [2].

THEORETICAL CONDITIONS

Let us consider a space-charge limited cylindrical diode. Space-charges concentrate not only around the filament but also in the vicinity of the anode if the longitudinal magnetic field is of the order of B_c,(slighty less, so that the cut-off regime is not reached).

The effect of the space-charge is to create a virtual cathode of radius r_o at potential $V_o = O(r_o \ll R)$; the electron cloud in the vicinity of the anode is rotating with the angular velocity $\frac{d\theta}{dt} = \frac{eB}{2m}$ $(R \cdot \frac{d\theta}{dt} \ll c$ so that relativistic effects can be neglected).

One electron leaving the virtual cathode ($v_o \simeq 0$) must therefore overcome a potential barrier prior to its reaching the anode; in the interelectrode space, this electron is moving under the influence of the constant magnetic field parallel to the axis, a symetrical radial external electric field and also a symetrical radial internal electric field due to the anode electron cloud.

The energy conservation law should be applied to the system consisting of N electrons located in the interelectrode space, (steady state regime) under the following form :

$$(2) \qquad \sum_i \frac{1}{2} m v_i^2 = N e V_a - \sum_{ij} \vec{F}_{ij} \, d\vec{r}_{ij}$$

where $\sum_i \frac{1}{2} m v_i^2$ is the total kinetic energy of the N electrons when they reach the anode, N e V_a is the total potential energy and $\sum_{ij} \vec{F}_{ij} \, d\vec{r}_{ij}$ represents the internal work due to electron-electron interactions.

The expression (2) is too complicated to be useful. Some simplifications have to be introduced here in order to be able to get some information on the effective cut-off magnetic field.

Let us assume that all electrons leaving the virtual cathode with $v_o \simeq 0$ arrive at the vicinity of the anode with approximately the same speed v and that the internal work done by the anode cloud on an individual electron moving towards the anode can be calculated from the coulombian interaction potential energy.

The energy conservation law applied to that individual electron is then:

$$(3) \qquad \frac{1}{2} m v^2 = e V_a - e \cdot \left[\psi(R^-) - \psi(r_o) \right]$$

where $\psi(R^-)$ and $\psi(r_o)$ are the coulombian interaction potentials at the vicinity of the anode (R^-) and at the virtual cathode (r_o).

It is obvious that the calculation of the coulombian interaction potential is not possible in the electron cloud itself nor at its boundary (since $\psi \to \infty$ according to Coulomb's law). At any rate, one would like to estimate the term $e \cdot \left[\psi(R^-) - \psi(r_o) \right]$ which represents the work done by the steady state electron cloud on a given electron on its way to the anode. Once the electron is very close to the anode cloud, it either becomes part of it or it has enough energy to go through the cloud and reach the anode. The distance R^- is therefore less than the internal radius R_1 of the electron cloud; the distance chosen in the calculations presented here is $R^- = R_1 - (R - R_1)$ where $R - R_1$ is the electron cloud thickness.

CALCULATION OF THE INTERACTION POTENTIAL

1 - Case of an electron planar ring

Let us consider first the following simple case : we want to calculate the coulombian interaction potential between an electron moving in a plane perpendicular to the axis of the magnetron, and the planar ring of the electron cloud located in the same plane.

This coulombian interaction potential is given (figure 1) by :

$$\varphi (r) = \frac{1}{4 \pi \varepsilon_0} \iint_{(S)} \frac{\sigma \, dS}{d} \tag{4}$$

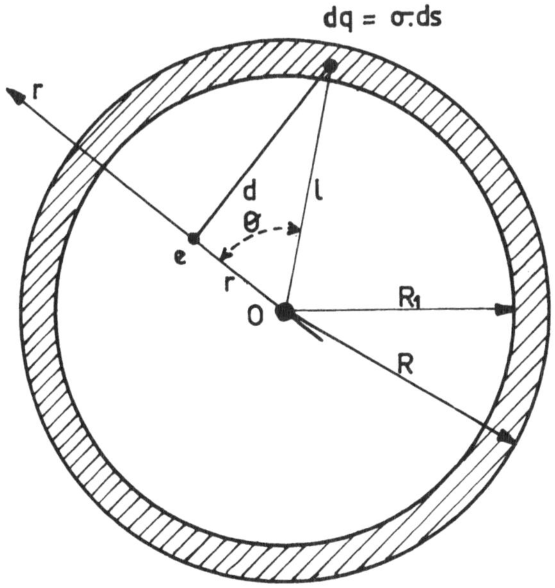

FIGURE 1 : Interaction between an electron (e) and a planar electron ring.

where S is the total surface of the ring and d is the distance between the electron (e) and the charge element dq. The surface integral can be calculated if one knows the spatial distribution of electrons over the ring; for simplicity reasons, and also because we want to have an estimated (or an order of magnitude) value of the interaction potential energy, we shall assume in this paper, that the charge density (σ, ρ) in the anode electron cloud is constant

(steady state regime, longitudinal magnetic field of the order of B_c)

In polar coordinates, the interaction potential (4) can be written:

$$(5)\quad \varphi(r) = \frac{\sigma}{2\pi\varepsilon_0}\int_0^{\pi}\int_{R_1}^{R} \frac{\ell\, d\ell\, d\theta}{\sqrt{r^2 + \ell^2 - 2r\ell\cos\theta}}$$

where the various symbols are easily understood from examination of figure 1.

The details of the integration will not be given here, the final result can be put under the form :

$$(6)\quad \varphi(r) = \frac{\sigma R}{2\pi\varepsilon_0}\left\{(1-\frac{r}{R})F_2 - (\frac{R_1-r}{R})F_1 + (1+\frac{r}{R})E_2 - (\frac{R_1+r}{R})E_1\right\}$$

where E and F are the classical elliptic integrals :

$$E\left(\frac{\ell}{R},\frac{r}{R}\right) = \int_0^{\pi/2}\sqrt{1-\sin^2\alpha\,\sin^2\varphi}\;d\varphi$$

$$F\left(\frac{\ell}{R},\frac{r}{R}\right) = \int_0^{\pi/2}\frac{d\varphi}{\sqrt{1-\sin^2\alpha\,\sin^2\varphi}}$$

$$\sin^2\alpha = \frac{4\ell r}{(\ell + r)^2}$$

and the indices 1 and 2 correspond to the internal (R_1) and external ($R_2 = R$) radii of the electron ring.

2 - Case of a thin electron ring

It is legitimate to make a few approximations to extend the domain of application of equation (6) if one deals with a very thin cylindrical electron ring. If h is the thickness of the ring, h must be small enough so that all distances d involved in the preceeding calculations do not depend upon h. In these conditions, one gets:

$$\varphi^*(r) = \frac{\rho}{4\pi\varepsilon_0}\int\int\int_{(V)}\frac{\ell\, d\ell\, d\theta\, dz}{d} = \frac{\rho}{4\pi\varepsilon_0}\int_0^h dz\int\int_{(S)}\frac{\ell\, d\ell\, d\theta}{d}$$

That is :

$$(7)\quad \varphi^*(r) = \frac{\rho h R}{2\pi\varepsilon_0}\left\{(1-\frac{r}{R})F_2 - (\frac{R_1-r}{R})F_1 + (1+\frac{r}{R})E_2 - (\frac{R_1+r}{R})E_1\right\}$$

For the calculation presented here, $R - R_1$ has been chosen equal to 0.01 R and h = 0.001 R. In the worse case, (electron e in the ring cross-section) the relative error on $\varphi^*(r = \frac{98}{100}R)$ is less than two percent.

3 - Case of a thin electron ring at altitude i

The same method could be extended to this more general case; however, one would find that the expression for the interaction potential involves not only elliptical integrals but also Legendre integrals of the type :

$$\Pi\,(\rho\,,k) = \int_0^{\pi/2} \frac{d\theta}{(1-\rho^2\sin^2\theta)\sqrt{1-k^2\sin^2\theta}}$$

where ρ and k are parameters related to ℓ (distance between the electron and the charge element in the cloud) and i (altitude of the ring with respect to the given electron).

If one wants to know these functions Π , which are not tabulated for obvious reasons (two series of parameters ρ and k), one has to use the method of numerical calculations; but since we intend to give only an order of magnitude of the interaction potential at r_o and at R^-, we have not used this direct but fastidious method.

We shall give, however, the expression of the coulombian interaction potential in these conditions:

$$(8)\quad \varphi^*(r,\,i) = \frac{\rho hR}{2\,\pi\varepsilon_0}\Big\{\sqrt{\left(1+\frac{r}{R}\right)^2+\left(\frac{i}{R}\right)^2}\,E_2^* - \sqrt{\left(\frac{R_1+r}{R}\right)^2+\left(\frac{i}{R}\right)^2}\,E_1^*$$
$$+\frac{1-\left(\frac{r}{R}\right)^2-\left(\frac{i}{R}\right)^2}{\sqrt{\left(1+\frac{r}{R}\right)^2+\left(\frac{i}{R}\right)^2}}\,F_2^* - \frac{\left(\frac{R_1}{R}\right)^2-\left(\frac{r}{R}\right)^2-\left(\frac{i}{R}\right)^2}{\sqrt{\left(\frac{R_1+r}{R}\right)^2+\left(\frac{i}{R}\right)^2}}\,F_1^* + \frac{i^2}{R}\,(M_2-M_1)\Big\}$$

with $E^* = \int_0^{\pi/2}\sqrt{1-\sin^2\gamma\,\sin^2\varphi}\;d\varphi$, $F^* = \int_0^{\pi/2}\frac{d\varphi}{\sqrt{1-\sin^2\gamma\sin^2\varphi}}$

$\sin^2\gamma = \frac{4\,r\,x}{(r+x)^2+i^2}$ $(x_1 = R_1,\ x_2 = R)$ and

$$M = (r^2+i^2+xr)\int_0^{\pi/2}\frac{d\varphi}{\left(i^2+r^2\sin^2\varphi\right)\sqrt{x^2+r^2+i^2+2rx\cos\varphi}}$$

Notice that when $i \ll R$, the term $\frac{i^2}{R}(M_2 - M_1)$ can be neglected and the simplified equation (8) gives a good value for $\varphi^*(r, i)$.

When $i \gg 0.2$ R, we have calculated the interaction potential $\varphi^*(r, i)$ only in the "limit" conditions, e.g. when the given electron is located at the virtual cathode (with $r_o \simeq 0$ since $r_o \ll R$) and when it is located very close to the anode cloud, ($r = 0.98\,R = R^-$, width of the cloud $R - R_1 = 0.01$ R).

In the first case, the given electron in its initial position on the axis of the magnetron), it can be shown that, for any value of i, we get :

$$(9)\quad \varphi^{*}(o, i) = \frac{\rho h R}{2\pi \varepsilon_0}\left\{\pi\left[\sqrt{1 + \left(\frac{i}{R}\right)^2} - \sqrt{\left(\frac{R_1}{R}\right)^2 + \left(\frac{i}{R}\right)^2}\right]\right\}$$

In the second case, (the given electron in its final location close to the anode cloud), the expression :

$$\varphi^{*}(r, i) = \frac{\rho h}{4\pi\varepsilon_0} \int_{R_1}^{R} \int_{0}^{2\pi} \frac{\ell d\ell d\theta}{d}$$

can be easily integrated when one assumes that the fluctuations of ℓ do not interfere with d which is then only a function of i and θ, (the approximation is sufficient when $i > 0.2$ R); the result is:

$$(10)\quad \varphi^{*}(R^-, i) = \frac{\rho hR}{2\pi\varepsilon_0}\left(1 - \frac{R_1^2}{R^2}\right) \frac{R}{\sqrt{i^2+4R^2}}\, F\left(\frac{2R}{\sqrt{i^2+4R^2}}\right)$$

$$\text{with}\quad F\left(\frac{2R}{\sqrt{i^2+4R^2}} = k\right) = \int_0^{\pi/2} \frac{d\theta}{\sqrt{1-k^2 \sin\theta}}$$

4 - Numerical results

Let us call $N_i(r)$ the factor of the common term $\frac{\rho hR}{2\pi\varepsilon_0}$ in formula 7, 8, 9 and 10; this factor is a numerical coefficient which can be calculated with the help of elliptical integral tables for various values of the parameters $\frac{r}{R}$ and $\frac{i}{R}$. Table I is a summary of $N_i(r)$ computed from these formulae for various values of i between 0 and 2R and for the values of r:

$$r = r_0 \simeq 0 \text{ and } r = R^- = 2R_1 - R \quad (R - R_1 = 0.01\ R).$$

TABLE I

$N_i(r)$	$r=r_0=0$ Equation 9	$r=R^-=0.98$ R Equation 10	$r=R^-=0.98$ R Equation 8	$N_i(R^-)-N_i(r_0)$
i = 0 Formula 8	0.0314	0.0612	0.0612	+ 0.030
i = 0.2 R	0.0309	0.0368	0.0366	+ 0.006
i = 0.25 R	0.0306	0.0351	0.0320	+ 0.004
i = 0.4. R	0.292	0.0292		
i = 0.5 R	0.0282	0.0276		- 0.0006
i = 0.8 R	0.0245	0.0222		- 0.0023
i = R	0.0222	0.0200		-0.0022
i = 1.5 R	0.0176	0.0160		- 0.0016
i = 2R	0.0141	0.0129		- 0.0012

As can be seen from Table I, the coefficient $N_i(R^-) - N_i(r_o)$ is maximum for $i = 0$, goes through 0 when $i = 0.4\ R$ and becomes negative for $i > 0.4\ R$ with a "minimum value reached around 0.8 R.

The term $e\left[\psi(R^-) - \psi(r_o)\right]$, which represents the internal work done by the anode electron cloud on a given electron on its path to the anode, is then given by equation (11):

$$(11)\qquad e\left[\psi(R^-) - \psi(r_o)\right] = \frac{e\rho h R}{2\pi\varepsilon_o}\sum_i\left[N_i(R^-) - N_i(r_o)\right]$$

where the summation has to be done over all possible thin electron slices in the cylindrical magnetron.

From experimental or theoretical results published in the literature [3, 4], we think that $\rho = 3.10^{-3}$ coul/m^3 could be a reasonable value for the volumic charge density in the anode electron cloud when the regime of the cylindrical magnetron is space-charge limited (with $B \simeq Bc$). As far as the other quantities are concerned, we can take the values which have been used in our experimental work:

V_a = 250 volts $\quad R = 1.7 \times 10^{-2}$ m $\quad h = 10^{-3}R,\ \lambda = 2\ R$

λ = height of magnetron $\quad \varepsilon_o = 8.85 \times 10^{-12}$

Assuming that the given electron is moving in the central cross-section of the cylindrical diode and taking care of the summation over all possible rings of the electron cloud, we finally get :

$$e\left[\psi(R^-) - \psi(r_o)\right] = 100\ \text{eV}$$

as compared to $e.V_a = 250$ eV.

Now that we have given an "order of magnitude" value for the expected interaction electron-anode-cloud potential energy, let us return to equation 3 and see what should then be the effective cut-off magnetic field. It should be pointed out at this point, that the angular velocity of an electron in a cylindrical magnetron (with $r_o \ll R$), with or without space-charges is nearly constant and is given by the relation :

$$\frac{d\Theta}{dt} = \frac{eB}{2m}$$

The "cut-off" condition corresponds to the situation whan the velocity of the electron at the anode is only tangential; therefore:

$$v = R\,\frac{d\Theta}{dt} = \frac{R\,eB}{2m}$$

Putting that value of v into eq ation (3), we easily obtain the effective cut-off magnetic field (when space-charge effects are not negligible) :

$$B_e = \sqrt{\frac{8m}{e}}\ \frac{\sqrt{V_a - \left[\psi(R^-) - \psi(r_o)\right]}}{R}$$

as compared to the classical critical field (valid when space-charge effects are negligible):

$$B_c = \sqrt{\frac{8m}{e}} \frac{V_a^{1/2}}{R}$$

By comparison of the two previous relations, we can easily see taht the effective cut-off magnetic field is shifted towards the lower values when space-charge effects cannot be neglected. With the numbers given above, we expect this phenomenon of relative displacement to be of the order of magnitude of :

$$\frac{\Delta B_e}{B_c} \simeq 22.5 \%$$

EXPERIMENTAL RESULTS

Several tubes have been studied, all conforming as closely as possible to the symetry conditions postulated above, namely straight filament in an axis of a cylinder (for more details see reference[5]). Only the results concerning the displacement of the cut-off magnetic field as a function of electron emission level will be given here.

Figure 2 shows a typical example of the experimental characteristics anode current versus magnetic field that have been obtained with a cylindrical magnetron (height of anode = 2R, R = 1.35 cm), at very low pressure ($p \simeq 2 \times 10^{-9}$ Torr). The various curves corresponding to different values of thermoionic emission levels (reference $I_{(o)}$ = current diode with B = O), are plotted on a dimensionless diagram $\frac{I_a}{I_{a(o)}}$ versus B (or I_b, this quantity being the current flowing in the solenoid). Figure 2 shows clearly that the current stays constant until the effective critical field is reached and then falls off rapidly; it can be seen also that the effective cut-off field at high emission levels is lower than the critical field at very low emission levels.

As one can see from figure 2, it is difficult to define exactly the experimental point where the anode current starts falling off (end of plateau); however, because of secondary effects, (mostly end effects), we noticed that the Hull value of the critical field (equation 1) corresponded to a drop of 5 % below the plateau value for anode currents lower than, or equal to 10^{-6} A and, therefore, the effective cut-off field B_e was experimentally chosen as the field able to reduce any anode current (high and low emission) to 95 % of its plateau value.

Figure 3 summarizes the results that we obtained in this way with another cylindrical diode, (R = 1.74 cm). On this diagram, the classical value of B_c given by the equation 1 and the experimental value of B_e are plotted against the anode potential V_a for various anode currents. It can be seen that all curves $B_e = f(V_a)$ are shifted towards the low magnetic fields: the observed relative shift increases when the electronic emission increases and when the voltage decreases, e.g. when space charge effects become important. When space charge effects can be neglected, (I_0 = 10^{-5} A, 250 < Va < 400 V), the effective cut-off field tends towards the classical Hull value.

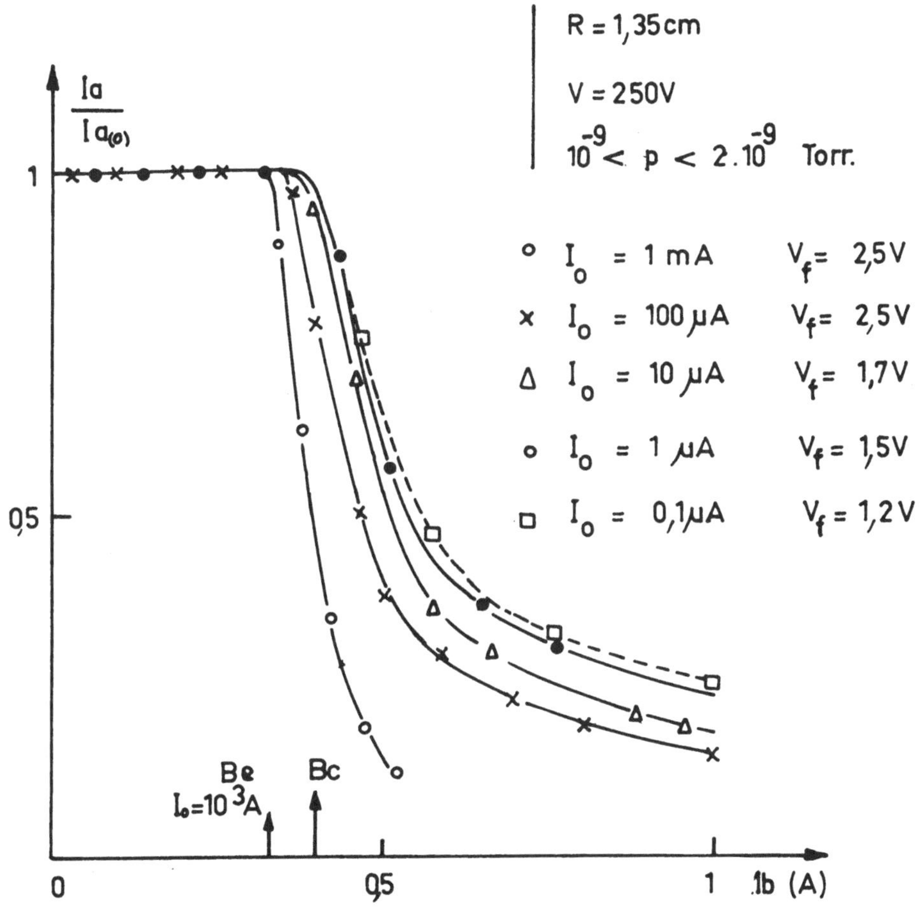

FIGURE 2 : Experimental characteristics of a cylindrical magnetron
(R = 1.35 cm V_a = 250 volts p = 10^{-9} Torr)

This shift of the critical magnetic field was predicted, at first we believe, by one of us [2] . The calculated and experimentally observed relative shifts are of the same order of magnitude. More work is being done in our laboratory to prove that this phenomenon is not caused by secondary effects such as end effects, filament heating currents, secondary emission etc...

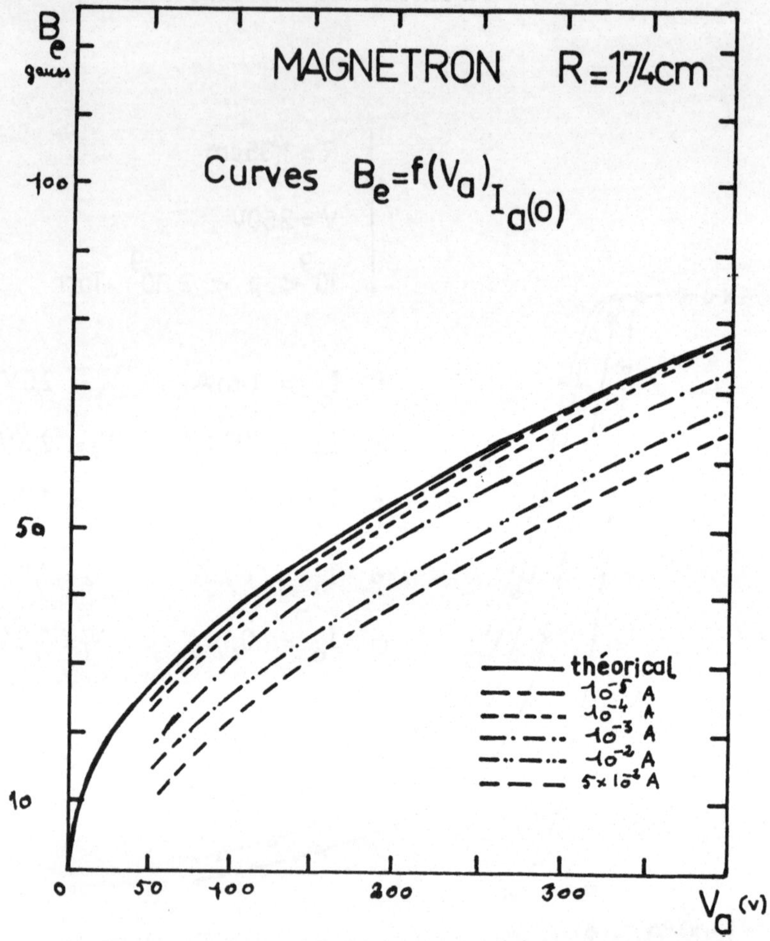

FIGURE 3 : Effective cut-off field versus anode potential. (parameter : electronic emission) (R = 1.74 cm)

REFERENCES

[1]- A.W. HULL Phys. Rev. 18, 31, 1921
[2]- J.M. ROCARD Rev. Phys. Appl. 2, 275, 1967
[3]- J.L. DELCROIX Thèse, Paris, 1954
[4]- J.L. REVERDIN Journ. Appl. Phys. 22, 257, 1951
[5]- A. CAPDEVILA Thèse 3ème Cycle, Toulouse, 1968.

DEVELOPMENT PROSPECTS IN ENERGY CONVERSION

J.P. CONTZEN

EUROPEAN SPACE VEHICLE LAUNCHER DEVELOPMENT ORGANISATION

ABSTRACT

Some of the most important conversion processes which have been developed during the last two decades to satisfy the growing needs of electrical power generation on board space vehicles are described. The impact of these initially space-oriented techniques in the terrestrial field is underlined.

The following systems are reviewed:
- photovoltaic cells.
- radiovoltaic, radiophotovoltaic and thermophotovoltaic cells.
- α and β cells.
- thermoelectric and thermionic systems.
- magnetohydrodynamic and electrofluid dynamic systems.

These systems are examined from the point of view of their current state-of-the-art, their development prospects in relation to different types of primary sources, nuclear, solar, chemical and their respective application areas with regard to the power range and the nature of the environment (land, space, sea and undersea).

A. INTRODUCTION-CLASSIFICATION OF ENERGY CONVERSION PROCESSES.

In the last fifteen years, activity in the field

of energy conversion has grown very rapidly. This growth, which has led to a wide diversification of processes, appears to be mainly due to two factors:

(a) the advent of a new form of primary energy, namely nuclear energy.

(b) the opening of a new area of applications - space - which has initiated or at least catalysed most of the research in this field.

This paper is concentrated on conversion into electrical energy and within this frame it is aimed at reviewing some of the most important conversion processes which have appeared in recent years. It analyses the current state-of-the-art and outlines the development prospects with reference to different types of primary energy sources, namely solar, chemical and nuclear energy.

Not only space applications of these techniques will be examined but also their impact on electrical power generation for land, sea, and undersea uses will be underlined: this impact is rather strong and energy conversion is certainly a field where initially space-oriented efforts have greatly helped to stimulate new solutions to classical problems on earth.

Before analysing any conversion system, it appears necessary to establish a clear classification of processes particularly as some confusion exists in the use of terms like direct, indirect, static or dynamic conversion. The following list together with the graph shown in figure 1, should help in this respect and also give a preliminary insight in the fundamental mechanisms which govern the conversion into electrical energy:

1. <u>One-step processes</u>
   - 1.1. Chemical energy - Electrical energy
     Batteries, Fuel cells.
   - 1.2. Solar (light) energy - Electrical energy
     Photovoltaic, Photoelectric cells.
   - 1.3. Nuclear energy - Electrical energy
     α and β cells, Radiovoltaic cells.

2. <u>Two-step processes</u>
   - 2.1. Nuclear energy - Light energy - Electrical energy - Radiophotovoltaic cells.
   - 2.2. Chemical, Solar, Nuclear energy - Heat-Electrical energy.
     Thermoelectric, Thermionic systems.

3. <u>Three-step processes</u>

3.1. Chemical, Solar, Nuclear energy - Heat - Light energy - Electrical energy. Thermophotovoltaic cells.

3.2. Chemical, Nuclear energy - Heat - Kinetic energy of a working fluid - Electrical energy. Magnetohydrodynamic, Electrofluid dynamic systems.

4. <u>Four-step processes</u>

4.1. Chemical, Solar, Nuclear energy - Heat - Kinetic energy of a working fluid - Mechanical (shaft) energy - Electrical energy. Turbines, Reciprocating engines coupled to conventional A.C. or D.C. generators.

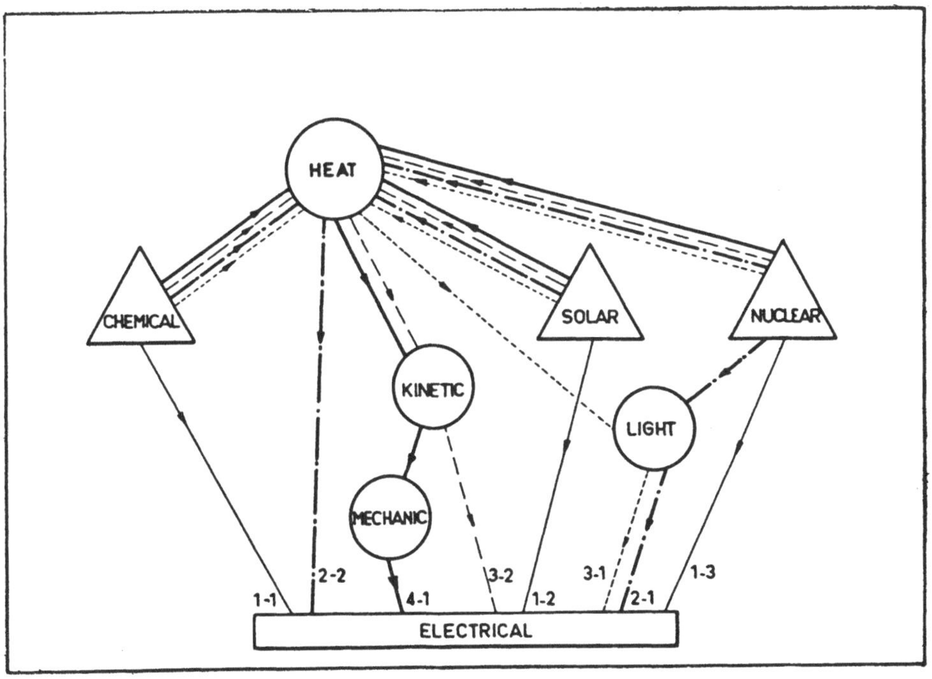

Figure 1. Energy Conversion Schemes

Processes under heading 1 are the only true direct conversion processes of primary energy into electrical energy but this description is generally extended to processes under headings 2 and 3 (systems under 2.2 are in fact converting directly heat into electricity). Furthermore, all devices corresponding to headings 1 to 3 are static converters as opposed to dynamic systems under 4.1 which are the only ones to involve the use of moving parts in the conversion section itself, the other devices possibly using moving parts in auxiliary components of the overall system such as pumps, blowers, etc.

The list permits also to determine whether a conversion process has its efficiency limited to the Carnot efficiency or not: if heat is used as an intermediate step (processes 2.2, 3.1, 3.2, and 4.1), this is the case and the device works then as a conventional heat engine. Having regard to the importance of efficiency, this factor has definite implications on the design and use of the conversion system.

Only chemical, solar and nuclear energy have been considered here as primary energy sources; such consideration is perfectly valid for space uses but it could be pointed out that electricity can be derived on earth from other forms of energy (hydraulic and geothermal plants, wind and tidal energy). Nevertheless, a close scrutiny on the process involved shows that all these systems are covered by the present list, their starting point (generally heat or kinetic energy of a working fluid) being merely an intermediate step in the classification used here.

In spite of the fact that only the most promising systems have been mentioned under each heading in the classification, - such devices as fission cells, ferroelectric or thermomagnetic systems have been dismissed in view of their questionable applicability -, it can be seen from the list that already many conversion processes are available for turning primary energy sources into electricity. None of these methods satisfies the whole spectrum of electrical power generation requirements and the discussion will be centered on the determination of the respective areas of applications of the various systems taking into account three parameters: the primary energy source used, the electrical power level required and the environment.

Systems listed under 1.1 (electrochemical) and 4.1 (dynamic) will not be reviewed here; they are the more conventional ones, except perhaps the fuel cell, and priority is given to the other processes which all rely on solid state or plasma physics mechanisms and which after being developed in the last decade for space needs, are now emerging as potential competitors for earth applications.

B. PHOTOVOLTAIC CELLS

The direct conversion of light into electricity has been for a long time solely utilised into light meters using the photoelectric effect (emission of electrons through light excitation). The requirements of electrical power generation in space vehicles have led to a very wide use of solar cells based on the photovoltaic effect, i.e. production by incoming photons of electron-hole pairs at the vicinity of a semiconductor junction.

Single crystal silicon cells of the n on p type with dimensions 1cm X 2cm, 2cm X 2cm or 2cm diameter are the most representative of the cells currently in use: they are made of a thin layer (0.25 - 0.5μ) of n-doped Si on top of a thicker layer (200 - 400μ) of p-doped Si, plated contacts covering the back of the cell and disposed as a grid on the illuminated surface collecting the current. A protective and antireflective coating e.g. 25μ thick is evaporated on the front of the cell. Solar cells are degraded by electrons and protons present in the space environment and n on p cells have been chosen in view of their better resistance to radiation damage compared with p on n cells: this resistance can be further increased by the addition of a transparent front cover e.g. quartz or sapphire, with a thickness of the order of 150μ.

Efficiency of a silicon solar cell at room temperature (efficiency decreases with an increase of operating temperature) is of the order of 10% and it can be estimated that in space (Air Mass Zero) approximately 100 W are produced by 1 m^2 of illuminated area: on earth this value is slightly smaller (60-70 W/m^2) owing to the lower solar radiation intensity and in spite of a higher efficiency due to a better spectral adaptation of the cell to the incoming light.

For space applications, silicon cells which are disposed either on the body of the space vehicle or on panels and paddles, are operationally producing electrical power from a few Watts up to 1.5 kW: specific masses for complete systems range from 50 kg/kW (Surveyor 4) to 650 kg/kW (FR1), the figure for recent developments averaging around 60-70 kg/kW.

Current developments for space are oriented towards improvements in the following areas: increase in radiation resistance, reduction of specific mass, increase in power range.

Cell restoration by annealing at elevated temperatures, use of lithium doped-Si cells and reduction in the thickness of these cells are considered as possible solutions to the first problem. Lithium doping appears to be rather promising but cells having received such treatment are exhibiting up to now an anomalous behaviour - a large drop in output has been observed both after apparent recovery from radiation damage and in unirradiated cells - and this problem requires further theoretical and experimental investigation.

Thin silicon cells (100 μ instead of about 400 μ) have a lower initial output due to reduced efficiency (8-9%) but their degradation with irradiation is less pronounced and after an irradiation equivalent to 10^{15} 1 MeV electrons/cm^2, their output is identical to that of thicker cells and their lower mass allows a better post-irradiation specific mass per unit power. Problems with thin cells are mainly related to their manufacture and handling in view of their fragile nature.

Thin film solar cells using <u>polycrystalline</u> Cadmium Sulfide, Cadmium Telluride or Gallium Arsenide have been under development for several years. The photovoltaic material (about 20 μ thick) is deposited on a thin plastic or metal substrate (e.g. 25 μ of Mylar or Kapton) and is covered by another film of transparent material (e.g. 25 μ again of Mylar or Kapton).

The aim is to produce a cheap, flexible cell of low specific mass; efficiencies are lower than for Si cells with typical values in the range of 3 to 6%, leading to lower specific powers per unit area (40 - 50 W/m^2 at Air Mass Zero). There are still uncertainties hindering their use on space vehicles: fabrication techniques need to be perfected (exclusion of moisture has been, for instance, a critical point), long term stability should be improved, radiation damage is largely unknown and operational experience is still lacking. Encouraging progress has been recently reported and thin film solar cells, and more particularly Cd S cells, may well appear in the future as strong competitors to silicon cells. As it will be seen in the next paragraph, the important potential reduction in cost provides the incentive for pursuing actively this line of development.

It is only a few years ago that solar cells have been considered for the generation of large quantities of electricity in space, up to 50 kW of power: this idea has now been generally accepted and hardware development is under way. Such systems require obviously large light collecting surfaces (up to 500 m^2) and the emphasis is put on structural problems and design of deployment mechanisms. Various technical options are envisaged, either multiple unfolding arrays of rigid or semirigid panels covered with conventional (200 μ) silicon cells (JPL/Boeing approach) or rollup arrays using flexible thin film Cd S solar cells (NASA/Lewis) or thin (100 μ) silicon cells fixed on a flexible substrate (RAE design).

Progress made to date permits us to envisage specific masses in the range of 20 to 30 kg/kW and costs of the order of 200,000$/kW for large silicon cells arrays (compared to 600,000$/kW to 800,000$/kW for smaller silicon cells systems) going down to possibly 20,000$/kW -40,000$/kW for thin film arrays. The achievement of such target figures would certainly widen the already extensive application of photovoltaic systems in space.

Reduction in the cost of solar cells will certainly allow an extension of their presently limited role for land applications, particularly in such fields as rural telecommunications, telemetry systems, marine buoys and navigational aids, unattended weather stations, repeater stations, etc..

Their obvious advantage for such applications is that primary energy is free: investment and possible maintenance costs are only to be accounted for: their main drawback being the discontinuous availability of solar energy which in most cases implies storage of electrical energy. Energy costs of 1 to 10$/kWh have been quoted: these are rather attractive figures if one remembers that in such types of applications, cost of the energy provided by conventional chemical batteries can be as high as 35$/kWh.

In spite of the fact that solar terrestrial systems will be limited to low power generators, up to a few hundred Watts, due to considerations of investment cost and surface requirements, it appears that there is a definite future for solar cells on earth, mostly in geographical areas where solar energy is abundant, let us say above 1,500 kWh/m^2 for an average year; these areas fortunately correspond mostly to developing countries where there is a crying need for electrical power generators in the low power range.

C. RADIOVOLTAIC, RADIOPHOTOVOLTAIC AND THERMOPHOTOVOLTAIC CELLS

The principle of electrical power generation by the production of electron-hole pairs in a semiconductor junction can be applied in other ways than the straightforward photovoltaic effect.

A first idea is to use another incoming radiation, namely nuclear radiation (β particles) from an isotope source. The radiovoltaic cell is composed of a thin slab isotope source sandwiched between two suitably doped semiconductor junctions: particles emerging from the source produce electron-hole pairs in the semiconductor and the minority carriers diffuse across the junction as in a photovoltaic cell. Thus, the radiovoltaic effect is also a one-step direct conversion process.

Main problems are selection of a suitable long-life isotope (the use of Nickel 63 could be advantageous), self absorption of the particles in the source itself - a thin layer of isotope must be used - and radiation damage in the semiconductor, β particles of sufficiently high energy causing lattice defects in the crystal hence reducing the minority carrier life-time and by consequence the efficiency of the device. Specific power outputs of 0.5 to 1 mW/cm^2 could be achieved and conversion efficiencies of 0.1% to possibly a few % are considered. In spite of these low figures, it seems that there is a possibility of applying these devices - attractive by their simplicity and compactness - to power generation in the μW-mW range, and more effort than has been devoted in recent years should be applied to this field.

Another possibility is to use a two-step process, namely transforming nuclear energy by means of a phosphor into light energy and then converting this light into electricity through the photovoltaic effect. This is the principle applied in the radiophotovoltaic cell where a plane or cylindrical layer of phosphor mixed with a radioisotope is surrounded by photovoltaic cells. The problems connected with the development of such devices are essentially the preparation of high specific activity luminophors (like Promethium 147/Zn S paints) and the matching of the energy distribution of the light emitter with the sensitivity distribution of the receiving cell.

Expected specific power outputs are rather low, between 15 and 45 $\mu W/cm^2$, (current experiments are only yielding 2 $\mu W/cm^2$) and efficiencies should not exceed

0.1%. Work is proceeding on this conversion process, notably in Germany, to attempt to increase the performance to a level which would allow the use of such devices as generators in the very low power range, viz. below 10 mW.

Finally, a three-step process has been suggested by P. AIGRAIN, still along the same basic lines but with the introduction of an additional intermediate step. In the thermophotovoltaic cell, the primary energy - nuclear energy (isotope source) or chemical energy (flame heater) - is converted into heat, heat is by incandescence turned into light, this light exciting photovoltaic cells which produce the electrical power. It could appear disadvantageous to degrade primary energy into heat but this additional step offers in fact some advantages:

(a) the power density of the light emitter is greatly enhanced: a black body radiates at 2000° K approximately 50 W/cm^2 compared to 0.14 W/cm^2 at best for solar energy.

(b) the spectral distribution of the emitter can be adapted either by the choice of the emitter material or by the use of optical filters, to the maximum sensitivity of the receiving photovoltaic cell.

Furthermore, in this device, the unconverted light energy can be reflected towards the emitter which minimises losses. Although the efficiency of this system is subject to the Carnot limitation, high overall efficiencies could be expected: 10 to 16% are potentially available with power densities in the range of 3 to 30 W/cm^2. Such figures are rather promising and it is rather surprising that this process has not been investigated beyond the stage of laboratory experiments.

D. α AND β CELLS

Before leaving the area of energy conversion in the low power range, one word should be said about charged particle collection in α and β cells. These devices which are in some way self charging capacitors are probably using the simplest, most direct conversion process. They consist of two concentric or parallel electrodes separated by vacuum or dielectric-filled space. The emitter electrode is coated with a thin layer of radioisotope material which ejects α or β particles towards the collector electrode where they are intercepted and

cells. Such generators will certainly continue to be used for space missions, at least until high performance thermionic generators appear on the scene: the SNAP 27 generator for a moon package and the SNAP 29 system for earth orbital stations are examples of future uses. Nevertheless, they will be limited to applications where independence from the environment is a predominant factor, because they are not competitive with solar cells from the point of view of specific mass (200 to 500 kg/kW) and investment cost (20,000 to 40,000$/W).

A much brighter future is offered by earth applications. Several types of isotopic thermoelectric generators are currently produced for terrestrial uses not only with government support but also on a pure commercial basis. Some have already experienced extensive operation powering fixed and floating automatic weather stations, fixed marine navigational aids and buoys, undersea beacons and this list could be extended to submarine cable repeaters, aircraft ground radio marker beacons, etc. These terrestrial applications correspond to power levels from Watts to a few hundred Watts. Investment cost of these generators is significantly lower than for space applications: 5,000 to 10,000s/W at the 1 Watt level going down to perhaps 1,000$/W int the 100 Watts range. Energy cost lies between 25 and 50$/kWh.

Thermoelectric conversion coupled with isotopes has recently penetrated in a new field of applications, namely the production of very low powers in the range of microwatts to milliwatts. Miniature isotopic batteries are considered for use in the medical field, particularly in implanted heart pace-makers, in the watch industry or as power supplies for special electronic equipment. The advantages of these devices over chemical batteries, the sole systems considered up to now, are reliability and life-time. As such factors are of paramount importance for medical applications, a large development can be expected in this field.

2. Reactor thermoelectric systems

It has also been attempted to apply the thermoelectric conversion process to the generation of larger power levels: in this use, the nuclear source is a reactor. The SNAP 10A system is an example of such an application in space: it has been tested in orbit in April 1965. This test though terminated abruptly was very successful and showed the validity of the concept. It should nevertheless be pointed out that SNAP 10A was merely an experimental system exhibiting a limited power output (500 Watts), a high specific mass (approximately 1,000 kg/kW) and a low efficiency (1.43%). All these parameters need a significant improvement to make the thermoelectric reactor attractive for space: target figures should be power range 10 to 25 kW, specific mass 100 to 200 kg/kW, efficiency 5%, life-time up to 5 years. They are within reach in view of the current technological progress but require more effort than is presently deployed.

Thermoelectric reactors are also considered for undersea applications in the kilowatt to 30 kW range using the same concept as SNAP 10A. Advantages claimed for this type of power generator are essentially compactness and long unattended, maintenance free operation (5 years, potentially 10 years).

Both for space and terrestrial uses, thermoelectric reactors appear to be limited in power level to about 25 - 30 kW due to the increasing size of the heat rejection radiator (for space), to the relatively high cost of thermoelements (10 to 100$/W) and to assembly problems of a large number of individual conversion units.

3. Solar thermoelectric systems

The combination of thermoelectric conversion with solar energy heating does not seem very promising. Efficiencies of these systems are lower than those of solar cells, which are furthermore cheaper than thermoelements. The only application which could be envisaged is possibly the supply of electrical power to solar probes in the immediate vicinity of the Sun where the high environment temperature would largely decrease the efficiency of solar cells.

4. Chemical thermoelectric systems

The use of thermoelements with chemical energy (combustion) is much more attractive than with solar energy and this explains that flame-heated generators are currently manufactured on a commercial basis and have seen operational land uses. They compete favourably with solar cells, chemical batteries and isotopic generators in the power range of Watts to a few hundred Watts for earth applications such as microwave repeaters, buoys, radio beacons, etc. Available data indicates investment costs of 50 to 100$/W with energy costs in the range of 1 to 3$/kWh.

This brief review shows that, in spite of a brilliant start in space, future applications of thermoelectric conversion will probably be mostly on earth.

F. THERMIONIC CONVERSION

Thermionic conversion is presently in a less advanced technological state than thermoelectricity but it will probably in the future supersede the latter in several fields, particularly in space.

In a thermionic cell, heat is brought to the cathode or emitter where it raises the energy of some electrons to a level higher than the work function of the cathode. This allows them to leave the cathode surface and to travel in the space between the electrodes. When they enter the anode, or collector, they loose an energy equivalent to the work function of the anode, this loss appearing as heat which must be removed through cooling of the anode: the thermionic diode is in fact a heat engine using electrons as a working fluid. The remaining potential energy of the electrons is available as electrical energy to the external circuit connected between the anode and the cathode. The flow of electrons in the interelectrode gap creates a space charge which acts as a barrier for the emitted electrons: this space charge barrier must be suppressed or at least reduced. Many ways have been suggested to achieve this aim but it appears that nearly all designs are now based on space charge neutralisation by cesium vapour at "high" pressure ("high" for thermionic diodes, i.e. 0.1 to 1 Torr.): cesium is easily ionised by contact on the cathode and cesium ions thus produced neutralise the electrons in the gap. Furthermore, cesium adjusts the work function of both cathode and anode.

Various materials are used for the electrodes: Tungsten, Rhenium, Molybdenum are typical emitter materials while the collector is generally made of Niobium, Molybdenum or Rhenium. Designs using the same material for both electrodes have recently been favoured as the deposition of emitter material on the collector does not, in this case, alter the work function of the latter. The gap between electrodes is small, going from 0.05 mm to 0.2 mm. Typical temperatures for emitters are in the range of 1700 to 2100°K, for collectors between 800°K and 1200°K. Power output densities currently attainable are in the range of 5 to 20W/cm^2 and cell efficiencies in the region of 15%.

Such figures give thermionics a place as one of the static conversion processes with the highest performances. Nevertheless, it should be remembered that a thermionic cell requires a high temperature source and this limits somewhat its scope.

The use of chemical energy (combustion) with thermionic diodes is still uncertain. The problem is to design converters compatible with an oxydising flame; this requires the use of coatings which presently exhibit a life-time limited to a few hundred hours. Despite all current efforts, it may well happen that more resistant coatings will never be developed and in this case, there will be no practical flame-heated thermionic generator. On the contrary, if these efforts are successful, the terrestrial applications of such systems will be numerous, covering a larger field than flame-heated thermoelectric generators in view of the possibility of extending their power level to the kilowatt range.

Solar heated thermionic diodes have been successfully designed and tested but still their future is doubtful. To obtain the required emitter temperature, solar energy must be concentrated which implies continuous and accurate pointing of the concentrator towards the Sun. This is not very practical and solar thermionic generators will probably not receive much attention in the future.

Nuclear energy appears as the most promising primary source for thermionic conversion.

Isotopic thermionic generators, in the range of some Watts to a few hundred Watts are under development for space applications. Compared to thermoelectric generators, they potentially offer for the same missions, a twofold increase in overall efficiency (going up to 8 - 12%) and a reduction of nearly a factor of 2 in specific mass (down to 100 kg/kW) at approximately the same investment cost. It is thus probable that thermionic isotopic generators will replace the thermoelectric ones in future space missions.

For terrestrial applications of such generators, the picture is not so promising. The requirement of high temperature sources implies the use of high power density isotopes which are unfortunately the most expensive: hence, for earth applications where cost is a governing factor, economics are playing against the use of thermionics. However, a possible exception to this statement has recently emerged with the development of a thermionic generator in the microwatt to milliwatt range. This generator operates in an unconventional range of temperatures for thermionic conversion, i.e. with an emitter temperature of 600 - 800°K and a collector temperature around 300°K. At 4 mW level, efficiency is of the order of 0.3% rising to 3% at 100 mW level. Such a device could compete with other miniature generators for applications as cardiac pace-makers, electronic clocks, etc.

From the reactor side, thermionic conversion shows great promise as the thermionic reactor potentially covers a wide range of powers from kilowatts to Megawatts. Two main concepts are investigated: the in-core thermionic reactor where the nuclear fuel element is directly in contact with the emitter, heat being removed from the collector by radiation or by circulation of a coolant in the reactor core, and the out-of-core concept where thermionic diodes are situated outside the reactor, heat being transferred from the core to the emitter by radiation, conduction (heat pipe) or by a circulating fluid. In-core concepts appear to offer greater promise of covering the whole spectrum of power levels.

The critical problem with thermionic reactors is life-time: this is also important for other thermionic systems but not to such an extent. To our knowledge, the longest in-pile test of thermionic diodes has a duration of about 8,000 hours, i.e. less than one year. To be really attractive, operational reactors should in most cases, have a life-time of about 5 years, e.g. for direct broadcast satellite applications. This means an increase in life-time of a factor of more than 5, not taking into account the increased complexity of an operational system compared to an in-pile experiment. Such an achievement is still questionable. If this objective could be met, thermionic reactors would then have a definite future in space with attractive specific masses of about 40 kg/kW at the 40 kW level down to perhaps 10 kg/kW at the megawatt level, reduced volumes and moderate investment costs of 15,000 to 25,000$/kW. Possibilities would also exist for such reactors in the hydrospace field.

Finally, if the cell efficiency of thermionic diodes could be further increased from the present 15% to 20 - 25%, there would perhaps be an application for such systems as topping devices in nuclear power stations equipped with conventional turbines.

5. <u>Magnetohydrodynamic and electrofluid dynamic systems</u>

These two processes which introduce an additional step in the conversion scheme, are still considered as static conversion processes. This is true for the converter section itself which involves no moving parts but the practical overall generator system could include some dynamic devices such as blowers, compressors, etc. We are really here at the boundary between static and dynamic conversion. In both systems, the primary energy (nuclear, chemical) produces thermal energy used to raise the enthalpy of a working fluid. This working fluid is expanded converting thermal into kinetic energy and passes through the converter section where the kinetic energy is partly transformed into electrical energy: it is then exhausted or recycled.

1. Magnetohydrodynamics (MHD)

In the case of MHD, the working fluid is electrically conductive and moves along a duct with a velocity v through an applied magnetic field B perpendicular to the duct. The charged particles (charge e) experience a Lorentz force $F = e\ (v \times B)$ which is mutually perpendicular to B and v.

In a gaseous conductor, both electrons and positive ions move under the action of the Lorentz force but because of the greater mobility of electrons, nearly all the current flow is due to their motion. If two insulated electrodes are placed on the walls of the duct, one of these electrodes will collect the drifting electrons and provided an external path through a load is established, current will flow across the duct between the electrodes. An electrical field E will be created between these electrodes, its magnitude depending on the internal resistance of the gaseous conductor and on the resistance of the load. This introduces a further phenomenon, namely the Hall effect which is the tendency of charged particles to gyrate about magnetic lines in the presence of an electric field: in this particular case, this means that electrons tend to drift in a $E \times B$ direction along the MHD channel rather than across it. This phenomenon leads to two basic generator configurations:

- the Faraday generator where current is extracted perpendicular to the gas flow, the Hall effect being minimised by reduction of the Hall parameter or by segmentation of the electrodes,
- the Hall generator where the Hall effect is fully used and current is extracted along the gas flow.

Both generators give direct current power. Another configuration, the A.C. electrodeless induction generator allows alternating current to be produced immediately. Its electrical part is analogous to that of a conventional induction machine, polyphase windings producing a travelling magnetic field: the working fluid replaces the rotor and currents induced in the fluid generate the electromotive force in the static windings.

MHD systems offer the possibility of extremely large power densities of the order of 10^9 W/m^3; this makes attractive the use of MHD conversion for the generation of very high powers in a limited volume. On the other hand, MHD devices are not suitable at low power levels, namely below a few hundred kilowatts. This lower limit which rules out MHD conversion for most of the applications described so far, is due to increasingly high losses with decreasing power level, heat losses through the duct walls as well as Joule losses in maintaining the magnetic field (the last being eliminated by the use of superconducting magnets). Qualitatively, this can be explained by the fact that power output varies with the volume of the generator, i.e. the cube of a linear dimension while heat losses to wall surface and magnet power vary only as the square of this dimension: hence, there exists a minimum dimension below which the generator becomes completely inefficient or possibly, does not even achieve self-excitation.

The basic requirement that the working fluid should be electrically conductive has oriented, from the beginning, the design of MHD systems. A gas to be conductive must be ionised though not to a large extent: when a gas is only 0.1% ionised, it is very nearly as good a conductor as it would be if it were completely ionised. Conductivity can be obtained by heating a pure gas to a very high temperature (equilibrium ionisation), by seeding the gas with a small amount of easily ionisable material (alkali metals like potassium) or by non equilibrium ionisation, i.e. bringing the electron temperature to a higher level than the gas temperature.

Chemical MHD systems are generally envisaged as open cycle systems with or without the addition of conventional conversion equipment (gas or steam turbines) after the MHD conversion section (mixed power plants). Combustion of fossil fuel in air or rocket propellant exhaust provides the working gas at temperatures reaching 2500°K - 3000°K. Ionisation is assisted by preheating the air, working in an oxygenated atmosphere, seeding the gas or by high velocity frozen flow where the electron temperature is maintained at the level reached in the combustion chamber while the gas has already decreased in temperature.

Dominant problems with chemical open cycle systems are:

- electrodes and insulating walls life-time which at present does not exceed a few hundred hours,
- efficiency which generally requires some recuperation of heat at the exhaust, the design of a heat recuperator working at or above 2200°K being no simple problem,
- economics of the overall process which do not appear to be attractive a this time in spite of expected high overall conversion efficiencies of the order of 45 to 50%.

Large scale experiments, in the megawatt range, are required for perfecting the process but apparently nobody, with the exception of the USSR, wants to launch into such a venture. The only chemical MHD system which seems currently applicable and in fact already reached operational status is the short duration (seconds to a few minutes) MHD converter driven for instance by solid or hybrid rocket exhaust gases: obviously, this type of system is limited to special applications (military field, power input to hypersonic wind tunnels, etc.).

Closed cycle MHD generators coupled with nuclear reactors have been mostly envisaged for space applications. Pure alkali metal vapours and seeded (Lithium, Cesium) noble gases (Helium, Argon) are fluids under consideration. Even with the recourse to non equilibrium ionisation by auxiliary RF or DC discharge or fission product ionisation, the required operating temperatures appear to be beyond those obtainable with current or even projected reactor technology. Furthermore, the low pressure level and pressure drop in the MHD duct do not seem to match the nuclear reactor characteristics. An alternative is to use a liquid metal as working fluid; alkali metals in the liquid state are good electrical conductors and such systems could work at much lower temperatures, in the 1300°K range, which is more adapted to future reactor possibilities. The important problem, not to speak of material compatibility problems which are in no way negligible, is to transform the thermal energy of the reactor into kinetic energy, a liquid not having the expansion property of a gas. To achieve this transformation, various systems have been devised using either a single alkali metal or two distinct ones undergoing in both cases a phase change, the vapour phase being used to increase by momentum exchange the kinetic energy of the liquid phase before the latter enters the MHD conversion section. Efficient operation of these devices, sometimes called thermodynamic accelerators, as well as of diffusors which convert the kinetic energy remaining after the converter section into potential energy (pressure) has still to be demonstrated. If this objective can be reached, liquid metal MHD systems could have some prospects for application to space power generators above a few hundred kilowatts, their potential specific mass being attractively low (10 to 20 kg/kW) in spite of limited efficiencies (8 - 10%).

In brief, the overall picture for magnetohydrodynamics does not seem at present very bright and even should any technical break-through occur in the current research, operational status would not be reached quickly. MHD appears to lag definitely behind other conversion processes.

2. Electrofluid dynamics

In the same class of processes as MHD, electrofluid dynamics has appeared more recently and is still in a relatively unexplored state.

In this process, primary energy is converted into heat which in turn is transformed into kinetic energy of an electrically insulating working fluid, usually a gas. This fluid flows through a dielectric duct bounded by electrodes which are connected through an external load. Upstream of the inlet electrode, unipolar charges are injected into the moving fluid. These charges are generally produced by a corona discharge method which requires only a minute fraction, of the order of 10^{-4}, of the possible useful electrical power output.

The charged particles are transported through the duct by viscous interaction with the neutral gas molecules (or sometimes by a pure ballistic process), this work being performed against the electrostatic field generated by the potential on the downstream collector electrode. Charges accumulate on the collector, causing a neutralising current to flow through the external load, thus completing the conversion into electrical energy.

This description shows that the EFD converter is in fact a fluid Van de Graaff generator, the gaseous working medium replacing the mechanically driven belt of the conventional machine. Two essential differences should nevertheless be pointed out: the gas speed is far higher than the belt speed and charged particles are distributed throughout the volume flow rather than being concentrated on the belt surface. These considerations should allow power densities orders of magnitude higher than those obtainable with Van de Graaff generators, making the EFD converter a power generation device and not merely a high voltage source.

The main characteristics of the EFD converter which influence both its design and its applications, can be summarised as follows:

- it produces high voltages, hundred kilovolts and higher: this is, in most cases, a disadvantage, as most applications require lower voltages. This also introduces an electrical breakdown limitation in the EFD converter which not only influences its geometrical configuration but also dictates the use of a gas with a high dielectric strength. This last requirement, combined with the need for low mobility and low molecular weight coming from power and efficiency considerations, orientates the choice towards such working media as air, helium, hydrogen or mixtures of the last two with breakdown inhibitor gases like freon 12 or sulfur hexafluoride, at relatively high pressures of the order of tens of atmospheres.

- as the conversion process uses unipolar charges, it is affected by space charge limitation: as a result, the number of neutral molecules per elementary charge is very high, of the order of 10^8. Hence electric current densities are low, a few mA/cm^2. The charged particles can be atomic or molecular ions as well as charged aerosols (liquid droplets, dust particles, etc.).

- the combination of very high voltage and low current produces reasonable power outputs per unit area of flow channel, of the order of 1 kW/cm^2, only an order of magnitude less than MHD systems. This factor added to high stage efficiency (possibly 75%) and the absence of heavy magnetic coils, could yield favourable power to mass ratios.

- there is no stringent temperature requirement in the EFD process: temperatures much lower than in MHD systems allow the use of a wider variety of heat sources and promise longer life-times. Furthermore, thermal losses do not limit the power range as they do in the MHD process.

Open cycle EFD systems using chemical energy appear to be the most directly applicable, particularly short duration systems in the kilowatt to megawatt range. These systems would satisfy the needs in an area not covered by MHD systems although some overlapping might exist, for any special application in space and on earth where instantaneous production of appreciable amounts of power is required. The next possible step for open cycle systems would be entire fossil fuel power stations based solely on the EFD process or on a combination with a gas turbine cycle. For such an application which obviously implies the achievement of efficient multichannel converter configurations, the production of high voltage DC power inherent to the process would be a great advantage as far as power transmission is concerned.

The situation for closed cycle EFD systems is less clear; basically these systems are attractive in view of the broad power spectrum they could cover (from a fraction of kilowatt to several megawatts) and of the possibility of using all types of primary sources, chemical, solar and nuclear. Immediate production of high voltages would be attractive for electric propulsion applications in space, particularly for colloid propulsion.

The difficulty lies in inserting the EFD converter within a favourable overall thermodynamic cycle. The very low total pressure ratio through the EFD section - which means a poor conversion efficiency of heat into kinetic energy - requires either multistaging techniques which raise numerous problems or the use of a special two component - two loop system which deviates from the fundamental simplicity of the process. Considerable engineering effort is needed before practical systems could be reached.

The EFD process is still in a too early stage of technological development to permit a definite statement on its future but due to its good theoretical performances and mainly to its wider flexibility both in usage and source requirements than MHD systems, it certainly warrants further investigation.

H. CONCLUSION

The paper has shown that a growing variety of conversion systems has been available for the generation of electrical power. No single system is applicable over the whole power spectrum with all possible primary energy sources in all conceivable environments. Each system has a certain area of applications and the graphs, given in Figures 2, 3, 4 and 5 which summarise the whole discussion, are an attempt to outline these areas; for terrestrial applications, electrical power level has been considered as the most significant parameter while for space applications, an additional parameter, duration, has been introduced.

The graphs not only show the conversion systems which have been reviewed but also their competitors, electrochemical generators (batteries and fuel cells) and dynamic systems (turbines, reciprocating engines) which exhibit no signs of disappearing. This competition is and will continue to be beneficial both in space whose new challenges have initially stimulated most of the research in the field and on earth where electrical power generation is an important factor in raising the standard of living of mankind.

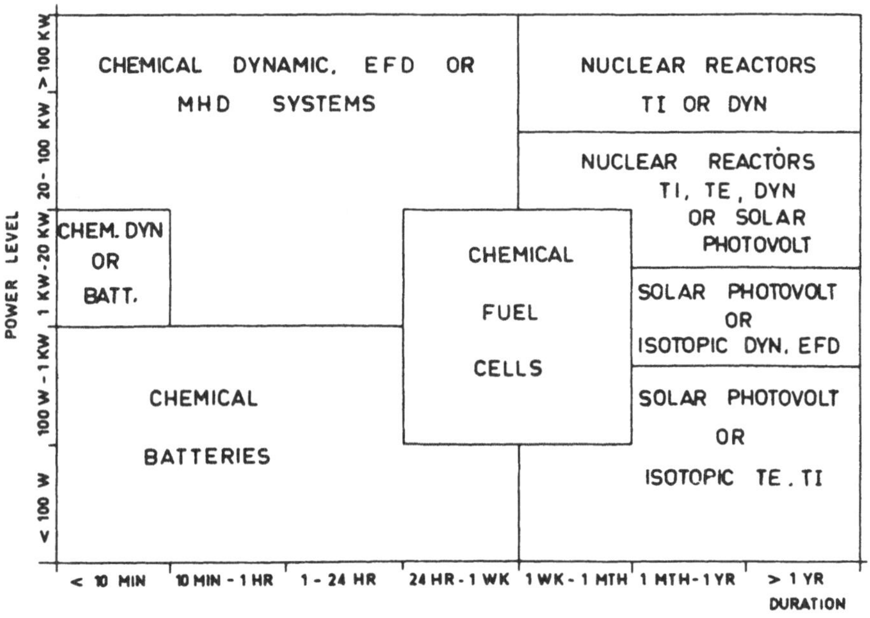

Figure 2. Space Power Generators Matrix

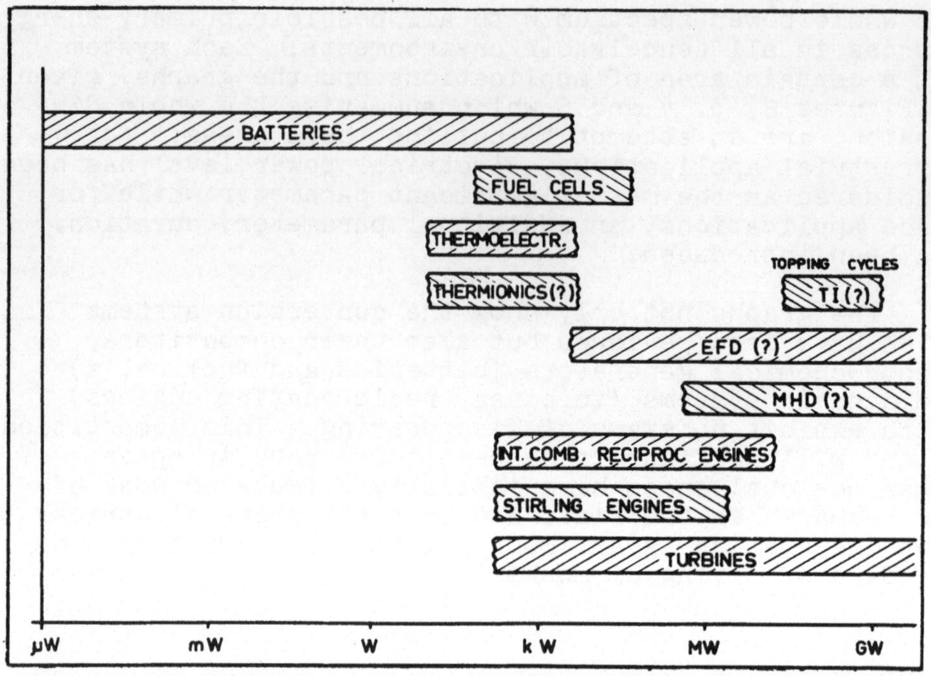

Figure 3. Chemical Sources - Terrestrial Applications

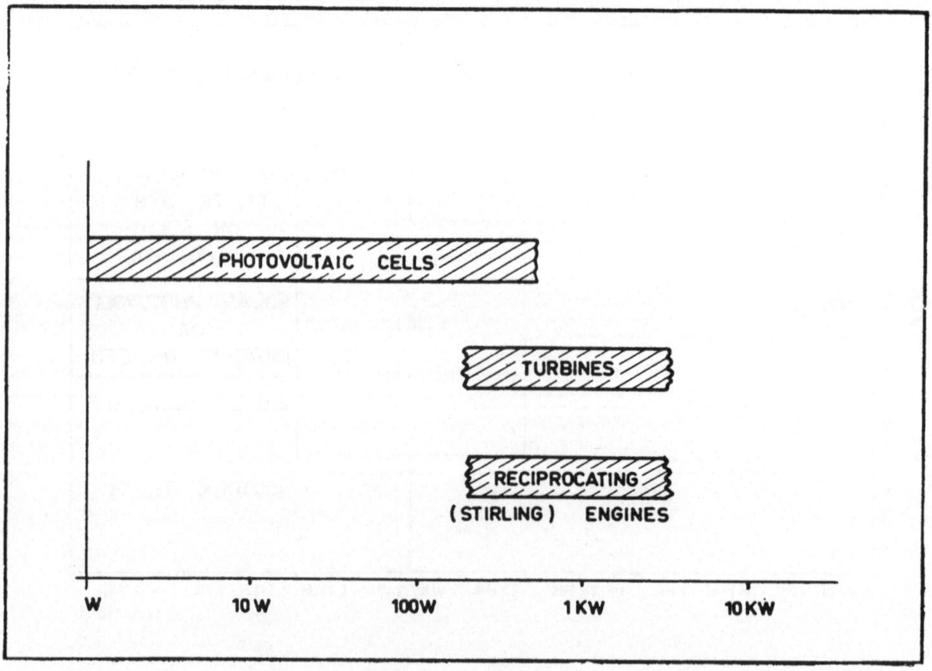

Figure 4. Solar Sources - Terrestrial Applications

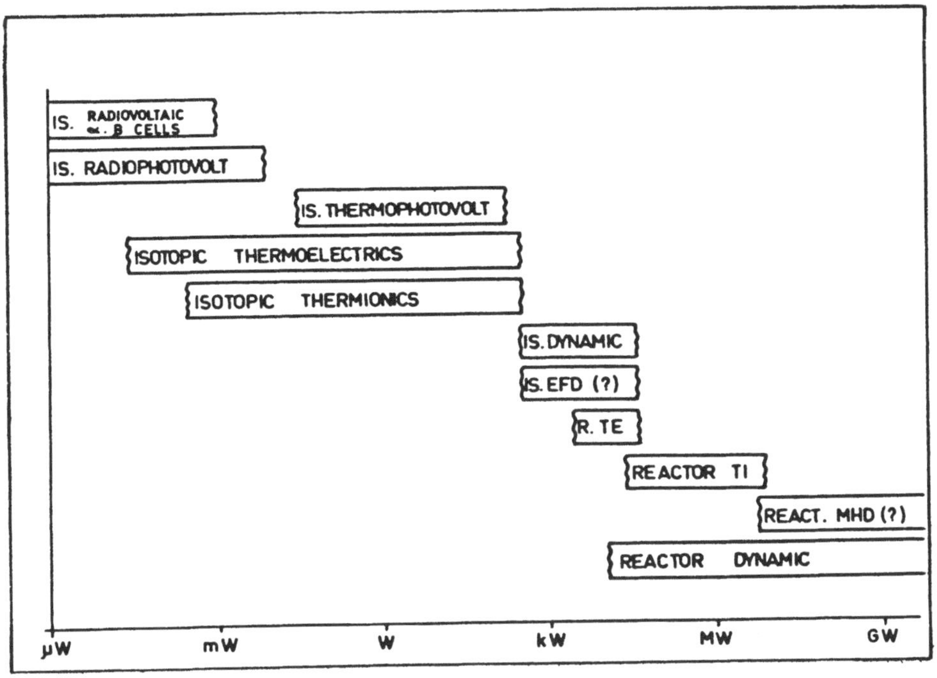

Figure 5. Nuclear Sources - Terrestrial Applications

Finally, it appears from this review that all the conversion processes which have emerged during the last two decades are much closer than conventional ones to fundamental physics. The realisation of practical devices rely heavily on theoretical and experimental work in various branches of this discipline and energy conversion is certainly one of the fields where the transfer from fundamental research to technological hardware has been the most efficient. It is hoped that this bridge between science and engineering will be strengthened in the future.

BIBLIOGRAPHICAL REFERENCES

A. Literature on Energy Conversion in General

- H.M. de GROFF et al.Ed.- Combustion and Propulsion. 6th AGARD Colloquium. Energy Sources and Energy Conversion. Cannes, France, March 16-20, 1964. Gordon and Breach Science Publishers.

- J. KAYE and J.A. WELSH, Direct Conversion of Heat to Electricity, 1960. J. Wiley and Sons, Inc.

- N. - Proceedings of the Intersociety Energy Conversion Engineering Conference 1966. Los Angeles, Cal., September 26-28, 1966. Published by the American Institute of Aeronautics and Astronautics.

- N. - Advances in Energy Conversion Engineering - Papers presented at 1967 Intersociety Energy Conversion Engineering Conference, Miami Beach, Florida, August 13-17, 1967. Published by the American Society of Mechanical Engineers.

- G.W. SUTTON, Ed. - Direct Energy Conversion, Inter University Electronic Series, Vol. 3, 1966 - Mc Graw Hill.

B. Literature on more specialised topics

- R.A. COOMBE Ed. - Magnetohydrodynamic Generation of Electrical Power, 1964, Chapman and Hall, London.

- W.R. CORLISS and D.G. HARVEY - Radioisotopic Power Generators - 1964 - Prentice Hall, Inc.

- H.M. DIECKAMP - Nuclear Space Power Systems, September 1967 - Atomics International.

- M.O. LAWSON Ed. - Selected Topics in Electrofluid Dynamic Energy Conversion - AGARDograph 122 - Published by Aerospace Research Laboratories.

- N. - Electricity from MHD - Proceedings of a Symposium, Salzburg, July 4-8, 1966, jointly organised by IAEA and ENEA - Published by the International Atomic Energy Agency, Vienna.

- N. - Industrial Applications for Isotopic Power Generators. Joint UKAEA - ENEA International Symposium, AERE Harwell, September 1966 - Published by the European Nuclear Energy Agency, Organisation for Economic Co-operation and Development.

- N. - Proceedings of the second International Conference on Thermionic Electrical Power Generation, Stresa, Italy - May 27-31, 1968; to be published by EURATOM, Brussels.

- G.C. SZEGO Ed. - AGARD Lecture Series N° XXVII on Energy Sources for Space Power - Brussels, Belgium, October 2-6, 1967, AGARDograph 123, to be published.

- F.C. TREBLE - Large Solar Cell Arrays - ELDO Technical Review, Vol. 3 N° 2, 1968.

- A.M. ZAREM and D.D. ERWAY Ed. - Introduction to the Utilisation of Solar Energy - 1963 - Mc Graw Hill.

A NOTE ON SCIENCE, MATERIALS AND THE REAL WORLD

N. E. Promisel

Executive Director, National Materials Advisory Board

National Academy of Sciences, Washington, D. C.

Because these Institute meetings are almost exclusively devoted to certain fundamentals and basic phenomena of solid materials, yielding a climate rather remote from the practical application of solid materials and thereby perhaps creating in the minds of some external observers illusions of an unreal world, the writer was asked to attempt to bring into better focus the interaction between science such as discussed at these meetings and the materials problems of the real world.

In a field as broad as materials, encompassing as it does probably all scientific and technological disciplines, it is not difficult to illustrate the relation between solid-state science and materials science and technology. Even going back only a relatively few years, one can cite, as examples: the theory and work that led to masers and lasers and their application to materials processing such as welding; the refinements in dislocation and imperfection theory which have contributed so much to our understanding of the properties and behavior of metals; the work on single crystals which has given us unique performance in single crystal turbine blades for aircraft engines; and these are but a few of many illustrations. This would appear to be only the beginning, however, because the growing recognition that materials are often the major roadblock to progress forces increased exploitation of the full spectrum of knowledge, from basic science through applied engineering. This will be so despite the fact that usually major advances in engineering come directly from engineering, although sometimes assisted by the oriented or directed research which the engineering problems stimulated. Often, however, a major contribution comes from basic research in entirely unrelated areas. Certainly the research leading to transistors and semi-conductors was not done 25 years ago with computers or aerospace in mind!

It is important to recognize, also, that the time lapse between fundamental science discovery and its practical application has tended to shrink, at least in many cases. Note the following table, based mainly on data published in Science Journal (1966):

Development	Approximate Time Period	No. of Years: Discovery to Application
Photography	1730-1840	110
Telephone	1820-1875	55
Radio	1865-1900	35
Radar	1925-1940	15
Television	1925-1935	10
Transistor	1945-1950	5
Gunn Oscillator	1963-1967	3

In recent years, a new and powerful force has been demanding contributions from materials and, through materials, from science and technology. I refer to the increasing emphasis, in many countries, on the social problems of mankind: pollution, health, low cost housing, transportation, etc. A recent U. S. report[1] draws attention to the opportunities in solid-state research in relation to such national needs.

In pollution, for example, the chemistry of fuels and their combustion, expansion in catalysis theory and the development of cheap catalysts, the technical development of economical treatment of waste materials and their conversion to useful products - all of these are important considerations which depend on materials and related science and technology. In health, there is a need for fundamental characterization of plastics to serve as containers and tubes for blood and its constituents. The control and specification of millions of metallic implants and other devices placed in the human body, as well as devices like artificial hearts and kidneys, require the attention of the materials scientist and engineer working closely with the surgeon. Solid-state concepts have contributed directly to the analysis of excited electronic states of ordered macromolecular structures, leading to calculations of optical properties in the fundamental genetic bio-polymer, the DNA molecule, which in turn has permitted quantitative deductions on structure from experimental results.

1. Research in Solid-State Sciences, by a Committee of the Solid States Sciences Panel, Publication 1600, National Academy of Sciences, Washington, D. C., 1968.

For low cost housing, composite materials developed for the aerospace industry may prove to be an effective answer, but there remains much to be understood here in terms of solid-state chemistry of fibers and polymers, the fundamental mechanisms of load transfer between fibers and matrices and the nature of crack initiation and propagation, surface effects on a microscale, etc. In transportation, high speed trains are handicapped (in their brake systems) by the need of better theory of surface wear and erosion. On the other hand, the excavation of tunnels through rocks may be aided by the application of the physicist's beam of electrons as a drilling tool. --- And so we must add to the points made earlier the relevance of solid-state science to social problems, through the medium of materials and processes, even though this route may not always be clear or directly visible to the casual observer.

Nevertheless, in spite of encouraging progress, we are still confronted by the challenge to accelerate the translation of scientific knowledge into engineering application. In this connection, the Materials Advisory Board of the National Academy of Sciences/National Academy of Engineering (U.S.) completed a study[2] about three years ago, the salient points of which are summarized below. The two major objectives of this study were: (1) to analyze representative examples of research-engineering interaction through 10 case histories; (2) to extract common denominators or clues which resulted in success or failure, particularly the success factors. The case histories covered a broad sample of the materials areas: metals, polymers, electronic materials, and others. They were analyzed systematically by uniform techniques.

It was possible to select some general findings from the above analyses, as follows:

1. Flexibility of support. This refers to having a broad enough basis of support to give the investigator adequate discretion to change direction of research when appropriate.

2. Close and frequent communication between the various independent groups. For example, one group could be primarily responsible for basic research and another group for engineering development. They might be geographically separated but in any case the communication between them had to be very good indeed.

---

2. Report of Ad Hoc Committee on Principles of Research-Engineering Interaction, by a Committee of the Materials Advisory Board, MAB-222-M, Washington, D. C., 1966

3. It was necessary that there be one or two key individuals who bridged the gaps in geography, organization and functions of the various groups. In some cases, these individuals served as coupling agents between the different groups to insure that the information flowed as necessary. However, in other cases, the individual might be one who adopted a leading and dominant role and who tackled all obstacles and overcame them. This type of individual might be called a "champion" of the cause. Sometimes the "coupler" was also the "champion".

4. Recognition of important need. This factor appeared most frequently in all of the cases studied and was an essential factor in bringing about the research-engineering interaction. In these cases, it was very rare that basic research by itself produced a new and unexpected opportunity which then stimulated an engineering application. By far, the most common situation was when there was an urgent need, which stimulated a search for a solution to this need and a search for prior basic knowledge. This appears to substantiate the commonly made statement that research serves as a reservoir of knowledge to which technology can turn to satisfy defined needs, rather than as a source which floods the engineer with solutions.

5. One of the most significant findings was that in only four of the ten cases did a majority of the events require the development of a new solution to a major technical problem. More often, technical approaches were available but had not been pursued and it was the act of timely recognition of existing knowledge that resulted in the final solution. This leads to the suggestion that there is much research knowledge that has yet to be tested, tried and applied to solution of current critical problems. It suggests that more effort could be profitably applied by the engineer in searching through research knowledge as he attempts to develop solutions to his problems.

Although some of these findings might not be new to research managers, this study made a contribution by providing the necessary documentary support for the above observations and at the same time disclosing no support for other generalizations frequently made about the transfer of research to engineering.

It would not be difficult to continue with specific examples of the dichotomy of science and engineering in many types of materials problems: in composites (already mentioned), in structural ceramics and glass (relation between composition, structure, surface diffusion, stress state), the theory of beryllium brittleness, in fabrication (superplastic behavior, unique role of ultrasonic energy on stress-strain behavior), nondestructive evaluation (using electron scattering, laser holography, acoustic emission and other physical phenomena). Perhaps enough has already been said, however, to demonstrate that not only is there a good linkage between solid-state science, materials and the real world (proceeding in that order) but increasingly the real world will have to look back in the reverse order for eventual answers to its problems and, in return, provide stimulation for further basic research.

CONTRIBUTOR INDEX